Frontiers of Complexity

THE SEARCH FOR ORDER IN A CHAOTIC WORLD

PETER COVENEY and
ROGER HIGHFIELD

Foreword by Baruch Blumberg

faber and faber

First published in the USA in 1995 by
Ballantine Books, Inc., New York

First published in Great Britain in 1995
by Faber and Faber Limited
3 Queen Square London WC1N 3AU
This paperback edition first published in 1996

Printed in England by Clays Ltd, St Ives plc

Peter Coveney and Roger Highfield are hereby identified as
authors of this work in accordance with Section 77 of the
Copyright, Designs and Patents Act 1988

A CIP record for this book
is available from the British Library

ISBN 0-571-17922-3

1 2 3 4 5 6 7 8 9 10

C 08/07/04

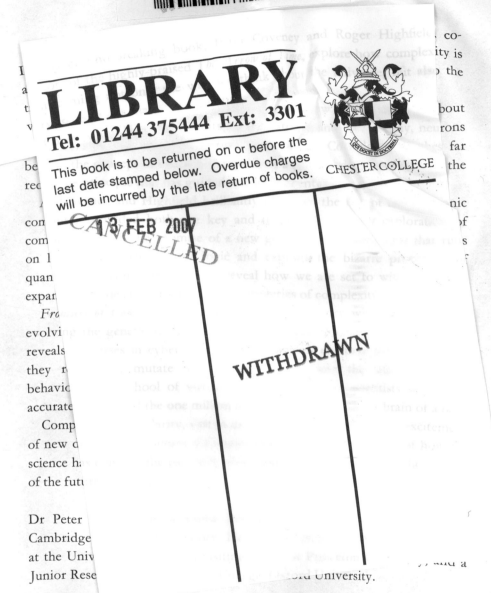

Dr Peter
Cambridge
at the Univ
Junior Rese

Dr Roger Highfield is the science editor of the *Daily Telegraph*. He was the
1987 Medical Journalist of the Year, the 1988 Science Journalist of the
Year, winner of the Specialist Correspondent category of the British Press
Awards in 1989 (commended 1991) and a winner in the first Commonwealth
Media Awards, 1994.

ALSO BY PETER COVENEY AND
ROGER HIGHFIELD:

The Arrow of Time

Glory be to God for dappled things—
    For skies of couple-colour as a brinded cow;
        For rose-moles in all stipple upon trout that swim;
Fresh-firecoal chestnut-falls; finches' wings;
    Landscape plotted and pieced—fold, fallow, and plough;
      And all trades, their gear and tackle and trim.
All things counter, original, spare, strange;
    Whatever is fickle, freckled (who knows how?)
      With swift, slow; sweet, sour; adazzle, dim;
He fathers-forth whose beauty is past change:
        Praise him.

—GERARD MANLEY HOPKINS
Pied Beauty

CONTENTS

FOREWORD

BARUCH BLUMBERG, NOBEL LAUREATE

In Molière's *Le Bourgeois Gentilhomme* the hero, after learning the meaning of the word "prose," realizes that he has unknowingly been speaking it for forty years. Similarly, on reading *Frontiers of Complexity*, I realized that I had been practicing complexity for decades without recognizing the exalted company I was keeping. My experience is that, in medicine, where observational science is crucial, the complexities of a phenomenon can be understood, at least in part, by repeated observations of a whole organism or a population of organisms under a wide range of circumstances; all the variables are retained and as many as possible are examined. For example, in studies of disease, it is possible to build up a knowledge of the effects of a large number of variables on the host, genetic susceptibilities to the disease, and outside factors which interact with each other, the host, and the environment. By contrast, in the reductionist approach, traditionally found in physics, chemistry, and molecular biology, experiments are designed to simplify the study of a natural phenomenon by the elimination of all but a few variables, and explanation is in terms of the most fundamental units.

The study of hepatitis B virus (HBV) and its interactions can be used as an example of observational science. HBV causes primary cancer of the liver, one of the most common cancers in the world. But not everyone who is exposed to HBV becomes chronically infected and not all infected persons develop cancer. The internal and external factors that determine infection and outcome are interactive and time dependent. There are interactions between HBV and other viruses which attack the liver, the AIDS virus, malaria, and probably other micro-organisms which influence both the probability of infection and the outcome. Genes, gender, and age have a significant effect. In addition, environmental agents such as aflatoxin—a carcinogen which is elaborated by fungi that contaminate foodstuffs—iron, arsenic, and probably others increase the risk of cancer and chronic disease. As a further complication, the environmental effects are influenced by products of the host genes.

The virus has developed remarkably clever strategies to maximize its persistence without killing the host prematurely, in order to be transmitted, primarily through sexual intercourse and by passage from mother to child during childbirth and infancy. Its strategies for the evasion of the human immune protective system are particularly clever; they include "smoke screens" of surface antigen produced in excess to befuddle the immune response; the development of tolerance to an antigen of the virus which allows it to reside for years within the liver cells of its host, replicating and spreading, but not immediately disabling its innocent bearer; and many other "intelligent" schemes.

All this data cries out for a model that will reveal the complex and time-dependent interactions of known and unknown variables. But conventional model makers find it difficult to deal with such complexity since so many assumptions must be made. Does the science of complexity offer a hope for the construction of dynamic and evolutionary models that provide a more comprehensive and complex image? *Frontiers of Complexity* encourages us to believe that this may be possible, since other very complex problems are now being addressed: *real* problems, as they exist in the world, not constructions of the experimentalist shorn of the richness of actual events. The book harbors a promise that complex problems will be solvable, and the forms of synthesis and analysis described are beginning to make an impact on biological and medical scientists who must deal with an embarrassing richness of variables.

The authors tell us that reductionism has its seductive attractions. Traditional science, particularly the Greek ideal, worships simplicity, harmony, symmetry, and other attributes of pure beauty. In some interpretations, Plato taught that the observed world was not as real as its essence which may be concealed by the complexities that cloud it, and it is this essence we should desire to know. Experimental science attempts to approach this essence by creating a world inferred from experience but with a minimum of complications. Reductionist science, for example, concerned with the elementary particles and forces of physics, or with the molecules and genes of biology, has propelled science forward, particularly in the past several decades with the availability of machines and methods which were not even conceivable in previous times. But, for nonlinear systems the whole is greater than the sum of the parts, and they can only be understood by examining "global" behaviors in addition to a detailed examination of the individual agents of which they are comprised. In interventional science, such as medicine, it may be the complexities themselves that provide the key to the problem. In a biological system gone wrong—due to disease—the more complexity we are aware of, the more numerous are the sites where intervention can take place. It is valuable to know the "real" cause of a disease, that is the one essential element that is causative, but this is often not knowable and may be a fiction of the investigator's desire to have simple answers to complex problems. But, as you will read, successful engineers and other applied scientists are practical and don't allow the perfect to drive away the good. Understandings that are derived at the border between chaos and order where, according to some, many of the problems of nature lie, may not provide exact solutions but rather those which can allow application and understanding to emerge.

You will read of heady things in *Frontiers*. For example, "The God Game" started as a mathematical amusement with simple rules, which has led to constructions that have attributes of living systems, such as virtual fish that scatter when a virtual shark approaches and virtual bees that show a preference for flowers similar to that of actual bees. There is the use of DNA itself to carry out complex computations. There is Artificial Life, computer-based systems which have attributes of life. And these attributes can be transferred to robots that operate independently in remote locations, such as the craters of active volcanoes, on the Moon, or on Mars, looking like insects devised by a logical mind, but designed to quest for the behaviors of natural beings.

Incorporated in these artificial life studies are evolutionary programs which allow the "biomorphs" within them to compete for survival—through access to memory space and/or processing time in the bowels of the computer. This differs from natural selection in that the rules for survival are provided by the scientist. Hence the creative scientist takes the form of a god in a universe which he or she has created. Indeed, Richard Dawkins, the Oxford zoologist, at the first (1987) conference on Artificial Life held in Los Alamos, presented his Blind Watchmaker, a program which allowed the evolutionary emergence of forms with a generated order where Dawkins selected those with aesthetic appeal. Now, Thomas Ray and others have developed programs that allow a menagerie of forms to emerge without the interposition of an intent-motivated creator.

This is a ripe area for the continuing concerns of the nonscientific public with the creative "arrogance" of science. The fears generated by Victor Frankenstein's Monster, who has terrorized the imagination of the world for nearly two centuries, can be rekindled by these imaginative possibilities. There are modern expressions of these ideas. Bernal, the crystallographer, is quoted as stating that humans will not be content just to manufacture a simulacrum of life; they will want to improve upon it. The molecular biologist in the movie and novel *Jurassic Park* was not content to leave the DNA as he found it when he was creating his dinosaurs. He added genes for more rapid development, attempted to produce single sex populations, and wanted to alter the locomotion of the dinosaurs to conform with the public's perception of dinosaur behavior. Yet these presuming aspirations were doomed to calamity and destruction in part because we understand too little of how the details of science interact with the environment and other vital agents within it.

The critic and essayist Gilbert Highet writes of Job, who, during a great storm, is made aware of the magnitude and greatness of the forces of nature. The continued study of complexity, even if it does not provide totally satisfactory solutions, should make reductionist science aware that no matter how many details are uncovered, no matter how comprehensive understanding may be, there will always be unknowns beyond the sum of current knowledge. Each time an experiment is performed to test a hypothesis, more questions are revealed; there is no limit to the mysteries of nature and to our desire to understand them. The study of complexity offers an opportunity to stand back and consider the global interactions of fundamental units—atoms, elementary particles, genes—to create a synthesis that crosses the borders of scientific disciplines, to see a grand vision of nature.

ACKNOWLEDGMENTS

Almost a decade ago we first thought of collaborating on a book, during a late-night conversation about the nature of time and its relationship with the evolution of complex systems. We then wrote *The Arrow of Time*, which was first published in 1990 and has since been translated into more than a dozen languages. *Frontiers of Complexity* continues that same mission, developing and extending concepts that we encountered in *The Arrow of Time*, notably complexity, self-organization, and artificial life.

The book is structured around those principal ideas, to ensure that they are not obscured by colorful accounts of personalities. We also attempt to put these ideas within a general context by sketching the historical development of relevant fields of science, technology, and mathematics. As far as possible, we have avoided the use of jargon, and we have banned mathematical notation altogether. Numbered notes aimed at the scientific reader are secreted at the back of the book, along with detailed references to the research literature and quirky, bizarre, or fascinating information that we could not bear to jettison. Anyone whose interest is kindled will be able to pursue the subject further by means of the Bibliography. Readers may also find the glossary of terms useful.

We owe an extraordinary debt of gratitude to the many people who helped us along the way. Many thanks are due to Max Hastings, editor of *The Daily Telegraph*, for allowing Roger Highfield to take up a visiting sabbatical fellowship at Balliol College, Oxford University, during Trinity Term 1994. In that term he received a great deal of support from Baruch Blumberg, the then Master of Balliol. The high-technology scientific environment within Schlumberger provided Peter Coveney with a great deal of stimulation during the writing of this book.

A large number of scientific friends and colleagues have given generously of their own time to provide information and explanations. We are particularly indebted to John Billingham, who read the entire manuscript more than once, and to Bruce Boghosian, for many valuable discussions and sound critical advice. We also wish to express gratitude to Baruch Blumberg for his encouragement, for his comments on the manuscript, and for writing the Foreword.

In terms of constructive criticism on draft chapters, we are especially grateful for comments made by Len Adleman, Shara Amin, Steve Appleby, Michael Arbib, Robert Axelrod, Per Bak, Gary Barker, Mark Bedau, Colin Blakemore, Tim Bliss, Gregory Chaitin, Dave Cliff, Francis Crick, Jim Crutchfield, Derek Denton, David Deutsch, Rodney Douglas, Gerald Edelman, Manfred Eigen, José Luis Fernández, Brian Goodwin, Geoffrey Hinton, Andrew Hodges, John Holland, Xiaoping Hu, Gerald Joyce, Stuart Kauffman, James Lake, Chris Langton, William Latham, Ralph Linsker, Seth Lloyd, James Lovelock, Luigi Luisi, Paul McIlroy, Misha Mahowald, Carlo Maley, Christof von der Malsburg, Norman Margolus, Mario Markus, Robert May, David Miller, Melanie Mitchell, Denis Noble, Martin Nowak, Leslie Orgel, Oliver Penrose, Roger Penrose, Edmund Rolls, Steven Rose, Riitta Salmelin, Antoine Schlijper, Terry Sejnowski, David Sherrington, Karl Sims, Olaf Sporns, Oliver Strimpel, Doron Swade, Harry Swinney, Jim Tabony, John Taylor, Roger Traub, Lotfi Zadeh, and Semir Zeki.

We would also like to thank the following for consenting to interviews or for providing information: Igor Aleksander, Anthony Arak, Wallace Broecker, Rodney Brooks, Marilyn Butler, John Conway, Malcolm Cooper, Elena Coveney, Richard Dawkins, Dan Dennett, Rodney Douglas, Tim Dowling, Karl Friston, Peter Fromherz, Hugo de Garis, Murray Gell-Mann, John Habgood, Danny Hillis, Peter Hilton, Rufus Johnstone, Julian Lewis, Robert Littell, Christopher Longuet–Higgins, Sidney Nagel, Tom Ray, Urs Ribary, John Searle, Hava Siegelmann, Tom Stoppard, Demetri Terzopoulos, Tom Toffoli, Giulio Tononi, Paul Verschure, Peter Walde, James Watson, Gerard Weisbuch, and Stephen Wolfram.

Many people read and commented on parts of the book, insisting that we improve the clarity and presentation in countless places. They include Samira Ahmed, Oscar Bandtlow, Julia Brookes, Jon Dagley, Richard Daly, Andrew Emerton, Allan Evans, Heather Gething, Ronald and Doris Highfield, David Johnson, Mehul Khimasia, Tony Manzi, Samia Nehmé, and Keir Novik. Mehul and Keir also helped to check the proofs.

Errors that remain are, of course, our responsibility alone. None of the above necessarily agrees with the opinions we express in this book.

We are very grateful to all those who provided graphics, in particular William Latham, Mario Markus, Karl Sims, Nick Waters, and Michael Whiteley. In addition, we acknowledge Daresbury Laboratory and David Stuart and coworkers at the Department of Molecular Biophysics, Oxford University, for use of the image of the foot and mouth virus structure.

We would like to thank the following for their permission to use graphics, pictures, or images: American Association for the Advancement of Science; J. Baldeschwieler; Gary Barker; British Telecom; Rodney Brooks; Peter Coveney; *Current Biology*; *The Daily Telegraph*; Richard Dawkins; Gerald Edelman; J. L. Fernández; Hugo de Garis; Joseph Harrington; Martin Harvey; Institute for Advanced Study; Gerald Joyce; William Latham; Norman Margolus; Mario Markus; Kevan Martin; Peter Newmark; Denis Noble; Martin Nowak; Przemyslaw Prusinkiewicz; Thomas Ray; Julius Rebek; Riitta Salmelin; The Science Museum, London; Kenneth Showalter; Karl Sims; Jakob Skipper (author of C-Zoo); David Stuart; Demetri Terzopoulos; Manuel Velarde; Nick Waters; Michael Whiteley; Andrew Wuensche; Semir Zeki. The graphics team of *The Daily Telegraph*, Alan Gilliland and Richard Burgess, provided the line drawings.

We also wish to thank Gulshan Chunara of *The Daily Telegraph* for her unfailing secretarial help; John Brockman, who first pushed us to write it; our editors, Joëlle Delbourgo and Andrea Schulz in New York, and Julian Loose in London. We particularly benefited from the close editing of successive drafts of the manuscript by Andrea Schulz. We would also like to thank Susanne McDadd, who played a key role in the very early stages of this book and in *The Arrow of Time*. Most important, we thank Samia and Julia for tolerating yet another long and exhausting project.

Peter Coveney, Schlumberger Cambridge Research
Roger Highfield, *The Daily Telegraph*

March 1995

PROLOGUE

It would be difficult to conceive of two more different individuals than John von Neumann and Alan Mathison Turing. "Good time Johnnie" loved to memorize jokes as much as he enjoyed the company of women.[1] He also had a fondness for smart suits, off-color stories, and parties. Alan Turing was an awkward, though approachable loner known as "the Prof." Habitually clad in a shabby sports jacket, he had a hesitant voice, yellow teeth, and poorly manicured hands.[2] The Christmas before the men first set eyes on each other, the twenty-three-year-old Turing had asked his mother for a teddy bear, which he named Porgy.[3]

These men, who are of central importance to our book, had little in common. Alan Turing was born on June 23, 1912, in a nursing home near Paddington, London. Von Neumann was then eight years old and enjoying private tutoring at the home of his wealthy Hungarian father. Their dazzling intellects would, however, share a common vision after they first met at Cambridge in 1935. During the next decade they independently laid down the mathematical, logical, and physical foundations of the electronic digital computer.

Alan Turing   John von Neumann

Both their names will be forever associated with this machine that now touches every transaction of daily life: electronic fund transfers, telecommunications, word processors, supermarket cash registers, programmable video recorders, movie special effects, the Internet—the list of applications is endless. Bits of digital logic now encircle the globe with a furious intensity, carrying technical data, commercial decisions, ultimatums, weapon-firing instructions, video conferences, late-night conversations, and messages of love.

Turing and von Neumann pioneered the theoretical study of complex systems, from turbulent fluids to thinking brains. They became figures of central importance in the science of complexity, the quest to see the forest for the trees, distill unity from diversity, and seek an account of the emergence of organization, whether a fleeting cloud or inspiring thought. It is no coincidence that their legacy—the digital computer— is today the most important tool used for conducting these investigations, since the mathematical equations describing complexity are usually too difficult to solve with pencil and paper. The foraging strategies of ants, convection patterns that form in a warmed pool of oil, and the irregular beating of a diseased human heart are among the vast number of complex systems that would have been impossible to simulate without a digital computer.

Their vision, however, went even further than computer simulation. Von Neumann invented a self-reproducing automaton, an abstraction that combined his interests in logic, computers, and biology, to show how a machine could perform the most basic function of life— reproduction. Von Neumann thus became the father of artificial life, one of the most extraordinary fields of contemporary research. Similarly, Turing's overarching ambition was to construct an artificial "brain," and he saw no reason why a computer could not one day realize that dream. In a now famous paper published in 1950, he formulated an operational definition of intelligence that was broad enough to encompass both biological and mechanical devices.[4] His work on computers and their relationship with brains made him the father of artificial intelligence.

Convinced of the importance of learning as an essential precursor to intelligence, whether biological or artificial, Turing spent a large part of his final years thinking about brain structure in living creatures and how vast assemblies of brain cells (neurons) act together during learning. These questions drove his interest in morphogenesis—the development of a biological pattern such as that found in a daisy. But this was to be the last flowering of an extraordinary career. As a result of police investigations into a burglary at his home, he was led to confess that he was a homosexual, a sexual preference not tolerated in Britain at that time. In 1952 Turing wrote to a friend that he faced a charge for sexual offenses. He felt his work was tarnished by this disgrace, signing off with the syllogism: "Turing believes machines think. Turing lies with men. Therefore machines do not think."[5] He pleaded guilty to "gross indecency," was put on probation, and sent for medical treatment—"organo-therapy"—in which hormones were administered to curb his "unnatural" sexual urge. Any chance of maintaining an effective collaboration with von Neumann evaporated, since Turing could no longer obtain a travel visa to America. The pressures of conformity became overwhelming. On June 7, 1954, the coldest and wettest Whit Monday for fifty years, the forty-two-year-old Alan Turing took several bites out of an apple he had dipped in cyanide.[6]

Turing was an unsung hero, a seminal figure in British wartime intelligence, the supreme code breaker, both of nature's secrets and the messages sent by the German naval Enigma machines. The British and U.S.

governments still prevent open discussion of the methods he used, though one former colleague maintained that Turing's work on crypt-analysis during the Second World War was the single most important individual contribution to the Allied victory.[7] This work had also been a vital stepping-stone in his mission to build a real computer. Yet his brilliantly individual, logical mind could no longer cope with the prospect of living in an uncaring and irrational world.

While Turing died branded a potential security risk, von Neumann had become a familiar figure on the many committees of the American military establishment, playing key roles in the development of the hydrogen bomb, the intercontinental ballistic missile, as well as the digital computer. Von Neumann dominated, and was not dominated by, his adopted country.[8] In 1956, he was awarded the Freedom Medal by President Eisenhower in a White House ceremony. It was one of his last official appearances, at a time when there was no longer any doubt about the outcome of a damaging illness.[9] Von Neumann remarked, "I wish I could be around long enough to deserve this honor." The president replied, "Oh yes, you will be with us for a long time."

Von Neumann attended the ceremony in a wheelchair. He had been invited to give the Silliman lectures at Yale in the spring of that year but, by then, had been admitted to the Walter Reed Hospital in Washington, D.C., with cancer. Two of his lectures, partly written on his deathbed, were eventually published in *The Computer and the Brain*. In that book, von Neumann expressed his belief that mathematics is a secondary language, derived from the primary language employed by the body's central nervous system.[10] For as long as he was able, von Neumann struggled to specify what that neural language was. But on February 8, 1958, his fight for life ended, leaving his mission, like that of Alan Turing, incomplete.

Yet Eisenhower was right: the vision of von Neumann, one that he shared with Turing, remains with us. It is to their memory that we dedicate this book.

Chapter 1

THE SECRET ART

God has put a secret art into the forces of Nature so as to enable it to fashion itself out of chaos into a perfect world system.

—IMMANUEL KANT[1]

When viewed in profound close-up, the universe is an overwhelming and unimaginable number of particles dancing to a melody of fundamental forces. All about us and within us, molecules and atoms collide, vibrate, and spin. Gusts of nitrogen and oxygen molecules are drawn into our lungs with each breath we take. Lattices of atoms shake and jostle within the grains of sand between our toes. Armies of enzymes labor to turn chemicals into living energy for our cells. Yet we think of the universe as a single harmonious system or cosmos, as the Greeks called it. Now a new branch of science is attempting to demonstrate why the whole universe is greater than the sum of its many parts, and how all its components come together to produce overarching patterns. This effort to divine order in a chaotic cosmos is the new science of complexity. It is weaving remarkable connections between the many and varied efforts of researchers working at its frontiers, across an astonishingly wide range of disciplines.

French scientists are studying how spots and stripes can spontaneously form in a soup of chemicals. These are uncannily similar to the markings found on an animal's coat, the wings of an insect, and on mollusk shells. Within the soup, cycles of chemical reactions interlock and turn so that, amazingly, countless numbers of molecules seem to know exactly what they are each doing. Colorful patterns emerge as the myriad molecules "communicate" with one another.

On the east coast of America, patterns of electrical activity are forming within a network of hundreds of thousands of artificial brain cells. No one tells the network what to do: all that is laid down are a few simple rules governing how one cell "talks" to another. Gradually, however, the cells organize so that different jobs are carried out by various cell groups within the network. Strikingly, the network integrates itself in a way similar to the way cells in the human body process vision.

In a suburb of San Diego, molecular biologists are aping the process that has allowed humans to evolve over the eons. Scientists are mutating and evolving trillions of molecular variants on a natural theme—a scrap of genetic code that nature has taken billions of years to optimize. This "evolution in a test tube" needs only a few days to alter the scrap so that it codes for an enzyme capable of speeding a novel chemical process, perhaps one that could help save a human life.

An ecologist is gazing at a series of multicolored bars on a computer screen in Kyoto. The rainbow of stripes evolves and changes, depicting the strands of mutating computer code that have survived a battle for memory space in the machine's core. After these competing codes have evolved for a few thousand generations, the result is a diverse menagerie akin to what seethes in a tropical rainforest.

And thousands of miles away, in Oxford University, a phenomenon called frustration is helping a theoretical physicist uncover the competing forces between atoms that endow metal alloys with odd magnetic properties. Real-life frustration arises when we run into unnecessary bureaucracy, nit-picking, or opposition. Similarly, magnetic frustration results when interatomic forces are at loggerheads. By understanding this conflict, we can not only make sense of the properties of alloys but also route traffic around global telecommunications networks and unravel the secrets of memory.

These scientists are all exploring manifestations of the same phenomenon: complexity, the "secret art" glimpsed by Kant. The *macroscopic* world abounds with complex processes and systems—religious rites and ephemeral emotions, musical musings and muddy meadows, global stock market crashes and wet Sunday afternoons. This complexity is intrinsic to nature; it is not just a result of the combination of many simple processes that occur on a more fundamental level.[2]

Within science, complexity is a watchword for a new way of thinking about the *collective* behavior of many basic but interacting units, be they atoms, molecules, neurons, or bits within a computer. To be more precise, our definition is that *complexity is the study of the behavior of macroscopic collections of such units that are endowed with the potential to evolve in time.* Their interactions lead to coherent collective phenomena, so-called emergent properties that can be described only at higher levels than those of the individual units. In this sense, the whole is more than the sum of its components, just as a van Gogh painting is so much more than a collection of bold brushstrokes. This is as true for a human society as it is for a raging sea or the electrochemical firing patterns of neurons in a human brain. A swirling vortex in a turbulent ocean cannot be expressed in terms of individual water molecules any more than a happy thought can be depicted in terms of events within a single brain cell. Conversely, the long-term behavior of only three billiard balls on a snooker table is unpredictable, even though the equations of motion describing the system are known precisely.[3]

Conventional science is frequently blind to the connections that can be drawn between frustrated metals, the rise and fall of stock prices, and a host of other complex phenomena. Most scientists today restrict themselves to the detailed study of one small aspect of a single subdiscipline within one branch of the tree of science: for example, the foamy structure of the universe; the decline of Partula snails on the Pacific island of Moorea; or the molecular structure of an enzyme from the Human Immunodeficiency Virus. This is inevitable, as more and more research becomes focused on ever smaller minutiae. The background expertise and sophistication of the techniques deployed in any field are so daunting that it is difficult to arrive at the cutting edge of knowledge without immense dedication. The specialization this entails brings with it, within every small area of inquiry, a unique methodology and jargon that are hard for an outsider to make sense of, let alone discover whether a common conceptual framework might be shared with other scientists working in different fields.

Yet the majority of real-world problems—and therefore most of those in modern industries and societies—do not fit into neat compartments. To solve them, people must be able to communicate across traditional boundaries, to approach issues in a collaborative, integrated way. Many scientists, who today are almost by definition specialists, may feel suspicious of, if not threatened by, this message. Unfortunately, our present education system hardly prepares us for such an approach. As the Nobel laureate Murray Gell-Mann has argued, we must get away from the idea that serious work is restricted to "beating to death a well-defined problem in a narrow discipline, while broadly integrative thinking is relegated to cocktail parties. In academic life, in bureaucracies, and elsewhere, we encounter a lack of respect for the task of integration."[4] In centuries past, it was possible for leading intellectuals to be genuine polymaths, making seminal contributions across the full range of human thought and ideas. This seems impossible today.

There is, nevertheless, a community of scientists—and philosophers—who are swimming against this tide. Motivated by a desire to establish connections across conventionally separate scientific disciplines, they want to show that there is an economy of concepts necessary for understanding the way the world works. Their ultimate goal is to come to grips not just with the complexity of any single phenomenon but also with the universal features of complexity itself, whether manifested by evolution in a rainforest or within the core of a computer, by the drip of a leaky tap, the spirals of color that can form in a chemical reaction, the magnetic properties of alloys, or the workings of a conscious brain. The quest is to find unity in diversity, to explain how order can emerge from a mass of evolving agents, whether atoms, cells, or organisms.

Conventional physics can predict large-scale events, such as the way starlight bends around massive objects, including black holes, and how spiral galaxies move. It can also probe events on the most minute scale, as when an electron hops between orbits within a hydrogen atom. Conventionally, however, physicists have steered clear of attempting to understand the workings of the brain or indeed any of the complexity that abounds on length and time scales that, ironically, are the most familiar to the human brain. The processes creating the exquisite macroscopic complexity we know so well from the primeval simplicity of atoms and molecules are part of nature's secret art. For the German philosopher Immanuel Kant, that "secret art" was a sign of God's hand guiding events in a propitious direction.

We are about to embark on our own quest for understanding the nature of complexity. In the following chapters we will investigate its basic mathematical description and seek out its manifestations in fields as diverse as chemistry, physics, biology, and computer science. En route, we will glimpse the awesome power of this new science and its ability to link the processes at work within our bodies and around us in the cosmos. We hope to demonstrate connections between such processes in ways never before possible.

CREATING COMPLEXITY

For complexity to emerge, two ingredients are necessary. The first, and foremost, is an irreversible medium in which things can happen: this medium is time, flowing from the past that lies closed behind us toward a future that is open. The reason for stating the apparently obvious is that the laws of motion traditionally used to describe the behavior of matter on the microscopic level do not distinguish one direction of time from another. Yet we know from the tendency of snowmen to melt and our skin to wrinkle that a preferred direction of time is singled out at the macroscopic level. This, the famous *irreversibility paradox*, arises from a discontinuity between these two levels of description, and is a problem we explored in our earlier book, *The Arrow of Time*.

The second essential ingredient is *nonlinearity*. We are all familiar with linear systems that have been the mainstay of science for more than three hundred years: because one plus one equals two, we can predict that the volume of water flowing down a drain is doubled when a tap drips for twice as long. Nonlinear systems do not obey the simple rules of addition. Compare the simple flow of water down a drain with the complex nonlinear phenomena that regulate the quantity of water in the human body, or the movement of water vapor in the clouds overhead. Nonlinearity causes small changes on one level of organization to produce large effects at the same or different levels. This is familiar to most of us through the example of positive feedback, which turns amplified music into a deafening howl, but the same effect is present in the propensity of plutonium atoms to fall apart during an explosive nuclear chain reaction. In general, nonlinearity produces complex and frequently unexpected results.

Irreversibility and nonlinearity characterize phenomena in every field of science: the complexity of the markings on a butterfly's wings, a leopard's spots, the shape of a spleenwort fern, the whorls of a little green alga called the Mermaid's Cap, and the rhythms in living systems such as the palpitations of a heart, the firing of nerve cells within the brain, and so on. Related, but more subtle, *chaotic* forms of complexity also arise from nonlinearity, including apparently random weather patterns, the outbreak of flu epidemics, and the spread of information and ideas.

One of the most striking ways to appreciate complexity is to chart the rich properties of a very simple nonlinear mathematical equation that describes how a population of organisms within an ecosystem varies in number from one generation to the next as a result of births and deaths. Using computer graphics, one can display the remarkable range of possible behaviors captured by this logistic equation in an evocative and eerie landscape (see color plate 1). Each graphic shows the results of computations carried out by Mario Markus and his colleagues at the Max Planck Institute for Molecular Physiology in Dortmund, Germany, for nearly a million possible combinations of environmental parameters. The images dramatically reveal the complexity that nonlinear equations possess.[5] Only with a computer are we able to explore such behavior.

ORIGINS OF COMPLEXITY

Matter has an innate tendency to self-organize and generate complexity. This tendency has been at work since the birth of the universe, when a pinpoint of featureless matter budded from nothing at all. From this putatively emphatic and utterly simple Big Bang, the universe ballooned. At first, matter consisted of only a soup of elementary particles and fields, but within about a billionth of a second of the Big Bang it began to coalesce into subatomic particles such as protons, neutrons, and electrons. These common building blocks obey what physicists believe to be simple mathematical laws, yet this matter generated structures of remarkable complexity as it agglomerated and curdled into lumpy features such as galaxies, stars, and planets. Energy and chemical elements produced by the stars have led to the emergence of intricate structures as organized as crystals and human brains. The most striking example we know of is the living economy of the Earth, in all its seething diversity.

But cosmology, astrophysics, and particle physics provide far from the whole story. In the case of life on earth, complexity in nature has been refined by competition for finite resources. Darwin popularized the notion of survival of the fittest, the struggle of every species—and of course every individual within that species—to adapt or *optimize* its ability to survive. As time goes by, an individual's environment—and therefore its chance of survival—is subject to changes as food supplies become exhausted, the local climate alters, or a killer virus spreads. As other species evolve and compete within that same environment, an organism must continually adapt in order to survive. A monkey is only one of the millions of patterns in space and time resulting from the evolution of a rainforest. The monkey embodies a vast range of complexity, from the chemical reactions within its cells to the electrical activity that crackles within its brain.

Comprehending the complexity of life is the biggest challenge facing modern science. What we stand to gain by succeeding in this endeavor is evident. A proper understanding of the living economy of the planet is the key to safeguarding its future. Understanding the living economy of the human body can help to treat it when it is sick. Could a similar understanding of the complexity of human societies aid us in anticipating riots, civil disruptions, and wars?

Many invaluable applications of this knowledge exist throughout science and technology. By copying the way living organisms handle the problems they face in the battle for survival, scientists have developed new tools that solve many complex problems. *Genetic algorithms*, computer programs that borrow ideas from biological evolution, and *artificial neural networks*, inspired by the structure of the brain, are two examples. Both play a central role in modern approaches to artificial intelligence; both are difficult to understand using conventional science alone.

SAVORING SIMPLICITY

For millennia, those seeking to understand the natural world have been seduced by simplicity. Their mission has been to boil the workings of the universe down to its component parts. So-called *reductionism* is a quest to explain complex phenomena in terms of something simpler. For a physicist, this means describing the properties of a gas in terms of the behavior of its constituent atoms or molecules. For a chemist, it means explaining a chemical reaction in terms of the changes occurring within its molecules.

Nowhere is the reductionist view more prevalent than in elementary particle physics, where the drive is on to find a "Theory of Everything" that would be expressed in one or a few equations describing the fundamental interactions between all forms of matter. What used to be the physicist's best game in town—searching for the simplest components of matter, the elementary particles, and their origins—is, however, looking a little passé given the dearth of discoveries since the W and Z^0 particles were found in 1983,[6] the yawn that greeted the putative identification of the top quark in 1995,[7] and, more seriously, the increasing lack of contact of many of its recondite theories with the observable world.

Within chemistry, the same philosophy adheres in the widely held view that all processes can be understood in terms of individual properties of the atoms and molecules involved. Similarly, in the life sciences it can be discerned as the "doctrine of DNA." This followed from the discovery, by Francis Crick and James Watson in 1953, of the structure of the DNA molecule. The science of *molecular* biology was born; thereafter, large parts of biology could be rationalized on the basis of molecular actions. No one can deny the tremendous impact on our lives of this triumph of reductionism: legions of scientists across the planet are using such knowledge to detect hereditary disease and correct it by gene transplants. So-called gene therapy heralds the ability to cure a host of hereditary diseases, such as cystic fibrosis and muscular dystrophy, and even offers the prospect of manipulating other traits influenced by our inheritance, such as athletic prowess or intelligence.

Because of its power, reductionism is all too often perceived as the universal route to understanding. Yet it has driven a wedge between science and other aspects of human life. Reductionism, used naively, offers an analysis of phenomena by splitting them up into their smallest possible pieces; but, as Alvin Toffler has remarked, modern science is so good at splitting problems into pieces that we often forget to put them back together again.[8] People are depicted as little more than survival robots who spread genes. Pain, suffering, and civil disorder are nothing more than manifestations of defective genes. And homosexuality is caused by a "gay brain," the product of gay genes.[9]

The reductionist view that anything and everything can be boiled down to atoms and molecules is widely viewed by nonscientists as a philosophy that also erodes our belief in "humanity" and the value we place on it: "After all, when you get right down to it, a human body is just a few dollars' worth of chemicals." Moreover, if mankind is ruled by natural forces, by deterministic mechanisms, we cannot develop a theory of human action based on free will. The vision of the world that naive reductionist science has proclaimed is a cold and solitary one that sets mankind apart from an unseeing and uncaring universe.[10] This is not an inspiring image, and it has led many to be critical of science and the scientific method, since they have little bearing on so much of human experience. As a result, this austere world-view has been instrumental in promoting a view of science as separate from the rest of human culture.

Complexity affords a holistic perspective and with it insights into many difficult concepts, such as life, consciousness, and intelligence, that have consistently eluded science and philosophy. For example, the question of whether viruses are living or nonliving is frequently debated. From the point of view of complexity, this question is meaningless, since life is a property of a large collection of entities undergoing evolution by natural selection, rather than a term that can be applied to any single entity within it. Indeed, there is growing support for basing a description of life on emergence and complexity, with some of that support coming from unexpected quarters.

"Life is not some sort of essence added to a physico-chemical system, but neither can it simply be described in ordinary physico-chemical terms. It is an emergent property which manifests itself when physico-chemical systems are organized and interact in particular ways." These are the words of the former Archbishop of York, John Habgood, a one-time physiologist who believes that the scientific world-view afforded by complexity is in many ways a more theologically comfortable notion than old ideas about vitalism.[11] In his address to the 1994 annual meeting of the British Association for the Advancement of Science, Habgood voiced the opinion that the creative work of God can be found in the growing complexity of organization during the development of organisms: "Indeed, there is a hint of this in the very first words of the first chapter of *Genesis* where God is seen as bringing order out of chaos."[12]

In the animal kingdom there is a similar debate over where to draw the line defining the place at which consciousness and intelligence begin and the actions of unthinking automata end, although it certainly depends on the complexity of the nervous system involved. Again, as the Archbishop of York put it, "One of the long-term implications of the acceptance of evolution is that we see all life as a continuum, therefore there is no precise break between other animals and ourselves. . . . The more we become aware of some very human-like capacities in animals, such as the higher apes, I think the more worried one is that they may have something at least beginning to approximate to a consciousness." For similar reasons, artificial life, artificial consciousness, and artificial intelligence cannot be assigned the simple sharp boundaries dictated by a reductionist world-view.[13]

The human brain is the supreme example of complexity achieved by biological evolution. Nowhere is the tension between reductionism and emergence more keenly felt. It is clear that the brain's functioning depends on a wealth of microscopic, cellular, and subcellular detail, yet it is equally evident that its extraordinary capabilities are emergent properties of the entire organ. Consciousness is one such property, and from it flow human emotions and spiritual values. Thus, through its emphasis on the study of the whole, rather than individual parts, complexity offers a means for transcending the materialistic limitations of reductionism and allows us instead to build a bridge between science and the human condition.

THE LANGUAGE OF COMPLEXITY

Just as we cannot understand any human language without reference to its grammar, we can only fully grasp and manipulate complexity by appealing to its own grammatical structure expressed in the language of mathematics. We gave our scientific definition of the term complexity at the start of this chapter. However, this word is often used in an infuriatingly vague sense by scientists, who may mean different things by it.[14] But within the realm of mathematics, the definition of complexity is unambiguous. There the complexity of a problem is defined in terms of the number of mathematical operations needed to solve it. Sizing up the degree of complexity of a given problem is the mission of mathematical complexity theory. It tells us whether the problem will be tractable—that is, whether it will be practical to attempt to solve it by systematic means. Because many aspects of nature's complexity are concerned with the solution of difficult problems (whether evolving the best enzyme to digest food or a visual system acute enough to recognize a nighttime predator), a deep connection exists between the concepts of mathematical and scientific complexity.

This increasing focus on complexity has shaken our touching faith in the power of mathematics, both pure and applied. The great French mathematician Henri Poincaré had already shown at the end of the nineteenth century that the motion of as few as three bodies was too complex to yield a neat, closed form of mathematical solution, foreshadowing the modern understanding of chaos theory. Many important real-world problems, such as that of a peripatetic salesman who has to devise the most economical way to visit a set of cities, can be formulated simply enough, but attempts to find their solutions by systematic means rapidly become impractical as the problem's size (for the salesman, the number of cities) increases beyond a small number. Other examples of mathematically complex problems include descriptions of how brains learn from their interactions with the external world and how evolution led to organs as complicated as the brain in the first place. Such problems lie beyond the scope of pen, paper, and analytical mathematics; because of their immense power, computers provide the only means of solving them.

To tackle complexity with a computer involves a combination of subtlety and brute force. Subtlety is needed for a precise mathematical formulation of the problem at hand; brute force entails feeding numbers and/or other symbols into this description and calculating from it, using a computer, the behavior for each and every set of circumstances desired. The essential role played by the computer explains in large part why such a rich field of investigation as the study of complexity was overlooked for so long by so many people. Before the advent of the digital computer, it was impractical for any person to sit down and feed thousands, even millions, of numbers into a set of equations describing a given complex problem—a whole lifetime would have elapsed without the likelihood of obtaining a single useful result. *The science of complexity is intricately entwined with and crucially dependent on computer technology.* The dizzying increase in computer power over the past fifty years has enabled scientists and mathematicians to model and simulate progressively more complex and more interesting phenomena.

SYMBIOSIS

The early "computer" envisaged by Charles Babbage was a machine designed to mass produce tedious arithmetical tables. Even today, many of its descendants are little more than dumb workhorses that carry out boringly repetitive tasks. This is now changing with a new generation of machine whose design and operation have taken a leaf out of nature's notebook. By using the techniques of *parallelism* and *massive parallelism*, computers come a little closer to working like the human brain.

Of all the sources of inspiration that computer scientists have drawn on, none can compare with the brain. It has long been a goal of those working in computer science to endow a machine with attributes resembling human intelligence, yet up to now all attempts at doing this have failed miserably. For while conventional computers are very good at performing tasks most people find difficult, such as arithmetic and algebra, people routinely deploy such skills as seeing and talking that even the most powerful machines cannot match.

But by mimicking the architecture of the brain—the key to its emergent properties such as intelligence and consciousness—neural network computers have managed to learn from and adapt to their experiences of the wider world. Nature's methods of adaptation and optimization, which refine the design of organisms through evolution, are now enacted by computers using genetic algorithms to tackle intractable problems. Like nature, such evolutionary computer programming techniques—which are so powerful, novel, and successful at dealing with problems of great complexity—all have *random* elements within them. This randomness leads to innovation—the discovery of smart and unexpected solutions to very hard problems.

The symbiosis between science and the computer is making it feasible to begin to understand and simulate some of the human brain's remarkable capabilities. This organ comprises a trillion cells, among which are one hundred billion nerve cells—the very stuff of thoughts, emotions, and the mind. The latter figure rivals the number of stars in the Milky Way. Yet, for the first time, scientists are now producing plausible models of certain aspects of brain function using artificial neural networks. In so doing they are lifting a corner of the veil that, since antiquity, has divided mind from matter.

These approaches to complexity are so successful that life itself is now gaining a new meaning. Neither actual nor possible life is determined by the matter that composes it. Life is a *process*, and it is the *form* of this process, not the matter, that is the essence of life.[15] As von Neumann sought to demonstrate, one can ignore the physical medium and concentrate on the *logic* governing this process. In principle, one can thus achieve the same logic in another material "clothing," totally distinct from the carbon-based form of life we know. Put another way, life is fundamentally independent of the medium in which it takes place. The implications of separating living complexity from its medium are stunning. Imagine creating an artificial world and planting in it the logical seeds of life. Given enough time, you could watch evolution in action, as a primitive self-replicating organism proliferates and mutates into a vast diversity of "offspring."

Such ideas are no longer wild flights of fancy, reserved for science fiction; attempts to create living complexity within computers are already taking place.

Chapter 2

THE ARTIST'S CODE

Go, wond'rous creature! mount where Science guides,
Go, measure earth, weigh air and state the tides;
Instruct the planets in what orbs to run,
Correct old Time, and regulate the Sun;
Go, soar with Plato to th'empyreal sphere,
To the first good, first perfect, and first fair;
Or tread the mazy round his follow'rs trod,
And quitting sense call imitating God;
As Eastern priests in giddy circles run,
And turn their heads to imitate the Sun.

—ALEXANDER POPE[1]

Mathematics is a language by which we can "Instruct the planets in what orbs to run, / Correct old Time, and regulate the Sun." It has freed us from the tyranny of magic and mysticism, allowing us to glimpse the secrets of complexity. It became the cradle of the Laws of Nature, providing algebraic patterns that could mimic the behavior of the real world. It is the sturdy undercoat that gives science its alluring gloss. Indeed, such is the power of mathematics that many have dreamed of constructing a machine that could represent any feature of the universe—no matter how complex—using only bricks of mathematical logic. This machine would provide the supreme tool for investigating and simulating the universe's complexity. Increasingly, however, mathematicians have challenged the deep belief in the infallibility of mathematics that underlies this idea.

Most people with a shred of scientific education have a surprising faith in mathematics. They believe that mathematics provides the foundation for capturing the essence of reality. One plus one equals two—whether the calculation is for one ounce of flour and one ounce of butter, or for the addition of one lead shot to another. Any circle, anywhere, has a circumference that can be expressed in terms of its diameter and the mysterious number *pi* (3.1415 . . .).

The English Franciscan Roger Bacon hailed mathematics as the door by which we enter all the sciences.[2] Mathematics is, after all, the only way we know of to carry out rigorous arguments, to extract infallible consequences from a set of statements. As such, it is a powerful language that scientists use to express relationships between measurable quantities. Most excitingly, we can also use it to anticipate real-world phenomena. From a single law of motion, we can predict the movement of a satellite through the heavens, or a baseball on earth. Using Schrödinger's equation, we can describe the shape of a negatively charged mist that cloaks the nucleus of a hydrogen atom. From Einstein's special theory of relativity, we can work out that a moving atomic clock appears to tick more slowly than one at rest.

Just as an artist employs clay, paint, and metal to create impressions of the real world, so scientists reproduce complex features of nature with pictures woven from mathematics. A painter may apply semi-opaque paint on top of a partially dry canvas to texture a tree trunk. Using rag, brush, or even fingers, a technique called scumbling can be used to depict skies, rocks, and fabrics. In a similar way, scientists can repeat a simple mathematical operation over and over to generate bifurcating structures that look surprisingly similar to those of trees or ferns.

Today, those who pursue the study of complexity are followers of a secret art in which the color, form, and motions of the universe are painted in atoms of logic. And, like their post-Impressionist forebears, these logical pointillists expect that a greater whole—the essence of the world—will emerge from the discrete elements of their mathematical code. Whether the results of these endeavors are superficial metaphors or potent insights into reality is an issue at the heart of the relationship between mathematics and the real world. For most scientists, it is enough that mathematical relationships can reliably predict real-world behavior. Recent developments, notably in mathematical logic, are increasingly testing our faith in mathematics, and making us question whether we can use it to describe all forms of complexity. Fortunately, we will show that these problems are more theoretical than practical.

THE RELIGION OF MATHEMATICS

One of the first significant cracks in the facade of mathematics came in 1936, when the twenty-four-year-old Alan Turing asked if it were possible to use purely "mechanical methods" to perform any and every mathematical and logical process. This question went to the very foundation of mathematics, for a negative answer would imply that there are profound limitations on our ability to distill real-world processes into mechanical logic and therefore simulate complex systems. We will see that his work on what came to be called the Universal Turing Machine would have bittersweet implications for the study of complexity.

Turing found that it was indeed possible to create an abstract universal machine, one that could take on the job of any other by working its way through every conceivable operation in every conceivable mathematical or logical calculus.[3] The physical embodiment of that machine was the digital computer, an object in the universe that could mimic any other part, from a supernova explosion to a surge of blood in the human heart, provided only that the process being simulated could be boiled down to a finite series of logical steps.[4] Turing himself argued that anything performed by a human "computer" could be done by a machine. In other words, a universal machine could perform the equivalent of the mental activity within that molecular machine we call the brain.

But Turing's work on the universal machine also contained profoundly bad news. He designed it to see if it were possible to use purely mechanical methods to perform any mathematical process. His answer—an emphatic no—came as a hammer blow to those who had spent decades attempting to put mathematics on firm logical foundations. Our efforts to erect a grand theory of complexity may be built on sand—certain processes in the natural world might exist whose complexity is beyond the simulation capabilities of Turing's universal computer. These processes would forever defy our attempts to model them.

Our faith in mathematics will be tested further. Turing was by no means the only mathematician to uncover its innate limitations. We will see that some of the simplest mathematics can—horror of horrors—contain randomness. And if that is not enough, we will learn that mathematics can sometimes be impossible to use. One class of problems can only be solved by someone with an eternity of free time. Other problems defy exact analysis: even Sir Isaac Newton speculated that finding some exact solutions to his laws of motion "exceeds, if I am not mistaken, the force of any human mind."[5] Indeed, we really do not know what mathematics is or why it works in our world. That begs a profound question: why is it that mathematics is so successful at describing nature? We don't know. It seems that the greatest engine of cultural change—the scientific world-view—rests on a mathematical foundation that, in many respects, is ultimately religious.[6]

THE DOOR TO THE SCIENCES

Early human mathematical activity was diverse and widespread because mathematics was found to be such a powerful language for describing the world.[7] The ancient Greeks believed mathematics had been invented in Egypt by bored priests with plenty of spare time for intellectual pursuits.[8] It was, however, in ancient Greece that abstract mathematical reasoning first fully emerged so that, for instance, the property of three-ness became independent of material objects such as cows, fingers, and spears.

The recognition that mathematics deals with abstractions may with some confidence be attributed to a religious, scientific, and philosophical brotherhood called the Pythagoreans.[9] The ancient Greek who lent his name to this order is a somewhat shadowy figure who was born on the island of Samos between 585 B.C. and 560 B.C.[10] Today his name lives on in a relationship between the sides of a right-angled triangle that holds regardless of whether the triangle is formed from a thread of silk, from three twigs, or by drawing it in the sand.[11] The Pythagoreans who first realized that such a relationship applied universally, regardless of context, must have felt that they were glimpsing a fundamental aspect of the world.

The origins of logic, a key ingredient of mathematics, can also be traced back to the Greeks. Aristotle, who was born in 384 B.C., developed a systematic framework for making logical arguments, and for a long time people assumed he had captured the laws of human thought. Aristotle's logical system contained three propositions: the law of identity (meaning "one is one," "two is two," etc.); the law of noncontradiction (both a statement and its negation cannot simultaneously be true); and the law of the excluded middle (any proposition is either true or false—it cannot be both true and false).

Logic is a peculiar subject in that it has no subject matter, since it is concerned with the study of the structure of arguments, the consideration of the validity of things that hold in virtual (logical) form but not in content. From the time of Aristotle until the mid-nineteenth century, logic was considered a branch of philosophy tackled only by those proficient in Greek.[12] Its subsequent fusion with mathematics into mathematical logic is due in great part to an astonishing individual, Gottfried Wilhelm von Leibniz (1646–1716), the philosopher, mathematician, alchemist, jurist, librarian, mining engineer, historian, archivist, wind-power pioneer, diplomat, and academician.[13] He was, as the Germans called him, an *Universalgenie*—a "universal genius."[14] In the 1670s, Leibniz put forward an ambitious scheme to formalize all human thought and mathematics into a universal language. His work was visionary, a quest that has a resonance for those who pioneered artificial intelligence 300 years later.[15]

Science, as we know it, was born a few years after Leibniz's vision of a universal language, on April 28, 1686, when Newton sent the first part of his Latin manuscript of the *Principia Mathematica* to the Royal Society in London. Some rate the *Principia* as the greatest scientific book of all time, a jewel in the crown of scientific literature. Built on the sturdy foundations laid by Galileo, it has been likened to a great edifice soaring about the ramshackle and temporary constructions around it. Newton regarded it as his greatest achievement in print. This magnificent work was rooted in a distant past, but even today is used to calculate the trajectory of various objects lofted into the heavens, from space shuttles to missiles. Its influence will undoubtedly extend far into the future.

No one who reads the preface to the first edition of the *Principia* can doubt the importance of mathematics to his enterprise. Written in

Trinity College, Cambridge,[16] it opens: "I have in this treatise cultivated mathematics as far as it relates to philosophy."[17] Newton revealed nature's apparently mathematical undercoat by laying bare the unity of the laws governing the heavens and the earth. He owed his success to a new mathematical tool he invented for the task (independently from but simultaneously with Leibniz) that today is called calculus. The core of the mathematics of his day included arithmetic, algebra, and geometry, which handled fixed and stable objects. Calculus allowed Newton to deal with change in a consistent and universal way, enabling him to describe falling apples and orbiting planets with one and the same mathematics. In effect, it allowed mathematics to be done on the fly.

Ever since his *Principia*, mathematics has been regarded as the most secure form of knowledge. Newton's successful mathematical description of motion transformed human perception of the structure of the universe beyond recognition. Galileo, on whose shoulders Newton stood, once remarked that the "great book of nature" is written in the language of mathematics. The physicist Sir James Jeans made the point equally in 1932 when he suggested that "the physical universe is constructed on mathematical lines. . . . It is then inevitable that the picture which modern science draws of the external world should be mathematical in nature; it could not be otherwise. . . . The secret of nature has yielded to the mathematical line of attack."[18]

REALITY AND MATHEMATICS

Science seemingly confirmed the position of mathematics as the most secure form of knowledge. Yet the nature of mathematics has long been debated along with why and whether the universe can be described by mathematical laws. Like the Pythagoreans, Plato (428–347 B.C.) attached great importance to mathematics. He asserted that the world we observe does not resemble the real world; according to him, our models of the universe should be manifestations of divine perfection.

The mathematical brand of Platonism holds that the foundations of mathematics itself are indifferent to the existence of our universe, indeed indifferent to our minds and evanescent existence.[19] Whereas the theorem named after Pythagoras has been independently discovered many times by different thinkers, Tolkien's *Lord of the Rings* or Handel's *Messiah* are unique products of transient human effort. Put another way, mathematics would exist even if mathematicians did not.[20]

Scientists, especially physicists, use observations of that Platonic cosmos to formulate often highly sophisticated descriptions of the real world. They browse through the range of goods offered in the mathematical supermarket, selecting those that fit with observations of real-world behavior. In the light of how successful the resulting mathematically based theories of nature are, particularly in physics, the mathematical physicist Eugene Wigner once wrote of the "unreasonable effectiveness of mathematics in the natural sciences."[21]

But there is no reason those believing in Platonism should expect the world to have a mathematical structure. "This is a great mystery for Platonists," says philosopher Bill Newton-Smith of Oxford University. "If mathematics is about this independently existing reality, how come it is useful for dealing with the world?"[22] An alternative view is that mathematics is a creation of the human mind, albeit its finest intellectual achievement. Indeed, there is no obvious reason to commit to the belief that mathematics should depict an independent, otherworldly reality. It has been argued that mathematics is an empirical science whose origins can be traced back to the rudimentary knowledge acquired by the perception of our distant ancestors, such as Egyptian bricklayers and Babylonian chandlers.[23]

The philosopher Nancy Cartwright denies that we can directly map Platonic objects and behaviors onto those in the real world. She argues in her book *How the Laws of Physics Lie* that, because the world is so complex, no system of laws can describe it: "There is no better reality than the reality we have to hand."[24] She accepts phenomenological laws, which describe what happens in the world, but rejects "fundamental" theories that are meant to furnish deeper explanations. She believes that what physicists tend to look for are simplified models that only provide highly idealized representations of the real world. "Paradoxically enough, the cost of explanatory power is descriptive adequacy. Really powerful explanatory laws of the sort found in theoretical physics do not state the truth."[25] With the recent advent of the study of complexity, these words have acquired renewed force.

The debate about the relationship between mathematics and the real world will rumble on. That felicitous symbiosis between mathematics and science will also continue for as long as it works well.[26]

But our unnatural faith in the power of mathematics may say less about its actual potential for expressing the complexity of nature in straightforward terms and more about the common scientific faith in the dogma of reductionism, which explains complex phenomena in terms of mathematical expressions about the behavior of simple units such as atoms and molecules.[27] For such highly idealized processes as compressing a cylinder of helium gas, studying the motion of mice sliding down frictionless pendula, and so on, simple mathematical expressions describing the behavior of the systems in question can be readily written down and then thoroughly analyzed.

However, the reason we study such simple systems is precisely because they admit facile analysis, and so can be easily taught, learned, and reproduced in examinations. In the past few decades, there has been a growing realization that such idealized relationships are the exception, not the rule.[28] They have the same relationship with nature as poetry does. Both elegantly provide insights into quintessential elements of reality—but in a rather mysterious and fleeting manner.

Scientists seeking to boil their observations of the world down into a final mathematical "Theory of Everything"—and even those who do not—must confront the limitations of mathematics, although this is a problem rarely considered significant by those engaged in the quest for such ultimate theories.

AUGUST 1900: HILBERT IN PARIS

Cracks within the edifice of mathematics first appeared during the early years of this century when mathematicians struggled to come to terms with a number of perplexing problems in diverse and, at the time, purely abstract fields such as logic, set theory, and the study of transfinite numbers (loosely speaking, infinities).[29] The hot Paris summer in 1900 witnessed one of the most influential bids to place mathematics on a firm logical foundation. Ironically, it ultimately led to exactly the opposite.[30] The venue was the second International Congress of Mathematicians and the hero was the representative of the *Deutsche Mathematiker Vereinigung*, David Hilbert (1862–1943), a German who had been professor of mathematics in Göttingen since 1895 and one of the truly great mathematicians of all time.

On the sultry morning of August 8, Hilbert approached the rostrum in the Sorbonne lecture hall. With slow deliberation for those who did not understand German well, he began to speak.[31] "Who of us would not be glad to lift the veil behind which the future lies hidden; to cast a glance at the next advances of our science and at the secrets of its development during future centuries?" To welcome in the new century, Hilbert listed twenty-three problems to inspire the mathematicians of the day.[32] The second problem on Hilbert's list was a challenge to demonstrate the consistency of the axioms of mathematics. In short, he wanted mathematics to be reduced to pure logic.[33]

Hilbert issued a rallying cry to the audience: "This conviction of the solvability of every mathematical problem is a powerful incentive to the worker. We hear within us the perpetual call: There is the problem. Seek its solution. You can find it by pure reason, for in mathematics there is no *ignorabimus*."[34] Hilbert's mission to stitch together a mathematical embroidery from a vast fabric of interwoven logical connections became known as *formalism*. By the winter of 1920–21, formalization of the foundations of mathematics had become Hilbert's chief goal.[35]

Renewed stimulus had come to him in the form of three papers totaling less than seventeen pages written by a young Dutchman named Luitzen Brouwer.[36] A so-called *Intuitionist*, Brouwer had challenged the formalists' belief that the laws of classical logic have an absolute validity independent of the subject matter they are applied to. Hilbert was alarmed by the gains that Brouwer's conception of mathematics was making among younger mathematicians. The source of his concern was the heresy advocated by Brouwer that the Aristotelian law of the excluded middle in logic should be rejected.[37]

Brouwer was eroding an important form of proof used by Hilbert: *reductio ad absurdum*. Instead of constructing a valid proof directly, step by logical step, Hilbert's method assumed that a logical contradiction would arise if at least one element in the proof was shown to be false. This, of course, assumes that statements are either true or false. By allowing some statements that are neither, Brouwer's scheme advocated direct proof. Many theorems of classical mathematics could be established by Brouwer's method, but in a more complicated and lengthy way than was customary. However, all pure existence proofs and a great deal of analysis would have to be abandoned. "If we would follow such a reform as the one they suggest," said Hilbert, "we would run the risk of losing a great part of our most valuable treasures."[38] He grumbled that forbidding a mathematician to use the principle of the excluded middle "is like forbidding an astronomer his telescope or a boxer the use of his fists."[39]

To counter Brouwer, Hilbert aimed to demonstrate that *finite* mathematical methods were free from contradictions. He wanted a limited set of axioms and rules of reasoning from which he would generate all mathematical truth. This procedure should be mechanical—in other words, so explicit that interpretation would not come into play. It would be impossible to use it to prove something that is false, and it was complete in that no truth was beyond its scope.[40] Step-by-step procedures for carrying out operations by blind application of specified rules are named *algorithms*, after the great scholar Abu Jafar Muhammad ibn Musa al-Khwarizmi.[41] Most are in the familiar form of computer programs. Thus, Hilbert's method was equivalent to attempting to find a *decision algorithm* for all mathematics, a process to decide whether any given mathematical statement was true or false. Hilbert's mission formed the basis for what is today known as computability theory—the study of the power and limitations of algorithms. Its consequences are of vital importance to the scientific study of complexity because they cast light on our ability to model and understand complex behaviors in nature using computers.

Hilbert continued his crusade until he retired in 1930 at the age of sixty-eight. The city of Königsberg, his birthplace, had voted to make him an honorary citizen that year and the presentation was made that autumn at the meeting of the Society of German Scientists and Physicians. Hilbert took his place at the rostrum during the opening ceremony looking, by one account, rather like Lenin. Again he reviled the "foolish *ignorabimus*," with his last words into the microphone: "*Wir müssen wissen. Wir werden wissen*" (We must know, we shall know).[42] But at almost the same time, a piece of work was completed that was to devastate the central objective of Hilbert's mission.

INCOMPLETENESS AND UNDECIDABILITY

On November 17, 1930, a journal called the *Monatshefte für Mathematik und Physik* received a twenty-five-page paper written by a logician working in Vienna. The author was twenty-five-year-old Kurt Gödel (1906–78), who was born in Brno, Czechoslovakia—at that time part of the Austro-Hungarian empire—and the paper was the first to demonstrate that certain mathematical statements can neither be proved nor disproved. There must always be statements whose truth value is undecidable. Thus, mathematical statements exist, even in arithmetic, whose validity cannot be decided *without using methods from outside the logical system in question*.[43] In effect, Gödel showed the inevitability of finding logical paradoxes in arithmetic that are the equivalent of the statement "this sentence is false."[44]

To make matters worse, he also showed that it is never possible to prove that a mathematical system is itself logically self-consistent. One must always step outside a formal mathematical calculus to determine its validity. At a stroke, Hilbert's goal of devising a decision algorithm for the whole of mathematics was shown to be a chimera; moreover, the logicist doctrine, according to which all mathematics may be deduced from the axioms of logic, was demonstrated to be incorrect. Gödel's achievement was "singular and monumental—indeed it is more than a monument, it is a landmark which will remain visible far in space and time,"[45] according to von Neumann, who was downcast by the finding.[46] He added, "The subject of logic will never again be the same." As John Barrow has written, "If we define a religion to be a system of thought which contains unprovable statements, so it contains an element of faith, then Gödel has taught us that not only is mathematics a religion but it is the only religion able to prove itself to be one."[47]

Gödel's faith in the Platonic worlds of mathematics seemed to reflect in some way his inability to deal with our transient physical existence. Though one of the greatest logicians ever, Gödel was a hypochondriac. Near the end of his life he became paranoid, almost deranged. Believing that there was a plot to poison him, he starved himself. On January 14, 1978, at one o'clock in the afternoon, he died at the Princeton Hospital in New Jersey. According to his death certificate, the cause was "malnutrition and inanition" brought on by "personality disturbance."[48]

THE UNIVERSAL COMPUTER

Despite being demolished by Gödel, Hilbert's ambitious plan for the foundations of mathematics gave us the crucial idea of computation and the notion of *computability*, originally abstract concepts that have led to modern computers and computer science, and the systematic study of complexity. A crucial aspect of Hilbert's system, hinted at earlier, can be summarized concisely in terms of his so-called *Entscheidungsproblem* (decision problem), fully formulated only in 1928 at an International Congress in Bologna where Hilbert had revived his call for placing mathematics on a firmer foundation.[49]

First, Hilbert wanted to demonstrate that mathematics was complete, in the sense that every statement could be shown to be true or false. Second, he wanted to establish that mathematics was consistent, in other words, that a statement like $1 + 1 = 3$ could never be arrived at by a sequence of valid steps. And third, he wanted to find out if mathematics was decidable, that is whether there was a definite method—a mechanical process or algorithm—that could, in principle, be applied to any assertion, and was guaranteed to produce a correct decision whether that assertion was true.

If mathematics had been consistent and complete, there would have been a mechanical way to run through all possible proofs.[50] But Gödel showed that formalized arithmetic must either be inconsistent or incomplete, ruling out this possibility. That left open the basic question posed by the third decision problem or *Entscheidungsproblem*: is there some mechanical procedure that can solve all mathematical problems? Given the formal definition of an algorithm as a recursive procedure for solving a given problem in a *finite* number of mechanical steps, it can in principle also be performed by "mindless" machines. Hilbert's *Entscheidungsproblem* was nothing less than an attempt to prove that all mathematics could be obtained from the mechanical action of algorithms acting on strings of mathematical symbols—an objective that one mathematician indignantly remarked would put mathematicians out of work.[51]

In the 1930s, the first step toward settling Hilbert's third problem was taken by Alan Turing, one of the greatest figures of twentieth-century mathematics and, from the perspective of complexity, also of science.

Turing's work on the *Entscheidungsproblem* was inspired by the nature of the mind and its relation to the physical world, which had fascinated him since his schooldays at Sherborne. Even as a ten-year-old, his eyes were opened to the possibility of machine intelligence by the book *Natural Wonders Every Child Should Know*, written by Edwin Tenney Brewster.[52] One striking passage reads: "For, of course, the body is a machine. It is a vastly complex machine, many, many times more complicated than any machine ever made with hands; but still after all a machine."[53]

Mathematicians had talked of mechanically applying rules to crack a problem. This was what Turing would do, though in an exceedingly general and idealized way. He set out to design an abstract machine able to tackle Hilbert's problem, to decide the truth of any mathematical assertion. To break down the concept of computation into steps, he came up with something akin to an old-fashioned typewriter, a device he had dreamt of inventing as a boy.[54] Just as a typewriter could type the works of Shakespeare, so Turing's abstract machine could type all of mathematics.

Turing proved that his device had the same capability as any machine devised to compute a particular algorithm. The actual feat would depend on the list of elementary operations written on the tape—in other words, the specification for a given machine. In today's language, a Turing machine would be equivalent to a computer program, for example, one that could add two numbers, or another that could determine their highest common factor. An infinite number of such Turing machines are possible, just as there is an infinite variety of computer programs (software).

Turing reasoned that to discover how an individual Turing machine would behave was itself a mechanical process or algorithm that could be carried out by another Turing machine. In other words it is possible to create one Turing machine to read through the instructions of another and carry them out. This led to the concept of a *universal* Turing machine. It can carry out any algorithm handled by a Turing machine in the sense that a modern computer can carry out any program.[55] "It all rests on a deep insight that programs and numbers are not distinct from one another," commented Andrew Hodges.[56]

Turing's work had remarkable consequences. First, the universal Turing machine was a theoretical blueprint of a computer. Second, it provided a tool for measuring complexity. Third, Turing saw that his step-by-step breakdown of what a logical operation is and of what it means to do a computation would be equally useful to describe the actions of the brain, making Turing the pioneer of artificial intelligence. Indeed, in his famous paper on computable numbers, Turing offered the bold proposal that his hypothetical machine could represent the "states of mind" of a human being. As Andrew Hodges put it, "The drift of his argument was obvious, with each 'state of mind' of the human computer being represented by a configuration of the corresponding machine."[57] Most immediate for Turing, however, was that the universal machine enabled him to tackle Hilbert's *Entscheidungsproblem*, which had preoccupied many mathematicians in the five years since Gödel disposed of the first Hilbert questions.

THE NOT-SO-UNIVERSAL COMPUTER

The universal Turing machine offered a splendid abstract device for exploring the theoretical limits of mathematics. Following consideration of Gödel's work, Turing showed that determining whether his machine would halt in a finite time after being presented with a given string of symbols, a crucial issue when tackling mathematical problems, was impossible.[58] To illustrate the halting problem, imagine a list of all possible numbers, written out one after the other as decimal expansions. This list will be infinitely long, and contains all the numbers that can be expressed as finite or infinite (i.e., terminating or nonterminating) decimals. Each number could be computed by an individual Turing machine and can thus be referred to as a *computable number*.

Turing employed the list to make a new number, which was formed by taking the first digit of the first number and changing it, then the second digit of the second number and changing it, and so on. The resulting number could not have been present on the original list. Since the list contained all computable numbers, the new number must be uncomputable. It followed, much to Hilbert's chagrin, that no "definite" method can exist for solving all mathematical questions.[59] Because of the existence of uncomputable numbers (in fact there are many more uncomputable numbers than computable ones), it is not generally possible to decide by mechanical means whether a Turing machine will halt in a finite time.

Independently from Turing, the American logician Alonzo Church had arrived at similar conclusions while working in Princeton in

1936–37, as did the Polish-American Emil Post.[60] All these approaches were soon shown to be equivalent: mathematics cannot be captured in any finite system of axioms. The world of mathematics is composed of two parts: the computable and the noncomputable, the finite and the infinite. As we will discuss later in this chapter and subsequently, it is debatable whether the noncomputable, ultra-complex elements of mathematics play a role within science. Yet even the computable aspects pose great challenges to science in *its* pursuit of the complex. Most of this book is concerned with the study of computable complexity; only rarely will we confront the noncomputable.

On September 4, 1940, Alan Turing joined the Government Code and Cypher School, where his deep understanding of mathematical logic played a pivotal role in the wartime code-breaking effort at Bletchley Park, Buckinghamshire. The cryptanalytic methods and machines he devised there, particularly in connection with breaking the German naval Enigma codes, provided him with the ideal yet entirely secret background from which to emerge into peacetime fully equipped to build a physical Turing machine—to turn the logical into the physical. We will return to this mission in the next chapter, after further exploring the limitations of mathematics.

DIOPHANTINE EQUATIONS

The demolition work on Hilbert's plan, initiated by Gödel and Turing, continues.[61] During the 1980s, Gregory Chaitin, who works on the east coast of America at the IBM Thomas J. Watson Research Laboratory at Yorktown Heights, New York, extended Turing's work on the halting problem by revealing that even the simplest version of arithmetic, which uses only whole numbers, contains intrinsic randomness. "I have found an extreme where you have no pattern, indeed complete chaos," he said.[62]

Ironically, Chaitin's inspiration came from the tenth problem listed by Hilbert: is there a systematic way of deciding whether a given Diophantine equation has a solution in whole numbers such as 1, 2, and 3?[63] Diophantine equations, named after a Greek mathematician, are those that only handle whole-number quantities, a restriction with important consequences.[64] If we want to find the solution of an equation in which quantity *A* squared plus quantity *B* squared equals one, we would have an infinite number of solutions. However, if we restrict *A* and *B* to positive whole numbers, there are only two: *A* is 1 and *B* is 0, or vice versa.

In 1970, an answer to Hilbert's problem was provided by Yuri Matijasevich at the Steklov Institute in Leningrad.[65] In a two-page paper that was the tip of an iceberg of mathematical endeavor, Matijasevich showed that no such method to find a whole-number solution to an algebraic equation exists.[66] In fact, he demonstrated that Hilbert's tenth problem was directly equivalent, in a deep way, to the halting problem for Turing's machine.

Chaitin exploited this profound relationship to use a Turing-type approach for tackling a variation of Hilbert's tenth problem. Instead of asking whether an algebraic equation has a whole-number solution, he asked whether it had a finite or an infinite number of whole-number solutions. To do this, he first specified a universal Turing machine capable of handling whole numbers. The result was a "universal Diophantine equation," although elephantine might be a more accurate adjective—it contains 17,000 variables, and can only be crammed into 200 pages of text.[67] The equation, in effect, represented a computer that, given an infinite amount of time, could calculate whether a program will halt.

Instead of exploring the halting problem one program at a time, Chaitin considered the ensemble of all computer programs and investigated the *probability* of any one such program halting for a family of universal Diophantine equations, each created by altering a single parameter. This probability is expressed by an infinitely long binary number called omega. Each equation will have finitely many solutions if a particular bit (binary digit) of the omega number is zero and infinitely many if it is a one. "Each equation in the family is perversely constructed. Whether it has a finite or infinite number of solutions is so delicately balanced that there is no reason it should come out one way or another," noted Chaitin.[68] Whereas Turing had found that the halting problem could only be resolved in an infinite time (in other words, it is undecidable by finite means), Chaitin found that the halting probability is algorithmically *random*: expressed as a binary number, its series of ones and zeroes is indistinguishable from a series of heads and tails obtained by tossing a coin.[69] That means that the answers to the questions about these elephantine equations must be random, too.

Chaitin expanded on the coin-tossing analogy to emphasize just how accidental and random "simple" integer mathematics can be. "This halting probability is maximally unknowable," he said. "Each outcome of a coin toss tells you nothing about any future outcomes or any past outcomes. It is exactly the same way with knowing whether each of my equations has a finite or an infinite number of solutions. The answer is an irreducible mathematical fact, not connected with any other mathematical fact."[70] Surprisingly, even arithmetic possesses random elements. As Chaitin put it, "God not only plays dice in quantum mechanics, but even with the whole numbers!"[71] This finding also has a deep resonance with the discovery in science of deterministic chaos, a seeming oxymoron we will return to later. It also has a corollary that alarms pure mathematicians. Sometimes the only way to explore mathematics is by trial and error—to conduct experiments.

TRACTABILITY

So far we have looked at tears in the very fabric of mathematics. There are, however, other, more practical, defects. It may be possible for us to solve a problem in principle but, even with a computer, impossible to do it in a realistic time frame. This is the issue of algorithmic "complexity," as opposed to computability, and it concerns the amount of time required to solve a problem using a Turing machine. While the work of Gödel, Turing, Church, and Chaitin highlighted the issue of computability, algorithmic complexity is very much a workaday issue. It turns out that among computable problems, certain classes are much more difficult to solve than others. The number of calculations, expressed in terms such as "floating point" operations[72] (the number of operations performed by an algorithm), indicates the amount of work needed to solve a given problem.

The algorithms used to describe computable problems can be divided into two classes, based on the length of time it takes to find the solution to a problem as a function of some number N that measures its size. The good news is for problems that are *polynomial* (i.e., an algebraic power of N, e.g., N squared, N cubed, etc.), when they are said to be *tractable*—the length of time required to crack them does not become unbounded as the size increases.[73] Problems solvable in polynomial time are said to be in the class P. Mathematicians and computer scientists blanch when the

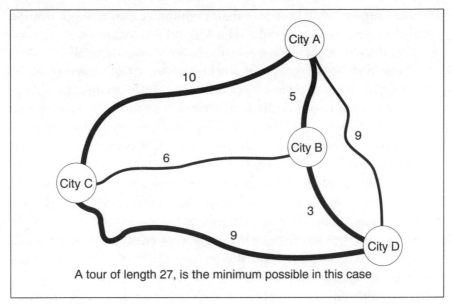

A tour of length 27, is the minimum possible in this case

Figure 2.1   The famous traveling salesman problem, for the case of four cities.

time required to solve a problem increases in an *exponential* fashion (something to the power of *N*). These problems are called *intractable* because the time required to solve them rapidly spirals out of control. Even the raw power of a computer has little effect. These problems, which are not solvable in polynomial time, are said to be in the class NP.[74]

Probably the most famous example of an NP problem is the traveling salesman problem.[75] This is the mathematical expression of the dilemma faced by a salesman who has to visit *N* cities once only in such a way as to minimize the total distance traveled: he has a penny-pinching boss and has to keep fuel costs as low as possible. (See Fig. 2.1.)

The problem is easy to formulate but, try as they might, no computer scientist or mathematician has come up with a well-behaved deterministic algorithm (i.e., one that is not random and allows only one outcome for any given set of circumstances) that can find solutions to it on a computer in polynomial time. For a handful of cities and roads it may be easy to determine the salesman's solution because not that many options exist. If the number of cities is five, say, a computer could easily calculate the twelve possibilities. With ten cities, there are 181,440 possibilities. However, even with the number-crunching power of the fastest available machine, the time required to solve the problem rapidly spirals out of control. For just twenty-five cities the number of possible journeys is so immense that a computer evaluating a million possibilities per second would take 9.8 billion years—around two-thirds of the age of the universe—to search through them all.[76]

A large number of other real-world problems are known to lie in this category, many of them concerned with similar optimization problems. For the owner of a printed circuit-board factory, for example, the function that needs to be maximized is manufacturing efficiency. For a pharmaceutical company, the function that must be maximized is the snugness of the fit of a drug molecule within a target protein found in the body. Frequently, elegant analytical mathematics is unable to provide us with a simple way of locating these optima, since hard optimization tasks are intractable (NP) problems.

But how do we know for sure that a given problem belongs in the NP class? Just as schoolchildren will always maintain that many mathematical problems are impossible to solve, the labeling of a problem as NP could say more about a mathematician's incompetence than anything about the problem itself. In fact, this question is one of the foremost open problems of contemporary mathematics and computer science. Little progress has been made to prove the conjecture that NP problems possess no solutions available by conventional—deterministic—algorithms in polynomial time. However, even without such a proof, at the very least the belief (supported by algorithmic experimentation) that a problem is of the NP variety implies that a significant breakthrough will be needed to solve it.

For those who want to reproduce the workings of the world within computers, these NP problems are at first sight rather depressing. Remarkably, however, as we will discuss in later chapters, nature has provided us with tools to tackle them.

COMPUTABLE MATHEMATICS

For more than 200 years after Newton, it was thought that a complete theoretical understanding of any mechanical process could be achieved by using sufficient mathematical ingenuity to analytically solve the equations describing those processes. Methods based on pencil, paper, and thought alone have always been regarded as the highest form of mathematical thinking.

It came as a shock when Henri Poincaré proved in 1889 that our supplies of ingenuity are severely limited—even for mathematical problems that are computable, in Turing's sense—as soon as one tries to analyze the motion of as few as three bodies, such as the Sun, the Moon, and the Earth. Mathematicians call this an intrinsically non-integrable system—a technical way of saying that an exact solution cannot be found by analytical methods. Thus, any motion involving more than three bodies, let alone the millions upon millions of molecules that constitute a gas-filled balloon or a glass of beer, would be even more difficult to describe. This limitation is a powerful reminder of the danger of the reductionists' attempts to render everything as simple as possible. By concentrating on oversimplified models—such as vicars sliding down frictionless bannisters—which yield to the seductive power of analytical mathematics, mathematicians easily missed the whole richness and complexity of the real world.

Fortunately, there is little danger of this occurring today. Thanks to the modern computer, it is possible to investigate the behavior of nonlinear systems, such as the motions of Poincaré's three bodies, where complex behavior begins to emerge. But once we leave the realms of compact mathematical formulas and resort to computers, new issues arise. How complicated must an algorithm (and hence computer program) be to produce a numerical solution to a problem? Of course, there will be clever algorithms and less smart ones that can all solve any given problem; the best (optimized) program will lead to the solution being found in the least computer time. But suppose, at least theoretically, that we confine our attention to the most efficient algorithms capable of solving all problems. Armed with our understanding of intractability, it does not take a moment's reflection to realize that different problems will be executed in different amounts of computer time. This is one of the most powerful, and least controversial, measures of complexity.

The complexity of algorithmic expressions can also be directly related to science: what is the degree to which behavior observed in an experiment can be algorithmically "compressed"?[77] Compression, a concept pioneered by Gregory Chaitin when he was fifteen, gives us a useful tool for measuring how easily we can reproduce natural phenomena and in principle offers a cast-iron definition of complexity.[78] In the words of Chaitin, "The complexity of something is the size of the smallest program which computes it or a complete description of it. Simpler things require smaller programs."[79]

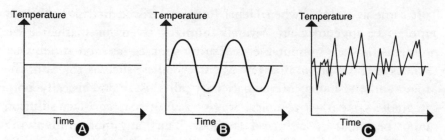

Figure 2.2   Increasing incompressibility. Three examples of time series, showing the variation of temperature with time. A is constant, B is periodic, and C is chaotic.

There are plenty of analogies in everyday life. One can tell unwanted guests that one is tired, has a great deal of work to do, or perhaps remind them that they are wanted elsewhere. Or one can shout "Get out," an effective compression that achieves the desired result. However, compression can be inadequate in more complex examples. The most interesting and extreme is that of incompressibility. "Boy meets girl, family intervenes, the couple dies" is a compression that falls far short of describing Shakespeare's *Romeo and Juliet*. The only way to enjoy the play is to go and see it yourself, not read a review or synopsis. Similarly, there are mathematical objects that are algorithmically incompressible. One good example is Gregory Chaitin's omega number, the algorithmically random quantity that we encountered earlier in connection with Diophantine equations.

We can cast compressibility in a form that is more familiar to scientists or mathematicians, placing emphasis on the time rather than information or program size required to tackle a problem. A simple set of physical examples is shown in Figure 2.2, which displays the values of some observable quantity—let's say temperature and its variation with time in three different situations.

The constant values shown in (a) can be compressed into a very straightforward algorithm: if the total length of the time series is N, the algorithm simply says "repeat the initial value N times." In (b), the regularly repeating sequence can be compressed by an algorithm that says "repeat the sequence from 0 to $(N/2)$ twice," which is a little more lengthy than for case (a). There are, however, examples of phenomena that look random but are in fact "chaotic" (c). Scientifically, this term has a quite specific meaning that we will explore later in more detail. For present purposes let us note that, although the sequence may appear entirely random, an underlying organization is implicit within its structure. It is possible to achieve a degree of algorithmic compression for a chaotic system. This is a relief for any scientist daunted by the complexity of the real world: some very complex phenomena can be captured in a small set of deterministic equations. Chaotic systems are exquisitely sensitive to initial conditions, and their future behavior can only be reliably predicted over a short time period. Moreover, the more chaotic the system, the less compressible its algorithmic representation.[80] Finally, if a property of the real world is algorithmically incompressible, the most compact way of simulating its behavior is to observe the process itself.

THEORIES OF NOT VERY MUCH

Inspired no doubt by the ubiquity of computers, it has become fashionable in some circles to argue that the view of the laws of nature as being mathematical in an abstract Platonic sense should be replaced with one based on natural processes performing computations. This perspective regards the laws of nature as a form of software running on the hardware of matter and energy. As Barrow points out, a modern definition of science could then be stated as the search for compressibility; a final "Theory of Everything" would be a law representing the ultimate possible compression.[81]

The elementary particle physicist and Nobel laureate Steven Weinberg is one scientist who dreams of a final compression. He remains optimistic, even though he admits that (reductionist) physics is today stuck in a rut. It possesses theories of only limited validity, which are tentative and incomplete, "but behind them now and then we catch glimpses of a final theory, one that would be of unlimited validity and entirely satisfying in its completeness and consistency. We search for universal truths about nature, and, when we find them, we attempt to explain them by showing how they can be deduced from simpler truths."[82]

This beguiling quest is predicated on the view that it should be possible to condense all natural phenomena into at most a handful of mathematical equations—everything that is amenable to scientific analysis should be explicable from this core of law. Yet not everything will be computable in practice: attempts to describe washing potatoes under running water or casting dice run into the problem of computational complexity. Tellingly, Weinberg admits that "wonderful phenomena, from turbulence to thought, will still need explanation whatever final theory is discovered."[83]

An even deeper problem is posed by Turing, whose work on computability is dismissed by Weinberg as belonging "more to mathematics and technology than to the usual framework of natural science."[84] Difficult issues are raised because in the Platonic realm of abstract mathematics problems exist whose solutions are not computable. At the very least, this obliges us to ask, even if we had a correct final scientific theory of our world couched in mathematical terms, whether we would be able to compute everything of interest from that theory.

THE CURSE OF TURING

How much is our mission to understand complexity hamstrung by the limitations on mathematics found by Gödel, Turing, and Chaitin? It would be deeply depressing if noncomputability was important in real-world processes, particularly the most complex phenomena, such as those associated with life and the intelligence that enables humans to do mathematics. Can computers simulate the moment when an inanimate process becomes a living one? Can a computer really simulate human "states of mind"? Is it possible to design computers that are living according to any reasonable definition of the term? Is it possible to produce computers that have intelligent properties, including consciousness? If life and consciousness do involve more than any single algorithm seems capable of achieving, is it nevertheless possible that the properties of sufficiently complex computing machines may display such attributes?

Very little work has been done to find out how the theoretical limitations on mathematics we have outlined in this chapter restrict our ability to describe real-world complexity. Perhaps the most serious suggestion that they do pose problems comes from the work of Marian Pour-El and Ian Richards at the University of Minnesota. They have shown noncomputable solutions do exist for certain well-known equations of mathematical physics, including the wave equation, which describes the way in which electromagnetic waves propagate—such as light, ultraviolet, infrared, X-ray, and radio waves.[85]

Yet it is difficult to see how this finding will restrict the study of complexity. The (computable) initial conditions necessary for yielding noncomputable solutions are somewhat unusual.[86] But as these mathematicians point out, an experimenter would observe a computable sequence of events and so would see nothing unusual occurring. Since science is about what we observe, even these peculiar mathematical results do not place significant limitations on the computer in its application to science. Even though pure logic has shown the limits of mathematical reasoning, it is significant that Turing himself never took these limits very seriously in his writings on artificial intelligence.[87] Chaitin also believes that the arithmetical randomness he discovered is a feature of the Platonic world, not the one we inhabit.[88]

The problems of mathematics all stem from the relationship between the finite and the infinite, or the discrete and the continuous. Turing's concept of a computer is based on a discrete, digital device with a *finite* number of internal states; with the possible exception of quantum computers (see Chapter 3), his universal machine defines the limits of what is mathematically computable. Nevertheless, physical systems that run on "classical," that is Newtonian, laws, are continuous rather than discrete. They therefore have access, in principle, to an infinite number of internal states. Such *analog* computers could therefore be expected to outperform universal Turing machines by computing results that digital computers cannot. Examples of this enhanced ability would include solving halting problems and generating noncomputable numbers.[89]

Despite the theoretical limitations mathematicians have found in digital computation, they have witnessed a profound change in their relationship with the computer. They are increasingly placing faith in computer proofs that are impossible to check and survey by hand. The most famous problem tackled in this way was put forward in 1852 by a student, Francis Guthrie, who found that four colors seemed to be all that were needed to color any map if neighboring countries are to be distinguished. In 1976, with the aid of a computer, Kenneth Appel and Wolfgang Haken at the University of Illinois solved this "four-color problem" by confirming that four colors did indeed suffice.[90] The computer not only functioned merely for the purpose of dealing with the very lengthy and tedious aspects of this problem; it also found ways of solving parts of the problem that were more ingenious than anything the human mathematicians could devise.

At the time, this use of computers in mathematical proofs was unprecedented. It has triggered heated debate among mathematicians because proofs are supposed to be free of empirical contamination: how do we know that the machine did not make a mistake? Is a proof really a proof if no human can check it with pen and paper? The four-color proof held up to scrutiny, despite rumors that it was faulty.[91]

That is not to say that computer proofs are free from error. For example, Clement Lam, with Larry Thiel and Stanley Swiercz of Concordia University in Montreal, provided a supercomputer "proof" of one problem that required checking 10^{14} cases. The computer was estimated to have generated undetected errors at the rate of approximately one every 1,000 hours. "Since we used several thousand hours of computer time, we should expect a few errors," said Lam. "I try to avoid using the word 'proof' and prefer to use the phrase 'computer result' instead. However, many mathematicians are willing to accept the result as a proof!"[92] And human mathematicians themselves frequently make mistakes, as the recent example of the announcement, subsequent retraction, and then reaffirmation of a proof of Fermat's last theorem demonstrates eloquently enough.[93] Eventually certain kinds of proofs will be accepted as watertight only if they have been checked by computer.

The reliance on computers has actually been strengthened by the discovery of profound intrinsic limitations within mathematics. These limitations by no means imply that mathematics is a failed enterprise, just that it is more like physics than many mathematicians would care to admit. As Chaitin puts it, "Mathematicians do not have a pipeline to absolute truth." He holds the treasonable view that mathematicians are increasingly having to conduct experiments to find the most appropriate principles and axioms in the light of their experience.[94] Physicists have already learned to accept this: Newton's laws work well in everyday circumstances, say when driving a car, but they know that Einstein's equations take over at speeds approaching that of light, and quantum mechanics operates on the scale of atoms and molecules.

By using the computer to expand their horizons, pure mathematicians are accepting its reliability. This is profoundly reassuring for anyone troubled by the implications of Turing's work in our quest to understand real-world complexity. Explorers of the Platonic realm will increasingly depend on an object in the physical world to make new discoveries on the frontiers of mathematics. It is to this object, the digital computer, that we must now turn, as the only tool capable of fully uncovering nature's complex splendor.

Chapter 3

THE ARTIST'S PALETTE

It is the age of machinery, in every outward and inward sense of that word.
—THOMAS CARLYLE

"The Great Wave off Kanagawa" is the best known of the many thousands of illustrations and prints produced by Katsushika Hokusai, the renowned Japanese master artist and printmaker (Fig. 3.1). The wave was one of a series of wood-block prints published in the early nineteenth century that marked the pinnacle of his endeavors as a member of the *Ukiyo-e* (pictures of the floating world) school. This print vividly depicts the inordinate mixture of many small- and large-scale liquid structures within a breaking crest of water.

In his own way, Hokusai spent a lifetime on a quest for mastery of the secret art of complexity. Fifteen years before his death in May 1849, he wrote: "From the age of six I have had a mania for sketching the forms of things. From about the age of 50, I produced a number of designs, yet of all I drew prior to the age of 70 there is truly nothing of any great note. At the age of 73 I finally apprehended something of the true quality of birds, animals, insects, fishes and of the vital nature of grasses and trees. Therefore, at 80 I shall have made some progress, at 90 I shall have penetrated even further the deeper meaning of things,

Figure 3.1   Classical turbulence. Hokusai's Breaking Wave.

at 100 I shall have become truly marvellous, and at 110, each dot, each line, shall surely possess a life of its own."

The study of fluid dynamics, the quest to understand the "true quality" of how fluids flow and shock waves propagate, obsessed John von Neumann as well. As one of the leading detonation consultants at the outset of the war, he wanted to understand each dot and line of the shock waves that occurred during explosions and implosions.[1] To portray what happened when a breaking wave of explosive energy compressed the critical mass at the heart of a nuclear weapon, von Neumann required a rich palette for depicting the immensely complex behavior that might ensue. In short, he needed the digital computer. Today, his name is almost synonymous with its birth.

As we have already seen, nature's complexity defies the capabilities of pencil-and-paper mathematics. The digital computer is essential for our exploration of universal mathematical principles underpinning complex phenomena. Using these logical machines, we can simulate, create, and control an enormous wealth of complex processes, from the natural to the artificial. We will see how the development of the digital computer has gone hand in hand with the exploration of complexity, and how the ever-increasing speed and power of these machines have enabled us to unravel progressively more challenging phenomena.

Even though the computer's impact has been overwhelming only in the last quarter of the twentieth century, the historical roots of this extraordinary tool date back many centuries. Indeed, its vast potential was glimpsed by such pioneers as Ada Lovelace and Charles Babbage, to the extent that there is a tantalizing hint that they conceived of the most exotic application of modern computers in a field called artificial life. However, it is unlikely that they could have imagined just how different the latest generation of computing technology is from their original mechanical machines. By basing new computer designs on flickering beams of laser light that rely on our most successful theory of physics—quantum mechanics—and exploiting exotic types of logic, such as fuzzy logic, we stand today on the threshold of a new dawn in our ability to apprehend complexity.

ORIGINS

The first primitive computing aids arose in the everyday transactions of ancient times. First were scratchings and other markings on wood or stone, and markers such as tokens, pebbles, knotted cords, and tally sticks.[2] These simple origins can be glimpsed in the word "calculus," which is Latin for pebble. Then came the abacus, derived from the Greek for board or tablet, abakos or abax. It started out as lines scratched in the ground, with pebbles representing the numbers of units, tens, hundreds, or whatever. In the Middle Ages came the wire and bead abacus such as the Chinese suan-pan, Japanese soroban, and the Russian stchoty.[3] Then, in 1614, a powerful method to ease the burden of calculation was developed by John Napier (1550–1617), the baron of Merchiston in Scotland. He found a way to reduce the arithmetical operations of multiplication and division to the simpler ones of addition and subtraction. His secret was the logarithm (often called log). Numbers could be multiplied by first adding their logarithms and then reconverting the resulting logarithm to give the answer. Division involved subtracting logarithms. In the 1620s, logarithmic scales were put to work by the country clergyman William Oughtred, who developed the slide rule that became synonymous with dreary mathematics lessons for generations of pupils.

The seventeenth century saw several attempts to mechanize calculation, notably by Wilhelm Schickard,[4] who outlined his Calculating Clock in 1623; Blaise Pascal, who in 1623 built a machine that carried out addition; and Gottfried Leibniz, who commissioned a craftsman in 1672 to build a device capable of mechanical multiplication. Their motivation was expressed by Leibniz, who had been studying astronomy and soon recognized how tedious it would be to calculate tables of celestial data by hand: "It is unworthy of excellent men to lose hours like slaves in the labor of calculation which could safely be relegated to anyone else if machines were used."[5] He was so proud of his invention that he thought of commemorating it with a medal bearing the motto "Superior to Man."[6]

These early pioneers of mechanized calculation faced a more fundamental limitation than reliability. None of the calculators so far described was automatic—each required an experienced operator to intervene during the process of number crunching. The first machine that was truly automatic, solved the "carry problem," and even printed the results, was conceived in 1821 by a dazzling polymath who went on to dream up a device strikingly close in concept to what Alan Turing would envisage a century later.

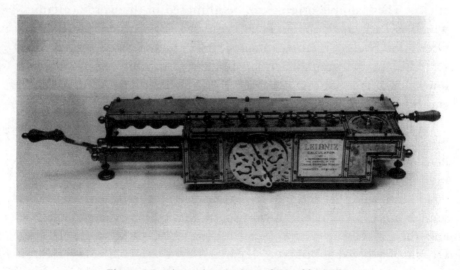

Figure 3.2   An early calculator devised by Leibniz.

THE STEAM-POWERED COMPUTER

Charles Babbage (1791–1871) is often hailed as the father of modern computing. He tried not once but again and again to build "engines" that would, he hoped, compute and print tables to meet the burgeoning demands of scientists, bankers, actuaries, engineers, and navigators, the last being particularly important for Britain, a leading seafaring country.

Computing was not the be-all and end-all of his interests. Babbage invented a cow catcher for trains, consisting of a "strong leather apron attached to a powerful iron bar,"[7] the flashing lighthouse, the speedometer, and footwear for walking on water,[8] and he devised techniques for operational research, the objective analysis and improvement of the behavior of complex systems and organizations. He was also an outstanding cryptologist, founder of the Royal Statistical and the Royal Astronomical Societies, and an overzealous campaigner against the playing of brass bands, bagpipes, and other "instruments of torture" in London, which he blamed for the loss of a quarter of his intellectual abilities.[9]

Charles Babbage was born on December 26, 1791, in his father's house in Walworth, Surrey, 500 yards from the famous hostelry of the Elephant and Castle. In his day, this spot south of the Thames had a rural feel that is a far cry from the faceless concrete and tarmac sprawl of what is now part of central London. Babbage's lifetime was to span the transformation of England from an almost entirely rural nation to by far the most highly industrialized country ever seen.

Babbage secured a place at Trinity College, Cambridge, in 1810. There he became involved in several student groups: a Sunday breakfast society, a Ghost Club, The Extractors (dedicated to extracting its members from the madhouse, if unlucky enough to be committed), a rowing club mockingly called Babbage's Tom Fools,[10] and a mathematics society, originally suggested as a parody of serious-minded religious groups, which marked the birth of Babbage's interest in computing.

Babbage's autobiography reveals that even as a child he had been bewitched by machinery, relating how he was entranced by two silver automata at an exhibition.[11] This fascination would grow into a vision for mechanizing mathematics some time in 1812 or 1813:[12] "One evening I was sitting in the rooms of the Analytical Society, at Cambridge, my head leaning forward on the Table in a kind of dreamy mood, with a Table of logarithms lying open before me. Another member, coming into the room, and seeing me half asleep, called out, 'Well, Babbage, what are you dreaming about?' to which I replied, 'I am thinking that all these Tables (pointing to the logarithms) might be calculated by machinery.' "[13] Babbage's actual "genesis episode" took place in 1821, during a meeting with the astronomer Herschel: "My friend Herschel, calling upon me, brought with him the calculations of the [human] computers, and we commenced the tedious process of verification. After a time many discrepancies occurred, and at one point these discordances were so numerous that I exclaimed, 'I wish to God these calculations had been executed by steam!' "[14]

It was 1823 before Babbage began work on his machines, however. He was aware of simpler calculators dating from earlier times but his first machine was far more advanced than anything that preceded it, giving new meaning to number crunching. Numbers were to be held in columns of toothed wheels in the Engine, which was worked by turning a handle. It was called a Difference Engine because it was designed to compute and then print tables of numbers by the mathematical method of "finite differences." Babbage used the word "Engine" to describe his brainchild because he planned to couple it to that other wonder of the age: steam.

Babbage thought the project to build Difference Engine No. 1 would take two or three years. The effort foundered in 1833, after a dispute with his chief engineer Joseph Clement, whom Babbage suspected of padding his bills.[15] The project was eventually dropped in 1842. Failure to complete the difference engine, by far the largest government-sponsored private research project of the time, was itself significant; had the engine been completed, other state-backed projects might have followed.[16]

There were, however, spin-offs. The project accelerated the development of advanced machine tools and machining techniques.[17] It also led to the construction of the first known automatic calculator. In 1832 Babbage had instructed Clement to assemble a portion of the Engine using about 2,000 of the intended 25,000 or so parts.[18] This finished part of the unfinished engine is an engineering triumph, one of the finest examples of precision engineering of the time that still works. It can be found in the London Science Museum.

In the autumn of 1834, Babbage had a vision of a computer that enthralled him for the rest of his days. This was to be Babbage's crowning achievement, albeit a purely intellectual one: the blueprint of the Analytical Engine. Unlike the Difference Engine, which tackled a restricted class of computations, the Analytical Engine was a general-purpose programmable computing machine. Like the first modern computers, it had a separate store for holding numbers and a central processing unit—the "mill"—for working on them;[19] a control and operations section; it even used punched cards as a program for input and printed output. Babbage believed it would be possible to skip or repeat the cards according to the results of the machine's calculations. This idea of "conditional branching"—if one condition is fulfilled, calculate the problem this way, if not continue as before—gave his Engine the ability to feed results back into the calculations, and thus the property of a universal machine of the kind Turing envisaged. Babbage himself said with some prescience, "As soon as an Analytical Engine exists, it will necessarily guide the future course of the science."[20]

His inspiration came from the automatic loom invented by the Frenchman Joseph-Marie Jacquard in 1801 that was so successful that 11,000 were built by 1812 in France alone.[21] As Babbage put it, the loom "is capable of weaving any design which the imagination of man may conceive."[22] Skilled weavers were replaced by using a string of pasteboard cards punched with holes attached to a rotating block over the looms. Where holes appeared, colored threads could be picked up by hooks to be woven automatically into patterns of any design.

"We may say most aptly that the Analytical Engine weaves Algebraic patterns, just as the Jacquard loom weaves flowers and leaves," commented the Countess of Lovelace, Augusta Ada Byron.[23] This charming, temperamental, and aristocratic hostess, who exuded all the romantic appeal expected of the legitimate daughter of the poet Lord Byron, likened the action of the Engine to musical composition, maintaining that it "might compose elaborate and scientific pieces of music of any degree of complexity."

Ada was an exceptional woman by the standards and values of the day, striving to work alongside Babbage rather than for him. When they first met at one of his parties in June 1833, she already knew something of his Difference Engines, having been taught mathematics by Augustus de Morgan.[24] She toured British industry in the Midlands and attended lectures by the popularizer of science Dionysus Lardner so that she could understand the Engines more completely. In time, she and Babbage became friends.

A turning point in their relationship occurred in 1840, when Babbage lectured in Turin on the Analytical Engine. An Italian military engineer, Luigi Menabrea (later prime minister of a united Italy) took on the job of writing a paper on the engine, which was published a year later in Geneva.[25] Following a suggestion from Charles Wheatstone, Ada translated Menabrea's paper from French into English. It was not so much the paper as her comments that impressed Babbage. Ada's insights ran to two and a half times the length of Menabrea's host material and, in Babbage's opinion, were good enough for publication in their own right, not merely as footnotes for an article in "Taylor's Scientific Memoirs."[26]

His suggestion implies that he realized Ada deserved more recognition for her contributions.[27] She is indeed credited with thinking about the abstract structure of the instructions that would control the machine, discerning the conditional branch we encountered earlier and the loop, a method used to repeat instructions.[28] Others, meanwhile, have noted the contrasting lack of evidence that Babbage thought about programming as a practical method for using his machines.[29]

If Ada had made an independent contribution to computing, it would be easier to judge the significance of her notes. This was not to be. After living for some time in seclusion, Ada began to seek excitement outside her marriage.[30] She became the mistress of Andrew Crosse, an inventor and gambler who was one of Babbage's wide circle of scientific friends. By Derby Day in 1851, she was £3,200 in debt and forced to pawn the family diamonds. She was also mortally ill with cancer of the womb, first detected the year before. Her mother, Lady Noel Byron, acted with a striking lack of compassion and tried to stop Ada from taking opium as a pain killer. She dismissed Ada's servants and forbade Babbage to enter the house. She even forced her drugged, dying, and helpless daughter to change her will.

Ada Lovelace died at the end of November 1852, yet her contribution lives on. We owe much of our knowledge of the Analytical Engine to her, including the only clear statement we possess of Babbage's views on the scope of his engines. A computer language is now named after her,[31] and, in the words of Noble laureate Arno Penzias, "The Countess may be said to have been the world's first computer programmer."[32]

Unfortunately for Babbage, successive prime ministers turned their backs on his project. He is often portrayed as a nineteenth-century victim of the "British disease"—the nation's chronic inability to exploit its ideas. However, Babbage may also have been a victim of his over-fertile mind. A few years before his death in 1871, the Cambridge mathematician John Moulton (later Lord Moulton) paid a visit and saw parts of the original calculating machine. "I have not finished it because in working at it I came on the idea of my Analytical Engine, which would do all that it was capable of doing and much more," Babbage told Moulton. "After a few minutes' talk we went into the next workroom where he showed and explained to me the working of the elements of the Analytical Machine," Moulton recalled. "I asked if I could see it. 'I have never completed it,' he said, 'because I hit upon the idea of doing the same thing by a different and far more effective method, and this rendered it useless to proceed on the old lines.' Then we went into the third room. There lay scattered bits of mechanism but I saw no trace of any working machine. Very cautiously I approached the subject, and received the dreaded answer, 'It is not constructed yet, but I am working at it, and will take less time to construct it altogether than it would have taken to complete the Analytical Machine from the stage at which I left it.' I took leave of the old man with a heavy heart."[33]

The bicentenary of Babbage's birth in 1991 produced conferences, exhibitions, plays, and commemorative stamps. The Science Museum in London completed part of Babbage's mission—the building of his Difference Engine No. 2, after a six-year effort that cost £300,000. The head of the project, Doron Swade, was gripped by the challenge Babbage's design posed: was it ahead of the abilities of Victorian mechanical engineering? "Our endeavor finally bore fruit in November 1991, a month before the bicentenary of Babbage's birth. At that time, the device—known as Difference Engine No. 2—flawlessly performed its first major calculation. The success of our undertaking affirmed that Babbage's failures were ones of practical accomplishment, not of design."[34]

The theoretical ideas that underpin the Analytical Engine have secured Babbage's place in the history of computing. The Difference Engine No. 2 that now stands in the Science Museum is a triumphant piece of engineering, a monument to the rigorous logic used by its inventor. Babbage even proposed an array processor for carrying out many simultaneous calculations[35] and invented a symbolic language for describing his machine's structure and behavior.[36] This laid the basis for, and gave a clear direction to, the invention of symbolic logic, the application of logical rules to the manipulation of symbols that takes place in every computer. Among the nineteenth-century mathe-

Figure 3.3 Number cruncher. Babbage Difference Engine No. 2.

maticians who pioneered this field, we now turn to the contributions of another Englishman.

BOOLE'S LAWS OF THOUGHT

Bertrand Russell once claimed that George Boole discovered pure mathematics.[37] Boole is better known for the algebras named after him, and as one of the pioneers of modern logic. His celebrated, though cumbersomely entitled treatise *An Investigation of the Laws of Thought on Which Are Founded the Mathematical Theories of Logic and Probabilities* opened up new possibilities, not just for mathematical logic, but also for the unborn subject of computing.[38]

Born in Lincoln, England, in 1815, the son of a shoemaker, Boole became a teacher as soon as he finished his schooling to help support his parents. By the accounts he gave in later life, it was early in 1833 that he first contemplated the ideas that were to grow into his major contribution to mathematics—the expression of logical relations in symbolic or algebraic form.

Although he had only rudimentary mathematical training, by the age of eighteen Boole was working his way through advanced treatises, learning his mathematics in the same way he learned his Greek and most of his Latin—by self-instruction. Publication in mathematics journals helped forge his reputation. His 1844 paper "On a General Method in Analysis" won the first gold medal of the Royal Society for the most significant communication in mathematics between 1841 and 1844.[39] Boole's achievement was all the more impressive since, by one account, the Council of the Royal Society had virtually rejected the paper until one member, Thomas Davies, argued that the fact that the author was poor and unknown was no reason to dismiss his work.[40]

Early in the spring of 1847, Boole's long-dormant interest in the connections between mathematics and logic was reawakened and inspired by a controversy between the great logicians Sir William Hamilton and Augustus de Morgan after Hamilton accused de Morgan of plagiarism.[41] Boole synthesized their approaches in his book *The Mathematical Analysis of Logic, Being an Essay Towards a Calculus of Deductive Reasoning*.[42] Though a somewhat rushed work, it marked the beginning of symbolic logic in the modern sense. Instead of focusing on shape and number, Boole enlarged the scope of mathematics by interpreting symbols as classes or sets of objects, concepts that, as we saw in Chapter 2, ultimately shook the very foundations of mathematics.

The effort launched by *The Mathematical Analysis of Logic* culminated in his book *An Investigation of the Laws of Thought*, published in 1854. In his own words, Boole wanted "to investigate the fundamental laws of those operations of the mind by which reasoning is performed; to give expression to them in the symbolical language of a Calculus, and upon this foundation to establish the science of Logic and construct its method . . . and finally, to collect from the various elements of truth brought to view in the course of these inquiries some probable intimations concerning the nature and constitution of the human mind."

Boole was proposing the following. First, logical statements, or propositions, should be expressed in the form of equations. Then, algebraic-like manipulations of the symbols occurring in these equations are carried out as the fail-safe method of logical deduction. Thus, reasoning is performed by calculating or, put another way, logic is reduced to algebra. For example, if the symbol x represents the class of all white objects and y the class of all round objects, Boole used xy to represent the class of objects that were both white and round. Similarly, if m is the class of all men and w the class of all women, $(m + w)$ is the class of all people.[43] If e represents the class of all Europeans, we see that class of European men and women is the same as the class of European men and European women, that is, $e (m + w) = em + ew$. Boolean algebra provides the basis for analyzing the validity of logical propositions, since it captures the binary or *two-valued* character of statements that may be either true or false.[44] Wittingly or unwittingly, Boole had discovered a new kind of mathematics, a variety ideal for the manipulation of information within computers.

Boole died in 1864. It was an untimely end,[45] perhaps hastened by his wife's homeopathic beliefs, and the wider significance of his work lay neglected for decades.[46] Although in 1867, the American logician Charles Peirce noticed that Boole's two-valued logic lent itself to a description of electrical switching circuits, it was not until 1937 that the American mathematician Claude Shannon, working at MIT, John Atanasoff at Iowa State College, and the German engineer Konrad Zuse independently showed that binary numbers (0 and 1) combined through Boolean algebra could be used to great effect in the analysis of electrical switching circuits and thus in the design of electronic computers.[47] Switching circuits are networks of relay contacts that occur in now outdated electronics, such as telephone exchanges, railway signaling, and the very first digital electronic computers. Boole's algebra is the natural language to describe the switching that computers depend on to do their work.

Switches are either on or off. We need to consider only these two values—there is nothing in between. In the case of the relays in old-fashioned circuitry, that meant the difference between the presence or absence of a current or voltage, for example, usually represented by the binary symbols 1 and 0, respectively. One can completely determine the function of a switching circuit by specifying how the states of a certain number of output switches depend on the states of the input switches. Hence it is possible to represent the circuit and how it operates by Boolean algebra, just as in the analysis of logical structures. The switches and circuits can be thought of as gates that open or shut in response to various bits of data according to simple logic, which is why they are called logic gates. By connecting logic gates together so that the output of one controls the input of another, complex tasks can be carried out.

In this way George Stibitz and Samuel Williams, at Bell Laboratories, were able to assemble logical gates that could add, subtract, multiply, and divide binary numbers. Similarly, in the living room of his parents' Berlin apartment, Konrad Zuse used binary mathematics to avoid Babbage's ten-spoked gear wheels and developed his simpler Z1 mechanical calculator, followed by electromechanical varieties that were used in the engineering of V2 rockets. Today digital computers and electronic circuits are designed to implement this binary arithmetic. Underlying all these diverse applications is a simple mathematical theory: the symbolic calculus of functions of binary variables, created by Boole.

BABBAGE AND BOOLE IN PERSPECTIVE

With the luxury of hindsight, it is tempting to meditate on what might have happened to the evolution of the computer—and complexity studies—if the hardware of Babbage's analytical engine had been married with the software suggested by Boole's algebra. It remains a mystery why, at the very least, they did not exchange views on mathematics, where the two men had a great deal in common.[48]

Babbage had speculated, even around 1820, on the type of symbolic algebra that Boole ultimately developed.[49] He also appreciated Boole's work: in 1847, on the publication of Boole's paper, *The Mathematical Analysis of Logic*, Babbage had scribbled in the margin of his copy: "This is the work of a real thinker."[50] Personal links might also have led to a friendship: Boole's wife claimed that Babbage was a friend of her father, while Augustus de Morgan and Sir Edward Bromhead provided other links. And in the autumn of 1862, Boole visited Babbage to view the Difference Engine.

The failure of Babbage and Boole to spark off each other's ideas is not the only puzzle in the development of the modern computer.[51] Although Babbage grasped the major principles of computing by the 1840s, the first electronic computer did not appear until more than a century later. Yet the pioneers of this computer made little use of Babbage's work; many were unaware of its existence. The lack of a direct line of development from Babbage to the present day has led some to suggest that Babbage should not be regarded as the father or grandfather of modern computing—he is more like a great uncle.[52]

Babbage did, however, leave a few fingerprints on the twentieth-century computer. His work was known to Alan Turing;[53] and Howard Aitken, who helped develop the American Mark I computer for the production of wartime ballistics tables, was particularly enthusiastic about Babbage's engines. The leading character in the genesis of the electronic computer, John von Neumann, was familiar with the inspiration for the Analytical Engine. His father, as director of one of Hungary's leading banks, Magyar Jelzalog Hitelbanka, had financed the introduction of Jacquard's looms into the country.[54]

VON NEUMANN'S INSPIRATION

By the late 1930s, von Neumann was established as one of the leading international mathematicians and mathematical physicists, though he had had no significant contact with computing. In the years that followed he would make such an impact that computers are now referred to as having a "von Neumann architecture," a token of respect for his lucid and masterful analysis of their structure and operation.[55] To be sure, others were involved in the development of the first machines. But, aside from Alan Turing, von Neumann was the dominant figure in defining the structure of the digital computer that now controls the industrial plant, the pilot's cockpit, and the video games arcade.

Von Neumann had several notable sources of inspiration for his move from mathematics to computing. First was Alan Turing, who had impressed him when they met in April 1935 during a visit to Cambridge.[56] Another inspiration was his correspondence from 1939 to 1941 with his friend Rudolf Ortvay,[57] director of the Theoretical Physics Institute of the University of Budapest, which drew parallels between the brain and electronic calculating equipment.[58] Finally, von Neumann was struck by a paper that aimed at providing a logical basis for brain function, which had been inspired by Turing's paper on computable numbers.[59] That paper, written by the Chicago neurologists Warren McCulloch and Walter Pitts,[60] impressed him with its potential for bringing mathematical regularity to the highly complex and ill-understood phenomena within the brain.[61]

THE TURBULENT BOMB

This interest in computation was given devastating purpose when von Neumann joined other scientists engaged in the huge effort of developing the first nuclear weapon. He soon drafted his friend Stan Ulam to help work out the hydrodynamics (i.e., fluid mechanics) of implosion involved in designing explosive lenses for compressing the plutonium at the heart of the weapon.[62] "The hydrodynamical problem was simply stated," said Ulam, "but very difficult to calculate—not only in detail but even in precise orders of magnitude."

Von Neumann initially used a traditional "top-down" method to work out the hydrodynamic properties, one that deals with overall fluid behavior rather than all the dizzying details of its component molecules. At directly observable scales of time and length, a fluid (in other words, a gas or liquid) appears to be continuous—its temporal and spatial properties are smooth, and most physical properties, such as its speed and local density, can take on any finite value. Differential equations are the standard tool employed by mathematicians and physicists since the time of Newton and Leibniz to describe how fluids—or indeed anything else—evolve in time. Von Neumann used a set of differential equations that describe the flow properties of a fluid, which included the well-known Navier-Stokes equations. Rather than dealing with the detailed behavior of an overwhelming number of molecules, these equations are derived by assuming that the bulk fluid is a continuous entity and then applying straightforward laws based on the conservation of mass, momentum, and energy that date from the time of Isaac Newton.[63] Although this amounts to an approximation, the differential equations accurately describe the profile of velocities within a flowing fluid, or an explosion, with all the complexity that it entails.

Von Neumann had become engrossed in the daunting task of solving these fluid-flow equations in the late 1930s. But, like Ulam and others, he found that mathematical solutions are hard to calculate, because many equations in fluid dynamics are nonlinear. A linear relationship between two quantities implies a directly proportional one: ten oranges cost ten times more than a single orange. But nonlinearity implies a disproportionate effect: in the case of a bulk purchase, for example, one crate of oranges may be free if nine are purchased, four with every sixteen bought, and so on. A concomitant effect of this nonlinearity is feedback—the outcome of an effect goes on to trigger more change. In this case, the size of the discount affects the number of crates one may decide to buy.

Such feedback between elements of the overall process—here between the cost and the person doing the purchasing—often leads to quite unexpected behavior. For example, if the discount is very great, you may decide to enter the marmalade business. Feedback comes in two varieties. One is the reinforcing power of positive feedback—the loop of amplification from microphone to loudspeaker that turns a whisper into a deafening howl. There is also the damping effect of negative feedback. A colony of rabbits can multiply so fast that it exhausts all available food, triggering a population crash. Within a fluid, a combination of positive and negative feedback effects intertwine and interact to create the endless swirls, eddies, and smaller vortices of turbulence.

While linear equations are rather easy to explore and can often be completely solved using pen, paper, and present-day mathematical methods, nonlinear equations are not. Until the computer, many people sought rough-and-ready descriptions of nonlinear processes by a familiar approximation route—they linearized them.[64] Such linear approximations are anemic, missing the very features of nonlinear systems that allow such fascinating processes as turbulence to occur. To tackle nonlinearity requires what appear to be hammer and tongs—numbers are fed directly into the equations and answers are worked out for each and every set of circumstances by sheer brute force.

Von Neumann recognized that a computer could act as an indispensable tool for exploring fluid dynamical complexity. Besides playing a key role in the genesis of the computer, he also pioneered and promoted its use for solving problems numerically, using the results as a "heuristic" guide to deeper mathematical theorizing. He would make a hypothesis about the equations under investigation, attempt to select some crucial special situations, employ the computer to solve these cases, check the hypothesis against the results, form a new hypothesis, and iterate the cycle. This use of computing to augment experimentation revealed physical and mathematical order in the mayhem of fluid dynamics.[65] As he remarked, "Really efficient high speed computing devices may, in the field of nonlinear partial differential equations as well as in many other fields, which are now difficult or entirely denied access, provide us with those heuristic hints which are needed in all parts of mathematics for genuine progress."[66] As usual, he was right: today people working in all areas of complexity use computers in an experimental capacity for gaining insights where no other routes are feasible.

Von Neumann was eventually drawn into the development of the infant electronic computer as the result of a chance encounter with the mathematician Herman Goldstine, an army officer with the Ballistic Research Laboratory in the late summer of 1944. Von Neumann learned of the existence of the Electronic Numerical Integrator and Computer, or ENIAC, while awaiting his train to Philadelphia.[67] "The conversation soon turned to my work," recalled Goldstine. "When it became clear that I was concerned with the development of an electronic computer capable of 333 multiplications per second, the whole atmosphere of our conversation changed from one of relaxed good humor to one more like the oral examination for the doctor's degree in mathematics."[68] Goldstine quickly arranged a visit on August 7 so von Neumann could see the progress on the ENIAC, a horseshoe-shaped affair that would eventually consist of 17,468 vacuum tubes, 70,000 resistors, 10,000 capacitors, 1,500 relays, and 6,000 switches.[69]

ENIAC was completed in November 1945, almost three months after the Japanese surrender that followed the first use of nuclear weapons. It was 200 percent over budget and became the star of a U.S. Army advertisement that trilled: "You should see some of the ENIAC's problems! Brain twisters that if put to paper would run off this page and feet beyond . . . addition, subtraction, multiplication, division— square root, cube root, any root. Solved by an incredibly complex system of circuits operating 18,000 electronic tubes and tipping the scales at 30 tons!"[70]

A serious drawback to the otherwise successful ENIAC was that it had to be rewired for each new problem—something that could only be carried out very slowly. Von Neumann helped the Moore School team refine their ideas for its successor, the Electronic Discrete Variable Computer, and in the spring of 1945 he offered to write an analysis of its logical design. Von Neumann's June 30, 1945, description of an early blueprint for a working computer, entitled "First Draft of a Report on the EDVAC,"[71] made a massive impact; indeed it influenced computer design for the next half century.[72] Goldstine called it the most important document on computing because it was the first written outline of the stored program idea. It stated: "While it appeared that various parts of this memory have to perform functions which differ somewhat in their nature and considerably in their purpose, it is nevertheless tempting to treat the entire memory as one organ."[73] By this emphasis von Neumann stressed the concept of storing both data and operating instructions—the computer's program—in a centrally located memory.

The importance of putting data and instructions in the same medium cannot be overstated. One could thus quickly alter a computer's function by rewriting its operating instructions (its software). This idea required a major conceptual leap, since computer scientists of the time regarded instructions and data as being different in nature.[74] Instead of operating by means of knobs, switches, and levers or indeed via continuously varying electronic signals, as the already available analog computers, the von Neumann machine simply followed the operating instructions located in memory.

Just as a mathematical algorithm is usually envisioned as proceeding in a sequential step-by-step fashion, so von Neumann thought of operations being carried out in series rather than in parallel.[75] His computer tackled one piece of digital information at a time. Von Neumann's stored programming method therefore falls far short of being an electronic analog of the human brain, where (as we will see later on) a great many different processing tasks whir away concurrently. Nevertheless, most computing devices today share a common architecture derived from the work of von Neumann.[76]

As the war drew to a close, the United States invited British scientists to see the ENIAC. The first to arrive was J. R. Womersley, superintendent of the Mathematics Division at the British government's National Physical Laboratory, NPL. In the wake of the briefing and the copy he received of von Neumann's report, he set out to organize a computing project. The first scientist he recruited was Alan Turing, who knew better than anyone the computer's full potential. The universal machine he had foreseen a decade before could be programmed to meet every challenge without the need of altering the hardware. In this vital respect, Turing was years ahead of anyone else. He wrote in 1945: "There will positively be no internal alterations to be made even if we wish suddenly to switch from calculating the energy levels of the neon atom to the enumeration of groups of order 720."[77]

Alan Turing studied von Neumann's EDVAC paper and then wrote one of his own, completing a comprehensive plan for a large computer called the Automatic Computing Engine (ACE) by the end of 1945. The name, coined by Womersley, was a nod in the direction of Babbage. The extraordinary implications of this enterprise were spelled out by Turing three years later when he wrote another report for the NPL with the evocative title *Intelligent Machinery*.[78] In that paper and at a lecture to the London Mathematical Society on February 20, 1947, Turing's central thesis was that the brain itself could be regarded as a computer. Although during these earliest stages of development the computer evidently had to rely on a human to program it to carry out instructions, he anticipated the possibility of his machine *learning* by a kind of supervised training procedure, much the way a pupil learns from a schoolteacher. He envisaged how computers could learn from experience, using a Darwinian analogy to express how a machine might explore solutions to problems by aping the way that nature combines genes to boost the chance of survival. He pointed out that "a human mathematician has always undergone an extensive training. This training may be regarded as not unlike putting instruction tables [programs] into a machine. One must therefore not expect a machine to do a very great deal of building up of instruction tables on its own. No man adds very much to the body of knowledge. Why should we expect more of a machine? Putting the same point differently, the machine must be allowed to have contact with human beings in order that it may adapt to their standards."[79]

Turing saw no reason why a machine could not thus become intelligent. And he argued that Gödel's theorems and his own answer to Hilbert's *Entscheidungsproblem* were irrelevant to this issue, because intelligent machines displaying "initiative" would not slavishly and infallibly carry out preprogrammed instructions. His commitment to the ultimate goal of artificial intelligence was made clear in a letter he wrote to a neurophysiologist: "In working on the ACE I am more interested in the possibility of producing models of the action of the brain than in the practical applications to computing."[80]

Unfortunately, ACE had little chance of fulfilling Turing's original vision, let alone the dream he outlined in "Intelligent Machinery." To achieve anything significant required a substantial team effort involving cooperation between mathematicians and electrical engineers. Lacking the wartime exigencies that had driven Turing's cryptanalytic work, the ACE project became bogged down in bureaucratic inertia and misunderstanding. A scaled-down version was constructed but by the time the "Pilot ACE" went into operation, Turing had left for Manchester University where a more promising effort was being led by his former Cambridge teacher and colleague Max Newman.

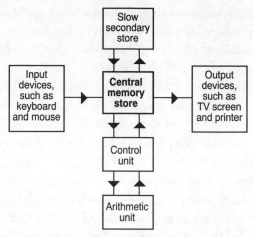

Figure 3.4   Von Neumann's computer architecture—the layout of a typical serial machine.

Though von Neumann had blazed a trail with his EDVAC report, the machine was not built until 1952. Its key hurdle was storing the programming instructions. The group at Manchester University, acting on an idea by its chief engineer Freddie Williams, decided to use three cathode ray tubes, similar to those bottle-shaped tubes in radar displays or today's TV sets, to store information in the form of dots without the need for expensive, custom-made equipment. On June 21, 1948, the Manchester team prepared and ran a test program, marking the creation of the first electronic, digital-stored–program machine, the Manchester Mark 1. The following May, another of Turing's former Cambridge colleagues, Maurice Wilkes, head of the University Mathematical Laboratory at Cambridge, put into operation the Electronic Delay Storage Automatic Calculator machine, the first stored-program machine capable of doing useful work. Around it Wilkes developed the world's first computing service.[81]

VON NEUMANN'S LEGACY

Most people who use computers have little time to worry about how they work. Their detailed architecture can vary considerably, depending on the manufacturer, though the majority are, at heart, von Neumann machines (see Fig. 3.4). These computers receive their instructions and data from an input device such as a keyboard or magnetic disk; these data are then stored in the computer's memory. The unit of memory is called a binary digit, or *bit*, that is, a 0 or a 1, on or off (hence the term "digital"); a *byte* is defined as eight bits of information.

All information is stored at specific memory addresses, rather like post office boxes. The shuffling around of data and arithmetical number crunching is performed by the *central processing unit* (CPU), which carries out instructions in its arithmetic/logical units to the beat of a clock. When the computer program requests some data, the CPU proceeds to locate, retrieve, and process them before reaching for the next specified bunch. Each transaction is slavishly carried out in a strictly serial—that is, sequential—manner by this postal-clerk CPU, hence the expression "serial" processing. It works exceptionally well for "mindless" numerical tasks; nevertheless, if von Neumann's brain had worked that way, he would never have designed the computer architecture that bears his name.

A computer programmer writes a program by first constructing an *algorithm* that provides a procedure for solving the problem of interest, as discussed in Chapter 2. The algorithm is then coded into a suitable programming language that will enable the computer to understand and successfully execute instructions. Usually, this involves one of the "higher" programming languages, so called because they are reasonably close to human language and do not require very specialized programmers.[82]

Since the CPU is composed of a set of Boolean logic circuits, it is capable only of performing elementary arithmetic operations such as addition, subtraction, multiplication, and so on. Thus, the original programming language fed into the computer must be first converted by means of a "compiler" into a machine-readable assembly language that employs the basic bit strings and arithmetic operations the computer can understand.[83]

FROM FLOPS TO TERAFLOPS

During the last half-century the speed of computers has increased more than a trillion times.[84] From around 1940 to 1955, the electronic switches that handled and processed all the data in computers were glass vacuum valves, which looked a little like lightbulbs, and com-

Figure 3.5 The walk-through computer. One way to explore the details of what is going on inside computers seen in universities, businesses, offices, and homes is to visit The Walk-Through Computer, a splendid exhibit in a former wool warehouse on Boston's historic waterfront, home of the world's only dedicated computer museum. Costing around $1.2 million, it is two stories tall—fifty times the size of a normal desktop personal computer. It also contains the biggest microchip in the world, measuring 7.5 feet square. The brainchild of Oliver Strimpel, the executive director, the exhibit was opened in June 1990 to show visitors how computers work in a way that would not alienate anybody. While walking inside the computer, past drooping ribbon cables, you can watch it operating in slow motion, from the whir of a giant spinning disk as it retrieves stored data to the frenetic electronic activity in banks of chips. Looking through a window on a chip, you can see an enlarged picture of the actual electronic circuit lines etched in silicon. That image fades and computer graphics take over, showing the step-by-step operation of the chip, from retrieving data located in the memory to projecting it onto the monitor.

municated with their masters through punched card and paper tape. The next five years saw the introduction of transistors—the pea-sized replacements for valves developed at Bell Telephone Laboratories— and magnetic core main memory, with new ways to communicate such as magnetic drums and disks. In 1959, Jack St. Clair Kilby of Texas Instruments invented the integrated circuit (IC). In his patent he declared that it was now possible "to achieve component densities of greater than thirty million per cubic foot as compared with five hundred thousand per cubic foot, which is the highest component density attained prior to this invention."[85] The first IC computers started appearing by the mid-1960s. At the beginning of the next decade, Marcian Hoff of Intel pioneered the idea of cramming a computer onto a chip. The microprocessor was born, a device that became the universal motor of electronics, whether computers, scales, or doorbells.

The result of Hoff's hunch was Intel's 4004, a chip containing 2,250 transistors able to process four bits of data at a time and around 60,000 operations per second. Then followed the 8008 and, in 1973, the first popular microprocessor, the Intel 8080, which was capable of eight-bit processing using the components crammed into its $1/10$" $\times$ $2/10$" chip. This formed the basis of the revolution in personal computing. The trend to put more components on chips has continued with VLSI—very large-scale integration of 10,000 or more logic gates within a single chip. This integration accelerated the development of advanced materials science and electrical engineering techniques that are aimed at controlled deposition of electronic circuitry on minute scales.[86]

In rough terms, every eighteen months or so new microprocessors appear that work at twice the speed of their forebears. Traditional and cumbersome "mainframe" computers have given way to the desktop "workstation" for many scientific forms of computing.[87] It is now possible to have on your office desk a computer far more powerful than the mainframes of the 1970s and early 1980s that had to be shared by the employees of a company or scientists in a research laboratory. Sun Microsystems pioneered this change in direction and the older companies such as IBM, DEC, and Hewlett-Packard have had to follow suit. Another development has also taken place: the construction of "supercomputers" designed to be dramatically larger and faster than ordinary mainframe computers.[88] We have now entered the era of the "teraflop" machine, one that can handle one trillion floating point operations per second.

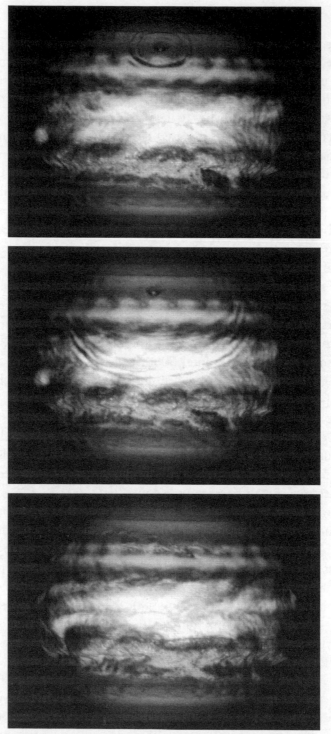

Figure 3.6 Cosmic impact. Computer simulations of the effects
of the comet Shoemaker Levy's collision with Jupiter.

TURBULENT TIMES

Despite this astonishing progress during the fifty years since the visionary work of von Neumann, simulating turbulent fluid flow in a realistic way is still largely beyond the capabilities of today's computers. Improving our understanding of turbulence remains an outstanding fundamental problem in modern science. The principal hurdle we must overcome is modeling the immense complexity of turbulent structures, where the behavior of the fluid at the tiniest of length scales has a dramatic effect on behavior over larger distances. As we saw earlier, the Navier-Stokes equations describe the flow of continuous fluids; digital computers are inherently discrete, however, so they necessarily *approximate* these equations by dividing up space and time into a grid and only take into account fluid behavior at points on this grid. Thus, the computational fluid dynamicist faces a dilemma: if she subdivides space too far, then the time taken to obtain a solution to the equations will be prohibitively long because she has a very great number of points to consider; but if she settles for a cut-off that is too coarse, then she will omit important details that affect fluid behavior such as eddy structures. In fact, the time taken to perform a fluid simulation increases as a high power of the Reynolds number, a measure of the propensity for the apparent mayhem of turbulence. At sufficiently high Reynolds values, the flow becomes turbulent and the Navier-Stokes equations are then a major headache to solve.[89] Even though this is not, technically speaking, an intractable (NP) problem, for any reasonably sized problem on any existing computer it is impossible to consider Reynolds numbers above around 10,000, a value corresponding merely to the onset of turbulence, rather than the fully developed form.[90]

Progress on this problem, whether through clever analysis or more likely with more powerful computers, will lead to immediate and very practical benefits, since fluid dynamics underpins so much of modern engineering and technology. We use it to model the flow of air around a car, of liquid sodium through the hot heart of a nuclear reactor, and of water around a stealthy submarine. It substantiates the meteorology used for weather forecasting, and not just on earth.

A group at MIT predicted a rippling effect on Jupiter's weather from the collision of comet Shoemaker Levy 9 three months before the first impact occurred in July 1994.[91] An expanding ring was indeed seen around the largest impact sites, moving at about 454 meters per second (plus or minus 20 meters); the team had predicted 400 meters per second. A different computer model gave a better account of the sharp structure of the observed rings but the wrong speed. Comparing these predictions to reality may provide valuable insights into the structure of the Jovian atmosphere.

Turbulence is regarded as one of the "grand challenge" problems in contemporary high-performance computing, along with such esoteric problems as the large-scale structure of the universe and lattice-gauge theories in elementary particle physics.[92] Meeting such challenges is a powerful spur for progress. "Electronic component speeds and densities have improved by a factor of more than 10^5 in the last half-century," according to computational fluid dynamicists Karniadakis and Orszag. "This development is unrivalled in other fields of human endeavor; if automobiles had undergone similar improvements, today a Cadillac would sell for less than a penny, or it would be capable of a peak speed in excess of 1% of the speed of light, or one gallon of gas would suffice for about ten trips to the Moon. Despite these remarkable advances in computer electronics, the motivating force behind computer developments has been (and will likely continue to be) the grand challenge applications."[93]

The grandeur of the challenge of fluid dynamics is easy to illustrate using the basic unit of machine speed called flops (floating point operations per second). A medium-sized calculation that today takes five hours on a Cray YMP running at 200 million flops (200 megaflops) would require twenty-eight years on an Apple Macintosh personal computer running at 0.004 megaflops, and would require less than four seconds on a teraflop (million million flops) machine. Similarly, the solution of an aerodynamic flow problem that could be expected to take about two weeks on a teraflop computer, would take several centuries on a Cray YMP and millennia on an Apple Macintosh.[94]

Developments in computer architecture, chip design, computer logic, even the fundamental physical basis and medium of computing, will greatly accelerate the pace of research by enabling bigger numerical problems to be tackled in greater detail. Today, our most powerful computers routinely process data and do simulations at the rate of a billion flops. In the next few years, we will see the introduction of numerous teraflop computers.[95] Among many tasks, they will be able to perform Navier-Stokes calculations of air flow past a complete aircraft at moderate to high Reynolds numbers, when turbulence sets in. In this way, it will be possible to put a prototype passenger jet through its paces within a computer, without the need for wind-tunnel tests.[96]

ARCHITECTURES

Connect a few thousand electronic circuits on a wafer of silicon and you have a microprocessor—the chip now at the heart of every conventional von Neumann computer. Because this single processor can perform only one job at a time, there are fundamental physical limits to the speed at which it can operate. So-called *scalar processing* requires a computer to complete one task before it can turn to the next. Most of its electronic circuitry stands idly by as one operation proceeds through the arithmetic functional units of the CPU. This gives rise to the von Neumann bottleneck, a restriction that, almost since its creation, has inspired the search for alternative computer architectures.

One significant early advance came through the efforts of Seymour Cray, who founded Cray Research in 1971. He was able to introduce substantially improved performance in his Cray-1 supercomputer, released in 1976, by moving away from the time-honored von Neumann architecture. The Cray-1 employed pipelined *vector processing*, which allowed a new arithmetic operation to be carried out immediately following the one in front without waiting for full completion of the first calculation. In this way, its speed is dramatically increased.[97]

Another innovation in computer architecture arose from an effort at the University of Illinois, that had been underway since 1949, producing the ILLIAC I, a computer based on von Neumann's EDVAC. Two decades later a team, under Dan Slotnick, designed ILLIAC IV, which was to become the first massively parallel computer. In contrast to the Cray-1's single vector unit, the ILLIAC IV had sixty-four identical scalar computers that operated in parallel, each with its own processor and memory. The principal attraction of parallelism is disarmingly simple: two processors should do a job twice as quickly as one. There is no limit to the number of processors a single computer can contain, so there appears at first sight to be no limit to the speed at which a single computer could operate.

Parallel processing has intrigued designers of computers since the days of von Neumann.[98] Early on, however, it was pointless to build a parallel machine while a single processor could tackle a problem at the same price or less. Between 1950 and 1965, electronics improved so fast that computer manufacturers were able to crank up the performance of their products over a thousandfold without resorting to parallel processing. The ILLIAC IV was perhaps ahead of its time, costing more than $100 million in 1990 money. But by the early 1980s, parallel processing was beginning to look both technically and economically attractive. By 1986 around two dozen companies were building parallel supercomputers in an attempt to meet the needs of the small but lucrative market for high-performance computing. A crowded bandwagon had begun to roll.

Probably the toughest job in designing a parallel computer is ensuring that it operates synchronously—that all its processors work in harmony. This can be achieved by singling out one processor as the "conductor" of the orchestra of processors, ensuring that they simultaneously carry out the same task on different pieces of the data in their local memories. Computer buffs call this approach SIMD (single instruction, multiple data). For example, if a parallel-data computer had to calculate the number $d = (x^2 + y^2 + z^2)^{1/2}$ for many different objects, it could store the coordinates (x, y, z) for each object in a separate processor. Each processor would then square, sum, and take a root for its own values. A SIMD computer thus produces as many results as it has processors in the same time that a conventional computer would take to produce a single result.

One example of such a computer was Thinking Machine Corporation's first Connection Machine, a black cube 1.5 meters on each side that was covered with winking red lights. The ambition of the corporation's founder, Danny Hillis, was to make a computer so sophisticated "that it could be proud of me."[99] To highlight the ability of the Connection Machine's 65,536 relatively unsophisticated processors, Hillis argued from the example of the brain, which could outperform a supercomputer despite its use of neurons, which are much slower than electronic transistors. The secret was parallelism. If, for example, a computer was forming an image from an array of 256 by 256 dots, then a simple imaging processing operation by a conventional machine would take 65,536 steps. "The Connection Machine, on the other hand, assigns a single processor to each point of the image. Since every operation can be performed on all the points simultaneously, a calculation involving the entire image is as fast as a calculation involving only a single point."[100]

An alternative approach, known as MIMD (multiple instructions, multiple data), allows each processor to attack *different* parts of the overall task concurrently. One processor's program may be quite different from those its neighbors are executing. In MIMD, a great deal of message passing occurs between processors in order to deal with the exchange of data and instructions concerning the relative progress made by each processor to ensure synchronization. The massively parallel machines from Intel and nCUBE have this architecture.[101]

Both SIMD and MIMD architectures have advantages and disadvantages. For example, in SIMD it is often possible to find many processors lying idle while only a few execute instructions, slowing down the overall program performance. In MIMD, one must program each processor separately, thus potentially vastly increasing programming complexity when using large numbers of processors.[102] Both approaches face a deeper problem. Consider by way of analogy the following two examples. Suppose it takes a man working alone one month to build a single house; it will then take him a year to build a dozen houses. But a workforce of twelve can achieve the same result in a month. However, even though it takes one woman nine months to gestate and give birth to a baby, it does not follow that nine women can produce a single baby in one month—morphological development and growth are inherently sequential processes. The pregnancy paradox shows that some tasks cannot be performed more quickly by sharing the work, as the designers of parallel computers discovered.

A parallel machine may in theory be able to tackle thousands of different tasks but the reality can be disappointing, depending on the internal communication between processors. "We have literally billions of active parts in our Connection Machines," said Danny Hillis. "They all have to work together properly for the machine to work."[103] When one processor needs to use another's results, it must talk to its colleagues to discover when that result will be ready and where it will be found. Dialogue between processors gets in the way of their sums. With only a handful of processors to worry about, it is easy enough to carve up the work among them so that bureaucratic chitchat is minimal. But as the number of processors increases, the chatter grows to a din as processors spend more time telling each other about their progress on a given task than they do solving it. Aside from the cacophony problem, parallel machines have problems with memory. When information is needed for a task, the processor summons it. With thousands of processors clamoring for attention, the machine grinds exceedingly slowly.

These problems are all aspects of Amdahl's law, named after Gene Amdahl who in 1967 argued that the boost obtained by adding more processors is limited by the eventual development of bottlenecks.[104] While this is the inevitable consequence of throwing more and more processors at a problem of fixed size, the law has been shown to be incorrect if one scales the number of processors in proportion to the size of the workload, since the speed then increases by close to the theoretical limit. This has led to the notion of scalability—scaling up the number of simple processors, for example, by networking together several desktop workstations—as and when a problem demands it, without any alteration to the software itself. Potentially it is a very practical concept, since it can be used to dynamically and inexpensively generate tailor-made supercomputer capabilities from a set of desktop processors.

RISC ASSESSMENT

Computing has also seen another revolution in the past few years, a move back to basics in chip design. Since the invention of the microprocessor, rococo chips have increasingly acted as the brains of ever more capable computers. The complexity of conventional chips rivals that of the infrastructure of a major city and offers programmers many thousands of different instructions for composing their software.

The instruction set is an essential component of a computer's operating system. Formulated in assembly language, it sends commands directly to the computer's microprocessor. Some instructions do such simple tasks as shifting information from memory into a register where it can be used in mathematical calculations. As a rule of thumb, however, around 10 percent of the chip's instructions perform at least 90 percent of the work. As long ago as the 1960s, manufacturers began to think about devoting more silicon-chip power to the numerically intensive tasks and, by 1971, an IBM team led by John Cocke realized that one only had to include a small portion of the full instruction set in the hardware:[105] the flashy accessories that reduce performance could be placed in software. Thus was born the reduced-instruction-set computer (RISC).

A debate among computer architects resulted,[106] hindered by the lack of valid comparisons between RISC and conventional CISC (complex-instruction-set computers).[107] During the past decade, RISC gradually took hold, however, in workstations introduced by IBM, notably the RS/6000 family, and those by Hewlett-Packard and Sun Microsystems. By cutting frills and the time needed to translate complicated instructions into simpler ones, RISC machines are many times faster than conventional computers. Simplicity also confers other advantages. Less circuitry is needed, so that a RISC microprocessor is easier to design and its components can be packed closer together to increase speed even further. The internal clock that keeps the various parts of a chip working in unison can also run faster on RISC machines.[108]

There can be drawbacks. To computer users the most important question is whether a machine can run their software, since overheads in terms of time needed to write code are heavy. A RISC chip must therefore look like its more baroque counterparts in preexisting software. Consequently, RISC machines must be able to emulate conventional processors without sacrificing that all-important gain in speed. Some computer manufacturers have avoided the compatibility problem by limiting the use of RISC chips to supporting roles. Others, including IBM, are aggressively pursuing this RISCy route in the belief that they can provide more computing power per dollar, given the right software. This sounds like a circular problem, but already many software companies are writing suitable code. And there are now many RISC machines available from companies such as DEC, Bull, and Apple as well as IBM.

FUZZY LOGIC

Just as the architecture and components of the von Neumann computer have been updated, revised, and changed, so there is now a similar movement in another apparently essential element. As we have seen, computer logic is traditionally restricted to the yes/no and one/zero, or true/false and black/white, of binary lore. Conventional computers depend on binary Boolean algebra for their problem solving and number crunching. But our brains often reason with vague assertions, uncertainties, and value judgments. Now there has been an attempt at constructing a logical model of human reasoning that reflects its approximate and qualitative nature.

Only in the twentieth century did logicians such as Jan Lukas-iewicz, Emil Post, and Alfred Tarski realize that they could formulate logical systems different from Aristotle's. They rejected the law of the excluded middle (which says that any proposition is either true or false), permitting statements that could be either true or false or unde-cided; later developments encouraged even more than three alternative options. There was no reason to prefer one logical system over any oth-ers—they all were equally consistent. Whether one is to be preferred over another in the description of natural phenomena is to be decided empirically, not by abstract philosophical argument. This point was not lost on Alonzo Church, who noted that it was akin to Einstein's search for real-world geometry among the varieties discovered by mathematicians as he formulated a new theory of gravity.

The last chapter showed how computers straddle the domains of the real world and the abstract one explored by mathematics. Computers are asked to solve complex problems, but can only draw on the purist resources of classical logic. Perhaps it was inevitable that they would eventually succumb to new forms of logic. One, in particular, is find-ing a number of applications. Called fuzzy logic, it was pioneered by Lotfi Zadeh of the University of California at Berkeley, who blurred the sharp boundaries of classical logic with his landmark paper "Fuzzy Sets."[109] In the sense in which we use it today, fuzzy logic is much broader than traditional multivalued logical systems.

"To some, fuzzy computing may sound like a contradiction in terms, since computing is usually associated with precisely defined op-erations on precisely defined sets," wrote Zadeh.[110] "Most of human reasoning, however, is approximate rather than exact. In a way that is not well understood at present, humans have a remarkable ability to make rational decisions in an environment of uncertainty and impreci-sion. We can understand distorted speech, decipher sloppy handwrit-ing, park in a tight spot, understand poetry and summarize complex stories. In so doing, we perform no computations in the conventional sense of the term. We do manipulate information, which is what com-putation does, but the objects of our reasoning are generally not num-bers but fuzzy patterns without sharply defined boundaries."

In mathematics textbooks, numbers may be grouped into sets in a precise fashion—the set of prime numbers or the set of even numbers, for instance. In the real world such sets are harder to find. If you try to classify your friends by height, you soon discover that it is easy to classify someone who is six feet as "tall" or someone who is four feet as "short," but what can one say about the majority who lie in between? And, like most other concepts, tall has a flexible meaning in everyday usage—taller than the average, tall for your age, taller than broad, and so on.

Fuzzy theory attempts to take account of these shades of gray by assigning each friend's claim to membership of a particular set as a number between zero and one. Thus in the case of tallness, one would assign the number zero to a vertically challenged individual and one to a bean pole,[111] while using other numbers within this range for different sizes within these extremes. Fuzzy classification evidently violates Aristotle's law of the excluded middle, according to which an object either does or does not belong to a given set. But, in fuzzy logic, the multivaluedness of truth is actually a point of departure. The key concept in fuzzy logic is a linguistic variable, that is, a variable whose values are words rather than numbers. For example, if age is treated as a linguistic variable, its values might be young, old, middle aged, very young, not very old, and so on, with each value interpreted as a label of a fuzzy set. This concept of a linguistic variable serves as a basis for computing with words. This is at the heart of most practical applications of fuzzy logic.[112]

Controversy is common whenever new ideas are introduced into science: people have to invest so much time and effort mastering a narrow field that the thought of having to start anew is repugnant. They therefore take up cudgels, motivated by considerations that are all too human. To survive such a battering, an idea must prove its worth. The skeptics claim that fuzzy logic is nothing more than a reformulation of existing techniques[113] and has nothing to do with the structure of logic. But the idea has already led to fruitful real-world applications and over 2,000 patents have been issued or applied for.[114] One of the earliest applications was developed by F. L. Smidth & Company, a Copenhagen-based manufacturer of cement kilns, large chambers where limestone and clays react at high temperatures to form small nuts of "clinker" that are converted into cement when ground with gypsum powder. Danish researchers interviewed kiln operators to build up a knowledge base of expertise and then developed a way to express and manipulate fuzzy notions such as "lime content high" or "kiln drive torque low." A mathematical model of what goes on inside the kiln would have been too complex to be manageable.

Fuzziness is spreading. In 1986 there were eight commercial and industrial applications of fuzzy systems. By 1993 the number rose to 1,500.[115] The turning point for the technology came in 1990, which is known as the "Fuzzy Logic Year" in Japan. As Zadeh commented, "In retrospect, the year 1990 may well be viewed as the beginning of a new trend in the design of household appliances, consumer electronics, cameras, and other types of widely used consumer products."[116]

At the Institute of Industrial Cybernetics in Sofia, Bulgaria, fuzzy computer vision is used to guide an arc-welder. In Shanghai, meteorologists have used fuzzy logic to determine the best regions in China for growing rubber trees. In Japan, fuzziness is widespread. Several cameras now use fuzzy logic to focus automatically. Researchers at Hitachi developed a fuzzy train driver that varies speed of travel to optimize such factors as safety, passenger comfort, and energy consumption in order to provide a smooth ride. A fuzzy washing machine made by Matsushita Electric can add the right amount of detergent after measuring the amount of grease in the water. Mitsubishi has developed a fuzzy vacuum cleaner that varies suction power according to dust flow. There are even fuzzy chips, microprocessors designed to store and process fuzzy rules. As Bart Kosko and Satoru Isaka remarked, "The next century may be fuzzier than we think."[117]

THE BRIGHT FUTURE

Another underlying assumption in our discussions of modern computers is that they are electronic.[118] But now scientists are developing a new medium for logical manipulations: light. The many advantages to running a computer on light are currently being explored by numerous laboratories worldwide. Light waves can cross, so that different channels of information pass through each other rather than short circuit, as in an electronic computer. The use of light in communicating with and between processors will pave the way to completely new classes of computer architecture. Imagine the difference to the traffic flow in a city such as London or New York if all the cars could pass through one another.

Optics avoids problems of "crosstalk" between wires at high-signal frequencies. It also circumvents signal loss at high frequencies. Vast numbers of interconnections can be spun and broken in complex topologies.[119] These advantages outweigh even the best known benefit of using light: that photons traveling *in vacuo* move about one thousand times faster than electronic signals propagate in the conducting and semiconducting materials used in microprocessors. This means that winks of light can be processed more quickly than electrical signals. Light thus offers the promise of machines that would be more versatile than the most powerful computers in use today.[120] Optical fibers are already being used to carry telecommunications data across oceans and continents, lasers help to store information on compact discs, and there are optical circuits that can carry out computing, though currently a mixture called *optoelectronics*, where optics is used for interconnection and electronics performs the logic, looks more promising. Indeed, no computer is complete without one crucial piece of optoelectronics—its display.

The properties of light can be exploited to carry out logical operations for computing. Light can be thought of as consisting of particles (photons) or waves. This dual nature can be described by quantum mechanics, as we will discuss later. For the moment, it is convenient to portray light purely as an electromagnetic wave. When a light wave passes from one medium to another, it changes direction. One can imagine that the equivalent of a Boolean logic gate could be achieved by bending a pencil beam of red laser light with a simple prism: "on" might correspond to the beam forming a spot on the wall; "off" might be the beam deflected on to the floor. Manipulating light beams with prisms would be a cumbersome affair but more subtle methods are at hand.[121]

A team of British scientists who relocated to the United States created the world's first digital optical processor, the forerunner of computers based on light. David Miller, Michael Prise, Nicholas Craft, and Frank Tooley did their pioneering work while at Heriot-Watt University in Edinburgh. In 1990, with support from twelve scientists at AT&T's Bell Laboratories in New Jersey, the team developed a simple optical counter that can be upgraded to make an optical adder or multiplier.[122] Although it was too small to be of any practical use, it did communicate all of its information internally using only light beams.

Like conventional electronics, optical processors require simple switches. The switch used in the Bell Laboratories device was called the S-Seed (symmetric self electro-optic effect device) and changes its refractive index and absorption properties to turn laser beams on and off.[123] In their first optical processor, four layers of units, each consisting of thirty-two S-Seeds, "talked" to each other using beams of near infrared laser light. The four arrays of S-Seeds were separated by lenses, mirrors, and masks that served the same function as wiring in a normal computer. The processor carries out calculations by alternating switches from on to off, and vice versa. Extensive follow-up work on this first processor has yielded experimental processors with a light show that would put a high-tech discotheque to shame: 60,000 light beams interconnecting 12,000 devices of a more advanced type, called a FET-SEED.[124]

Conventional semiconductor diode lasers employed for optical information storage and communication are too big to be useful in optical computers. Complementing the work on processors, dramatic advances have already been achieved in the manufacture of miniaturized lasers.[125] Individual neighboring lasers within the array can also be made to emit light at different wavelengths, offering technologists the possibility of sending multiple signals along optical fibers simultaneously. And microlaser arrays can also be switched on and off with pulses of light, a feature that will prove useful in constructing parallel optical computers.[126] These arrays, integrated optoelectronics, and optical processors will very likely be at the heart of future computers. According to David Miller of AT&T Bell Laboratories, "The difficulties in design of current electronic computers are so daunting, and the reality of computational problems we cannot solve so striking, that one would be a pessimist indeed not to believe in a future for optics in digital computing."[127]

QUANTUM COMPUTERS

Turing's work has provided the theoretical bedrock for all contemporary forms of computation. His universal machine is an abstract mathematical system in its own right, and the computers we have described so far in this chapter all represent its real-world embodiments. As Turing showed, his hypothetical machine can be programmed to perform any operation that "would naturally be regarded as computable." This statement, the original Church-Turing hypothesis, means that all real-world embodiments of the universal Turing machine—whether serial or parallel computers—should be capable of performing exactly the same set of tasks.

However, Rolf Landauer at IBM was among the first to emphasize that a real-world embodiment of a Turing machine opens up important new questions. He pointed out that certain properties of abstract Turing machines are not completely valid in the world we inhabit. In particular, Turing's concept of linking computing with "mechanical processes" in the spirit of Newtonian physics did not take quantum theory into account. And the difference between classical and quantum physics is stark indeed.

Any actual computer, whether it uses light or electrons, and whatever its design, has to work according to the laws of physics. This seemingly obvious point has profound consequences, which have only recently been recognized. In the case of Babbage's mechanical device, the motion of its toothed gear wheels is correctly described in terms of the physics introduced by Newton in the seventeenth century. And modern computers also run on the basis of classical physics. But the fundamental microelectronic components of modern computers—not to mention tomorrow's optical computers—have shrunk to the point where Newton's mechanics breaks down: their behavior can only be described using twentieth-century quantum theory.[128]

Equations of quantum mechanics are designed to describe the microscopic world of electrons; its laws thus underpin the entire microelectronics industry. Quantum theory provides a dazzlingly successful description of the subatomic world, not just of electrons in transistors and the movement of photons in fiber-optic cables but of chemical and nuclear reactions, and much else besides. Indeed, quantum theory was invented to overcome the failure of classical physics to describe the world of the atom. According to classical physics, atoms—the very building blocks of matter—should not even exist. An electron orbiting an atomic nucleus should radiate away its energy, slow down, and consequently spiral inward: atoms should collapse. The early pioneers of quantum theory, including Max Planck, Albert Einstein, and Niels Bohr, assumed—with little justification other than that it seemed to work—that quantities such as energy are not infinitely divisible but come in chunks, called quanta. The laws of quantum mechanics as subsequently formulated by Werner Heisenberg and Erwin Schrödinger explained these assumptions in a mathematically consistent way:

quantum rules prevent an orbiting electron from radiating energy continuously. In this way, we avoid the embarrassing collapse of the atom.

Quantum theory is good at explaining the results of scientific observations, but it puts our concept of an independent underlying Platonic reality in serious jeopardy. In our everyday world, we expect effects to have causes. Yet quantum mechanics seems to admit intrinsically unpredictable hops—"quantum leaps"—between electronic, atomic, and molecular states. There seem to be no limits to how accurately we can measure the properties of an object like an apple, such as its weight or dimensions. Not so in quantum theory. The uncertainty principle, enunciated by Werner Heisenberg in 1927, states that measurements of certain pairs of quantities, such as position and momentum, can be made only to a certain degree of precision and no further. This restriction is so small on the everyday scale that for all practical purposes it appears to be nonexistent, yet it dominates the microscopic world.

According to this unsettling and strongly counterintuitive theory, all physical objects are intrinsically ghostly. They exist in a twilight state—a "superposition"—of all possibilities of position and velocity. Only when a measurement is made on an object do we gain information about specific values of its observable properties.[129] Particles of matter are waves of energy, and waves are particles, appearing as one or the other depending on what sort of measurement is being performed in any experiment. Stranger still, a particle moving between two points in space simultaneously travels along all possible paths between them. Indeed, the behavior of particles that are at opposite ends of the universe cannot be described separately by quantum lore.

Without entering into the mathematical niceties, the main difference between quantum and classical physics is that the latter deals directly with *observable* quantities such as the position of a ball, its velocity, and its acceleration. But if we shrink this ball to the size of an atom, quantum mechanics replaces such continuously variable properties with discrete properties—quanta (chunks) of energy. And a deeper level of description is used, based on so-called wave functions, which provide the probabilities of making particular observations of these quantities at particular times and places. The wave function contains information on all possibilities that could befall a system. It is used to calculate, for instance, the probability of an excited atom spitting out a photon of light when a measurement is made. Although the wave function contains information on all such observable properties, it is not itself observable. When an actual measurement is carried out, the system's wave function is usually said to "collapse" to yield a particular value of the quantity we are interested in.

Since all current computers are embodiments of classical Turing machines, one might ask whether the bizarre quantum world affects computer science at all. Even though both quantum theory and the modern mathematical theory of computation have been with us for more than half a century, only recently have people begun to study the implications of one for the other.[130] Various milestones have been laid down along the route extending Turing's pioneering work to the quantum domain. In 1982, Paul Benioff described classical computers made of quantum components[131] while Richard Feynman introduced the idea of a "universal quantum simulator."[132] Two years later, David Albert described a "self-measuring quantum automaton" that could perform computational tasks that have no classical analogues.[133] The key development came in 1985, when David Deutsch of Oxford University first described a "universal quantum computer."

Deutsch reinterpreted the familiar Church-Turing hypothesis as a new principle of physics—"the Church-Turing principle." This principle states that "there exists [or can be built] a universal computer that can be programmed to perform any computational task that can be performed by any physical object." Atoms and molecules are physical objects; therefore it follows from this new principle that it must be possible to build a quantum computer—though even Deutsch recognizes that it would place extreme demands on present-day technology.

The issue then arises of what difference, if any, exists between computing with a quantum device and a classical Turing machine. A quantum computer has yet to be built, but there is good evidence that it would have quite different properties from contemporary machines. The essential ingredient that makes a quantum computer capable of performing tasks that Turing machines cannot is the quantum property known as "coherence." It is a consequence of the fact that quantum theory uses wave functions to describe dynamical processes rather than observables, as classical physics does. A computing example illustrates what we mean: a classical computer has information stored as "bits" in memory; these bits are either 0 or 1, as decreed by Boole. A quantum computer may also be found in a coherent (superposed) state of 0 *and* 1, that is, packets of intertwined quantities. This novelty arises because computational logic is applied to these packets rather than single quantities, marking a qualitatively new computational state. The result is massive parallelism. As Deutsch explains, "What the quantum computer does is to deal with these packages wholesale, rather than retail. A single bit of a quantum computer can contain an entire package of true/false values that are related to each other in complex ways by the wave function."[134] Deutsch went on to prove that all the computational capabilities of any finite machine obeying the laws of quantum mechanics are contained in a universal quantum computer.[135]

What could a quantum computer do that a classical one cannot? The subject is so new that many questions remain open at the present time. However, some comparisons have been made. As noted in Chapter 2, in classical physics, taken strictly, it would not be possible to build digital computers, only analog ones. "In an analog machine, a small error may increase exponentially in size, so there is an absolute limit on accuracy," said Deutsch. By comparison, numerical errors would tend to remain the same in quantum computers. Moreover, quantum computers should be able to solve some of the same problems that classical computer programs can, but faster. For some classes of problem, quantum computers would be very much faster: the superposition of states enables them to, in effect, carry out many calculations at once. Because of the inherently probabilistic nature of quantum mechanics, quantum computers should also be able to generate genuinely random numbers, as opposed to pseudo random numbers that are the best a conventional (classical) computer can deliver. After all, any number generated by a conventional computer is the result of the mathematical manipulation found in an algorithm. Any number generated by such an algorithm is therefore predictable in principle.

Because they do not obey classical two-valued logic, quantum computers also require new types of logic gates.[136] Moreover, they pave the way toward a novel form of quantum cryptography, which could provide a new, more secure type of data encryption than what Turing considered.[137] Encryption relies on a mathematical operation that is easy to carry out on a message but very difficult—intractable—to reverse. The classic example is factoring a large number: it is trivial to multiply two prime numbers together to form a product, but extremely difficult to reverse by the process of factorization. Peter Shor of AT&T Bell Laboratories has found that quantum computers should be able to factorize numbers much more quickly than before.[138] At the time this was announced, Deutsch warned that the advance, "a major event in the history of computation," would make the most secure *classical* encryption system—the RSA system—obsolete.[139] "This puts a question mark over the whole of cryptography, except quantum cryptography, and quantum cryptography has severe limitations."

QUANTUM PARALLELISM

If it is to operate effectively, the universal quantum computer does face one preeminent problem. It must preserve the coherence of in-limbo quantum superpositional states—those packages of interlinked logical quantities must be maintained during the intricate processes undergone in a typical computation. Loosely speaking, the outside world has to be isolated from a coherent quantum computer, rather than the reverse. Coherence is lost when an interaction with the environment effectively "measures" or acquires information about the state of the computer, as this leads to collapse of the wave function. Unfortunately, quantum theory is at its weakest at precisely this point because no agreed theory accounts for this collapse.

The need for quantum coherence is a consequence of the role played by wave functions in quantum computers: it is crucial that these wave functions are not collapsed by any measurement-like process. But for us to realize that for any macroscopic object is a very tall order. To quote from the late John Bell, one of the leading thinkers in the foundations of quantum theory, "If the theory is to apply to anything but idealized laboratory operations, are we not obliged to admit that more or less 'measurement-like' processes are going on more or less all the time more or less everywhere?"[140] The question is whether it is possible in principle to maintain a quantum system without collapse occurring.[141]

For Deutsch, the many-worlds interpretation of quantum theory offers a way out of the predicament, although for most people not in an entirely convincing way. Deutsch, whose background is in astrophysics and cosmology, is an advocate of this interpretation, like many others who work in these areas.[142] According to the conventional many-worlds picture,[143] there is no collapse of wave functions: with every measurement process, the entire universe buds off an uncountable multitude of new and completely disconnected universes—the "many worlds," each representing a different possible outcome of a measurement. Instead of wave function collapse, an infinite proliferation and propagation of totally separate, new, branching universes occurs. This bears a certain similarity to the teachings of Islamic theological scholars a thousand years ago that held that with every event the world is born anew. Indeed, the many-worlds interpretation has even been used as part of a wildly reductionist speculation, called the Omega Point Theory, which claims physics has absorbed theology to "explain" God, the afterlife, and resurrection.[144]

Though it is beyond the credulity of many people, Deutsch argues that the existence of a quantum computer would provide firm evidence in favor of many worlds as the only correct interpretation of the theory. Instead of maintaining an interpretation in terms of branching universes, however, he regards the calculations done by a quantum computer as being carried out in parallel in many universes. Each computer in each universe begins with the same program. The individual computers evolve by performing separate computations that interact with each other until, eventually, they converge on the same answer.[145] This is "quantum parallelism."[146]

Such parallelism is very different from the parallelism we discussed earlier, which was rooted in classical computation. While a set of N processors working in parallel in a classical computer can accelerate a computation by up to N times, quantum parallelism can be faster by an arbitrarily large amount, depending on the wave function. There are limitations: quantum mechanics allows only some information to be extracted, unlike classical parallel computing that places no such restrictions.[147] As noted before, this dramatic surge in speed can only be expected for certain types of problems running on appropriate hardware.

BUILDING A QUANTUM MACHINE

It is well worth attempting to construct a quantum computer for probing the physical limits on computation, but the task poses fiendish technical challenges as well as problems of principle concerning wave function collapse. Recently, however, a paper laying down a conceptual blueprint for the computer was published by Seth Lloyd, who works at the Los Alamos National Laboratory in New Mexico.[148] His paper was the first design for a computer operating in a completely quantum mechanical fashion that has any hope of being built. The paper included a discussion of how such devices could work, what materials they might be made from, and what techniques we could use to operate them, using today's technology.

A similarly pragmatic approach has to be taken in selecting the quantum system. A "nuclear computer," where the nuclei of atoms in a crystal are controlled by microwaves, is one possibility.[149] However, it is much easier to use visible laser light to manipulate the state of electrons in atoms. Quantum theory shows that electrons can exist in various energy states within an atom. Visible light at the correct frequency can knock the electron from one energy state to another. These two states may be thought of as corresponding to Boole's on-off logical alphabet that drives a conventional digital computer. Recent developments in laser technology provide a means for controlling such quantum mechanical states of electrons within an array of atoms. Specific "resonant" frequencies of light force each electron to an upper or lower energy state. By defining and applying the proper sequence of light pulses of different frequencies, the atomic array can be placed in any desired energy state. A second sequence of pulses transforms these binary states in a logical fashion so that the array can perform computations. A third sequence of pulses allows one to unload information from the array.

Lloyd claims that his design could use any material made up of repeating sets of atomic units. These include polymers, very long, essentially one-dimensional molecules composed of many thousands of repeating sequences of atoms; quantum dots, clumps of atoms, or imperfections that each contain a single electron and that behave, in effect, like one giant atom;[150] and suitable crystalline materials, whose three-dimensional structure comprises a vast number of regularly repeating atomic units.

Quantum computers do face the problem of random noise, caused by ever-present fluctuations in the ubiquitous electromagnetic field, which threatens to destroy the required quantum coherence. Nevertheless, at the time his paper was published in September 1993, Lloyd was investigating suitable materials for building his quantum computer. Other groups have also joined the quest for quantum gates, circuits, and machines. At Oxford University, David Deutsch and Artur Ekert put forward a scheme for a "Quantum Factorization Engine" based on the work of Peter Shor. Deutsch envisaged an array of quantum dots on a silicon crystal. The array in itself is not the quantum processor. What makes it a microprocessor is light: it is driven by

pulses of 60,000 different frequencies of light, altering once every pi-
cosecond (a trillionth of a second).

"That is a very, very complicated bit of light and it is that light
which makes the device a computer," said Deutsch. Each dot would lie
at the crossing point of a grid of wires. Voltages applied across each
dot would make electrons crowd over to one side of the dot, creating
the potential to interact with the electron in a neighboring dot. Any
interaction would alter the way laser light is absorbed by each electron.
"You might be able to induce a transition that only occurs if the
neighboring electron is in an excited state. That allows logic, in fact it
allows quantum logic." Providing the light and quantum array would
be an immense effort but one with the potential to deliver a machine
as significant as Babbage's Analytical Engine. According to Deutsch,
"It could factorize a number comprised of, say, 250 digits, which is totally impossible with a classical computer."[151]

Quantum computing could be ideal for modeling complex fluid behavior, the task that first inspired von Neumann to develop an interest
in computing and, as we have seen, one that still poses one of the
greatest computational challenges. One computational physicist interested in how far computer technology can be pushed, notably in tackling problems in quantum mechanics, is Bruce Boghosian, currently at
Boston University. He anticipates how each electron in a vast array of
quantum dots could be used to play the role of each individual molecule in a fluid. A splash of real water could be simulated by a similar
splash of activity in this quantum array of electrons. "In that case," he
said, "the physics and the computations begin to merge."[152]

Whether such attempts ever realize a quantum machine, let alone
obtain results from it that are beyond the reach of a conventional computer, the effort will not have been in vain. From a theoretical standpoint, the concept of quantum computation is already making an
impact on foundational questions in physics, mathematics, computer
science, and philosophy: it serves to emphasize that computation is a
physical process that can be used to simulate other physical phenomena. From a practical viewpoint, the developments in quantum cryptography have already underlined the constant symbiosis between
pure, basic science and technology. As noted above, these developments suggest a powerful new way for sending sensitive information
between computers without any risk of eavesdropping.

THE FUTURE OF COMPUTING

The scratchings and other marks that were used in the earliest days to count cattle and warriors evolved into a new mathematical language that enabled increasingly sophisticated societies to emerge. Mathematical understanding was power: on the stone head of the 3000 B.C. ceremonial mace used by the Egyptian king Menes, the founder of the first Pharaonic dynasty, you can count his property: 400,000 oxen, 1,422,000 goats, and 120,000 prisoners.

Today's proliferation of the global information network will have an equally profound effect, creating an electronic metasociety. Information of all kinds surges from one brain to another by a multitude of routes in this era of digital information. The cascade of electronic pulses flowing through computers since the days of the Manchester Mark 1 has been at the heart of a communications revolution that now enables new ideas to ripple across the planet in less than the blink of an eye.

The digital computer has empowered the human mind, enabling us to achieve a deeper understanding of certain limited aspects of nature's complex mysteries. Despite the colossal changes the computer has wrought, and the apparently superhuman things it can do, it is still at a nascent stage where, as Turing recognized, it depends on the human beings who program it. Today, the ability of the computer to model real-world complexity is beginning to increase dramatically, not simply because of the advances described in the latter part of this chapter.

The existence of the class of intractable problems we encountered in the last chapter, which remain obstinate in the face of all conventional computing methods, has encouraged a move away from the rigid step-by-step approach of systematic programming, through a renewed emphasis on how nature solves problems, the original inspiration of Boole, Turing, and von Neumann. While the conventional "deterministic" approach makes a computer slavishly carry out every step in an algorithm, computer scientists have found that by emulating features of evolution they can tackle intractable problems.[153]

Before we turn to this inspiration, we might ponder that a still deeper fusion of computing and nature may have even been glimpsed in those heady days of steam-powered computing by Charles Babbage. While discussing his engines with some friends, Babbage mused on what would happen if one set of calculations triggered another set, governed by a different algorithm, and so on indefinitely. He was not sure whether the results would be of much use, but added that such calculations "offered a striking parallel with, although at an immeasurable distance from, the successive creations of animal life, as developed by the vast epochs of geological time. The flash of intellectual light which illuminated the countenances of my three friends at this unexpected juxtaposition was most gratifying."[154] Today, this profound insight is at the heart of a field called *artificial life*. We shall return to it in Chapter 8.

Chapter 4

INSPIRATION FROM NATURE

The chess-board is the world; the pieces are the phenomena of the universe; the rules of the game are what we call the Laws of Nature.

—T. H. HUXLEY[1]

When the first computer was built, its architects ensured that it solved problems according to the strictly logical rules of classical mathematics. Today, scientists are looking to previously little-understood facets of nature to extend the computer's range and power, drawing among other things on the growth of crystals, magnetic properties of exotic alloys, and the annealing of metals for inspiration. Equally, our use of computers to gain an understanding of these processes and properties has enabled us to come to grips with more of nature's complexity than ever before. In essence, the common underlying theme linking nature's complexity with computation depends on the emergence of complex organized behavior from the many simpler cooperative and conflicting interactions between the microscopic components concerned, whether they are atoms, bits of logic, or spinning electrons.

A powerful way to reveal this synergy is to construct metaphors of the real world, wherein dizzying microscopic detail is sacrificed for the sake of apprehending the bigger picture; in short, to devise ways of seeing the forest for the trees. To be more precise, these metaphors can reveal how the complexity of the forest emerges from a tangle of trees, shrubs, and undergrowth. Many processes that create natural complexity provide a powerful means of understanding complexity in general. By incorporating them into the designs and programs of computers, scientists are able to simulate even more of the complexity of natural processes, most notably that of living, breathing creatures.

John von Neumann took the first step toward reproducing living complexity within a computer when he announced his bid to define life as a logical process, akin to what might take place in a machine, in a lecture on the "General and Logical Theory of Automata" given to the 1948 Hixon symposium in Pasadena.[2] His lecture drew on the "remarkable theorems" of Warren McCulloch and Walter Pitts on neural network theory, and Alan Turing's concept of computable numbers and universal computing machines. The result was a precise mathematical theory that allowed comparison of the complexity of computers and biological information processing systems, notably brains. Von Neumann pointed out that "Natural organisms are, as a rule, much more complicated and subtle and therefore much less well understood in detail than are artificial automata"; nevertheless, he maintained that some of the regularities we observe in the organization of the latter may be quite instructive in our thinking and planning of the former.

His mission was to make a machine complex enough to reproduce. To this end, he devised a "self-reproducing automaton," a reproductive robot afloat a sea of its own components; in so doing he united the apparently divergent worlds of nerve tissue and electronic valve through mathematical logic. To von Neumann, the ability of living things to reproduce seemed almost paradoxical, but he recognized that the amount of complexity involved was the key to understanding this paradox. At its lowest level, complexity is degenerative: every automaton that can produce other automata will only be able to produce less complicated ones.[3] However, above a certain minimum level, this degenerative characteristic ceases to be universal and automata that can reproduce themselves then become possible. "This fact," he wrote, "that complication, as well as organization, below a certain minimum level is degenerative, and beyond that level can become self-supporting and even increasing, will clearly play an important role in any future theory of the subject."

Von Neumann had made a leap in laying down the logical framework to explain self-reproduction and in so doing uncovered a profound truth. He had realized that there is a connection between complexity and life itself. Extending the work of Turing, he showed that in theory a universal automaton exists, that is, a machine of a certain definite complication that will do anything any other machine can do so long as it is equipped with correct instructions. Beyond a certain point, you do not need to make your machine any more complex to get complicated jobs done—all you need are more elaborate instructions. Thanks to the inevitable mutations that arise from logical errors within the automaton, this reproduction would be open ended, allowing the possibility of more and more complex organisms to emerge over time; in an environment with finite resources, a selection pressure would emerge, leading to something akin to Darwinian evolution.

CELLULAR AUTOMATA

The robot's proclaimed capabilities were too remote from chemistry, physics, and mechanics to be of widespread interest.[4] The mathematician Stanislaw Ulam advised von Neumann on how to make the self-reproducing machine more abstract, simpler, and elegant.[5] Born in Lwów, Poland, then part of the Austro-Hungarian empire, Ulam could recall speculating on the possibility of artificial automata while sitting in a Lwów coffeehouse twenty years before von Neumann's lecture.[6] When he became familiar with von Neumann's work, he suggested replacing the floating robot with "tessellation robots." This quaint term revealed how Ulam had been inspired by the growth of crystals, which occurs by the buildup of unit blocks, or tessera.

What Ulam proposed was a *cellular automaton*,[7] an abstract array of cells programmed to carry out rules en masse. This collection of cells performing computations in unison could be viewed as an organism, running on pure logic. Ulam envisaged a novel automaton structure comprising a lattice or grid of points—square for simplicity—with the squares referred to as cells and the points as lattice sites. Time is not continuous in the automaton, but discrete, so that it advances with each tick of a clock. The state of such an automaton at each instant in time is given by the state of all its sites at that moment. The automaton evolves through a set of simple rules that determine how any cell changes from one instant to the next. These rules depend not just on the state of the given site at a fixed moment but also on the states of its *neighbors*. At any moment in discrete time, a site takes into account

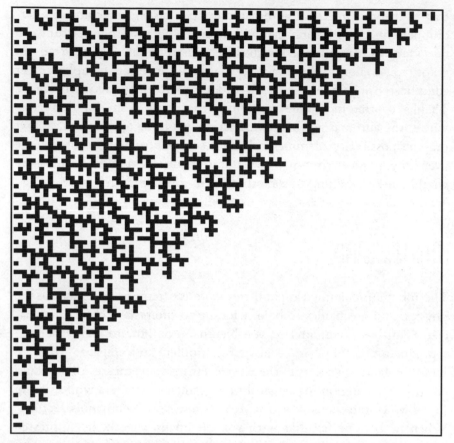

Figure 4.1   Ulam's automaton. A one-dimensional cellular automaton, showing how each site is updated from one time step to the next. The state of each time step is displayed by the pixels on each line.

this local information and consults a predetermined "look-up" table based on these rules to decide what to do next.

A cellular automaton operating along these lines carries out logical operations as von Neumann desired. He could see that anything achieved in his original automaton by means of motion could occur in Ulam's cellular system by transmitting information from cell to cell. Von Neumann was particularly receptive to Ulam's proposal, since at the time he was interested in accelerating the speed of computer-based calculations through parallelism.[8] Because the evolution of a cellular automaton is based on *local* rules, the information needed by any given site to determine what it will do next depends only on a small number of other nearby sites. This means that cellular automata are inherently parallel, since we can divide the lattice into separate regions in which computations can proceed concurrently, with problems only occurring at boundaries between a pair of such regions. Thus, cellular automata can be regarded as a parallel computing paradigm, in much the way a universal Turing machine fulfills that role for serial computation.

Von Neumann's cellular automaton was unveiled at the Vanuxem lectures delivered at Princeton University, New Jersey, between March 2 and 5, 1953.[9] His blueprint revealed an inelegant structure consisting of a basic box of 80 by 400 squares, a constructing arm, plus a tail comprising another 150,000 squares. In all, the self-reproducing object consisted of about 200,000 cells, each with 29 states. To reproduce, the machine used a combination of a brain and brawn: neurons to provide the logical control; transmission cells, which carry messages from the control centers; and muscles to change the surrounding cells. The machine also boasted a Turing tail, a long strip containing coded instructions. Under the control of the rules of the cellular automaton, the machine would extend an arm into a virgin portion of the universe, then slowly scan back and forth, creating a copy of itself by a series of logical manipulations. The copy could then make a copy, and so on. Since this work, it has been possible to greatly simplify his automaton. Nonetheless, von Neumann had succeeded in demonstrating the kind of complexity that was needed for self-reproduction. Indeed, these cellular automata could evolve in a Darwinian manner: as random mutations alter the "genetic makeup" of these computer-based "creatures," the more favorable ones are selected in a competition for the finite computer resources of memory and central processing unit time. There was "no conclusive evidence for an essential gap between man and machine."[10]

THE FALL AND RISE OF AUTOMATA

After von Neumann's premature death, the torch for cellular automata was carried by Arthur Burks, who had worked as an electrical engineer at the Moore School in Philadelphia while the ENIAC was being developed, and edited von Neumann's posthumous manuscripts on automata theory.[11] In 1949, Burks had been appointed to head a Logic of Computers Group at the University of Michigan; from there he began to systematically investigate the relationships between computation and biological processes under the sponsorship of the Burroughs Company.

Burks actively encouraged others to investigate cellular automata and soon his student John Holland started to apply them. In 1960, Holland outlined an "iterative circuit computer,"[12] related to cellular automata, that could mimic genetic processes: programs of arbitrary length were stored in a contiguous set of cells.[13] Such programs could move around, split, combine, control, or construct other programs, and duplicate themselves. Several years later he outlined ideas for automata that can adapt to their environment[14] and also be used to solve optimization problems.[15] Meanwhile, Ulam and Robert Schrandt showed that cellular automata simulations can generate remarkably complex, even life-like, large-scale patterns from very simple laws.[16] They constructed bizarre and ungainly three-dimensional structures of color blocks and, using computers at Los Alamos, even staged "fights" between automata.[17] However, despite these indications of the beginnings of a new science of complexity, the field largely languished, partly due to lack of computing power and partly because it did not sit easily within conventional demarcation lines defining academic disciplines.[18]

Nevertheless, because cellular automata are really like mathematical games, by the end of the 1960s, mathematicians who had heard of them started to show some interest. This game-like feature is the underlying appeal of their most widely known example, which had a major reinvigorating impact on the field in 1970. In that year, John Conway, then at Gonville and Caius College, Cambridge, invented the "Game of Life." He was familiar with the earlier work by Ulam and Schrant, which experimented with different neighborhoods, numbers of states, and rules. Conway selected his rules with great care so as to strike a delicate balance between two extremes: too many patterns that grow quickly without limit and too many that fade away rapidly.

The evocative name reflected Conway's fascination with how his combination of a few rules could produce complex global patterns that would expand, change shape, or die out unpredictably. Conway's brainchild was introduced to the public by Martin Gardner in *Scientific American* as "a fantastic solitaire pastime."[19] That article offered readers a way to reproduce real-life behavior, yet all that was required to play was "a fairly large checkerboard and a plentiful supply of flat counters of two colors." For a while it enjoyed a popularity close to a cult, and turned cellular automata into a household term for a generation of scientists.

Some people call it the "The God Game" in recognition of the toy universe it puts at the disposal of the player. Taking part is easy, requiring no miracles, religious followings, or holy books. Imagine a checkerboard with counters in a few squares. Then follow these rules. If an empty square has three occupied neighbors (including diagonal as well as adjacent sites), it, too, acquires a counter. The cell comes "alive," nurtured by its neighbors. If a square has two occupied neighbors, then it remains unchanged. Finally, if an occupied square has any other number of occupied neighbors, it loses its counter—to anthropomorphize, the cell dies because it is pining for neighborly love or is smothered by overcrowding.

The checkerboard rules represent the laws of physics (or life) and the patterns signify material objects. For several months in the late 1960s, the mathematics department at Cambridge University was consumed by Conway's early efforts to play the game. From a small table an assortment of counters and shells marked the mass of "living" squares that proliferated across the floor of the common room as the rules were enacted during tea breaks. Even then, Conway used a PDP-7 computer to study particularly long-lived populations.

The handful of people who took part discovered that very complex behaviors can emerge from a simple starting arrangement by applying basic rules to every square. Several patterns emerged, with such evocative names as Beehive, Loaf, Pond, Tub, Snake, Ship, Long Ship, Honey Farm, Pulsar, and the Glider, which, as its name suggests, was a group of counters that moved across the board. Conway conjectured that no initially finite population could grow in number without limit, and he offered fifty dollars for the first proof or disproof. The prize was won in November 1970 by a group in the Artificial Intelligence Project at MIT that discovered a "glider gun," a pattern that ejects a glider every thirty moves.[20] Since each glider added five more counters to the field, the population could grow without limit.

Streams of intersecting gliders were found to produce fantastic results, spawning strange patterns that in turn emit gliders. Sometimes the collisional configurations expanded to digest all guns. In other cases, the collision mass destroyed guns by "shooting back." Gardner reported: "The [MIT] group's latest burst of virtuosity is a way of placing guns so that the intersecting streams of gliders build a factory that assembles and fires a lightweight spaceship every 300 moves." The patterns produced by Life can be exceedingly complex; indeed the cellular automaton can be shown to be equivalent to a universal Turing machine—any computation

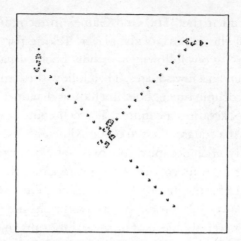

Figure 4.2   The Game of Life. A glider gun.

that could be carried out by any conceivable Turing machine computer could also be performed by a Game of Life.[21]

Using computers, endless variations on the Game of Life are still played today. They have been employed to simulate complex phenomena such as biological development, chemical reactions, crystal growth, the structure of snowflakes, and the meandering of rivers. Cellular automata have proved successful at reproducing many aspects of the complexity of the observable world, such as growth, aggregation, reproduction, competition, and evolution. In later chapters, we will encounter cellular automata used to explore some of these processes.

The attraction of cellular automata for studying complex emergent phenomena is summed up neatly by Tommaso Toffoli and Norman Margolus of MIT: "Cellular automata are stylized, synthetic universes defined by simple rules, much like a board game. They have their own kind of matter which whirls around in a space and a time of their own. One can think of an astounding variety of them. One can actually construct them, and watch them evolve."[22]

TURBULENT AUTOMATA

In the light of von Neumann's deep involvement with fluid dynamics discussed in the last chapter, it is ironic that the study of cellular automata has received further impetus in the past decade or so largely through belated recognition of their use in modeling and understanding complex fluid flows. So-called lattice-gas automata can model the flow of oil and water through porous rocks, and the swirl of sea water around obstacles such as propellers.

Previous methods of modeling fluid flow used a top-down approach based on the Navier-Stokes equation, as we mentioned in Chapter 3. When the existence of atoms and molecules was established at the turn of the twentieth century, it became theoretically possible to model fluid flow using the bottom-up approach called molecular dynamics. In molecular dynamics, one uses an atomistic description and applies Newton's equations of motion to determine what happens as all the atoms jiggle about in the fluid and collide. This sounds simple enough but is hopelessly impractical for studying fluid flow. Given the vast numbers of molecules in any macroscopic sample of fluid—of the order of 100,000,000,000,000,000,000,000,000—molecular dynamics is computationally overbearing.

The class of cellular automata called lattice gases enables laminar flow, turbulence, and other fluid flow phenomena to be computed on large space and time scales without the need for including all the microscopic information necessary in molecular dynamics.[23] Lattice-gas cellular automata thus offer a convenient halfway house. They are called *mesoscopic* models because they offer a grainy picture of a fluid that lies somewhere between the macroscopic Navier-Stokes and microscopic molecular dynamics pictures. They model a collection of particles moving on a symmetrical lattice in two or three dimensions, with several directions for motion. This universe of artificial particles is usually subject to two basic laws of physics: when in collisions with other particles, both energy and linear momentum are conserved.

Lattice-gas models may sound distant from the Navier-Stokes equations used in the classical top-down approach. However, Uriel Frisch, Brosl Hasslacher, and Yves Pomeau showed that the so-called mean-field or average behavior of such cellular automata do approximate the Navier-Stokes equations.[24] One can also argue that a lattice-gas model is a more realistic starting point for modeling fluids than the Navier-Stokes equations themselves. As we pointed out earlier, all fluids are really composed of discrete particles and cannot truly be regarded as continuous.[25] Indeed, for some types of complex fluids (e.g., those containing large polymer molecules), no known hydrodynamic descriptions of any kind may exist.

Cellular automata models do, however, have some drawbacks; and for some simple problems, the lattice-gas methods can be slower than methods based on the Navier-Stokes equations.[26] Nevertheless, there is general agreement that lattice-gas automata methods do come into their own for describing complex situations, including mixtures of oil and water, flows involving chemical reactions, and trickles over rough and random boundaries. One application where cellular automata excel is in the study of percolation through carbonate and sandstone rocks where oil is found, thereby helping to improve our ability to extract it.[27]

Given von Neumann's twin interests in computational fluid dynamics and cellular automata, it may seem ironic that he did not apply the latter to the former, but it is not surprising. Cellular automata were too hungry for computer power and ill suited to serial machines to catch on in the early days. To uncover the complicated macroscopic patterns that frequently emerge, one needs large numbers of sites; for such structures to evolve, we must allow a program to run for long intervals of time. A typical computation may require the execution of many billions of events (corresponding to updating the individual lattice sites), each taking some thirty or so machine operations, each in turn involving a few machine cycles (perhaps ten microseconds on a fast machine). It would have taken years to compute anything substantial on the machines available to von Neumann.

WOLFRAM'S CLASSES

Give a scientist a universe and he or she will play God. As our quote from Toffoli and Margolus states, one can define endless cellular automaton universes by choosing different rules that govern their evolution. This begs a larger question: does this mean that the general behavior of cellular automata are equally varied, or do any universal features emerge? If there are universal features, they would enable us to draw general conclusions about the kind of behavior possible within *any* kind of complex system simulated in such a way.

This question was posed by Stephen Wolfram, then at the Institute for Advanced Study at Princeton, who was interested in the link between local rules and global behavior. The beauty of cellular automata is that they offered Wolfram an exceedingly simple computer-based environment for studying the origins of complexity. Wolfram studied the simplest "universe," one-dimensional automata in which all the sites are arranged on a single line. The initial state of the cellular automata, defined by the states of all its sites, was set at random and shown at the top of a computer screen; the result of applying the local rules—which can alter the state of each site—could be conveniently displayed on the next line of pixels. This procedure could be repeated indefinitely. One glance at the computer screen would reveal any patterns and regularities. Figure 4.3 depicts an example of what can emerge.

Wolfram discovered that he could classify the longtime behavior of the cellular automata into four types, regardless of specific local rules employed (see Fig. 4.4): Class I, in which the pattern disappears with time or becomes a fixed, static, homogeneous state; Class II, in which the pattern evolves to a fixed finite size, forming structures that repeat indefinitely; Class III, which yield so-called chaotic states (i.e., structures that never repeat) with little semblance of regularity; and Class IV, in which complex patterns grow and contract irregularly.[28] Thus, four qualitatively distinct kinds of global dynamical behaviors can emerge from the infinite set of possible simple local rules—a dramatic condensation of complexity. Nevertheless, Wolfram's classification scheme is a purely phenomenological one, and does not explain how these classes arise, or indeed the relationships between them.

Class IV is the most mysterious, and its inhabitants possess the most complex behavior. Conway's Game of Life is one example from Class IV. Such automata exhibit considerable local organization, but little

Figure 4.3 The simplest universe. Typical behavior found in one-dimensional cellular automata. The update of each site is shown on the next line down.

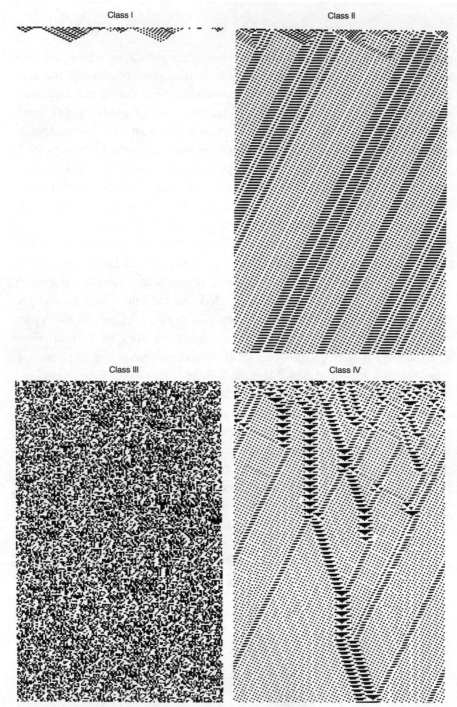

Figure 4.4 Examples of Class I, II, III, and IV cellular automata. (Images courtesy of Andrew Wuensche; generated using his PC software "Discrete Dynamics Lab," available as shareware on the Internet by downloading the file ddlab.zip by anonymous ftp from the directory pub/alife/ddlab at ftp site ftp.cogs.susx.ac.uk or from the World Wide Web at http://alife.santafe.edu/alife/software/ddlab.html)

long-range order. Noting that the Game of Life was already known to support universal computation, Wolfram suggested that cellular automata in Class IV would in general be capable of carrying out computational tasks, including possibly universal computation. This class was later shown to be located between periodic (Class II) and chaotic (Class III) behaviors.[29]

From using cellular automata as models of living systems in all their complexity, some researchers have claimed in recent years that there is a suggestive relationship between biological life and artificial systems capable of performing universal computation. We will explore this intriguing claim, which relies on the argument that a living system has to be able to perform feats of arbitrary complexity in its struggle for survival, in more detail in Chapter 8.

PROGRAMMABLE MATTER

Cellular automata not only offer a natural way for modeling the processes of nature, but have also provided the motivation for designing new computer hardware to enable them to be run more efficiently. In conventional serial computers, the duration of each cycle of calculations is limited by the speed of signals within the electronics—and therefore ultimately by the speed of light. If a system becomes larger, the paths over which signals must propagate grow, and so the cycle time increases. In a hardware realization of a cellular automaton, communication takes place only between *neighboring* parts of the hardware, so that the path length becomes independent of the machine's size. Thus, adding more lattice cells has, in principle, no effect on the time needed to update a large cellular automaton.

The first attempts at developing hardware implementations of cellular automata came in the early 1950s, when British scientists investigated how to combine a TV screen with a computer to analyze airborne particles. By the middle of that decade, simple neighborhood logic circuits were being used to count microscopic objects.[30] The descendants of that effort can be found today in hospitals, where these techniques form the basis of operation in machines that screen and analyze images of blood cells. But the effort continues to develop ever faster general-purpose cellular automaton machines. The Massachusetts Institute of Technology's Laboratory for Computer Science contains a rather unprepossessing bin-sized box of electronics, the latest in a line of dedicated hardware devices designed to harness and unleash the full power of cellular automata. When in full flow, the handful of circuit boards in the Cellular Automata Machine 8 can outperform a supercomputer, generating patterns that model phenomena as diverse as the evaporation of a liquid droplet, the spread of a forest fire, or the expansion of a ripple of sound wave energy.[31] According to its architects, it is a universe synthesizer: "Like an organ, it has keys and stops by which the resources of the instrument can be called into action, combined and reconfigured. Its color screen is a window through which one can watch the universe that is being 'played.' "[32]

The machine, completed in 1993, is the baby of Norman Margolus, a scientist who has struggled for almost a decade to develop it with his colleague Tommaso Toffoli. It was Toffoli who originally wanted to build a fundamentally different computer architecture suited to carry out cellular automata simulations. As physicists, they were interested in how closely computation could be mapped onto real-world behavior.[33] Their first advance was spurred when they became frustrated by the limitations of early 1980s workstation technology that could only update a million-cell universe once every minute—quite a miserable performance. Toffoli found he could speed the process by adding one piece of simple hardware, worth a few dollars. It scanned over the cells in the machine and was optimized to apply simple rules to neighborhoods of cells.

In this way, CAM-1 was born. Year by year saw improvements, from CAM-2 to CAM-6, as Toffoli and Margolus tweaked the hardware and the programming.[34] The logical extreme of their work would be the construction of a machine that contained one processor for every cell. But this is impractical. "The harder you try to cram processors on a chip, the worse your problem will be in connecting these chips," Margolus observed. The latest generation, CAM-8, continues the trend to share each processor over many cells but adds a new trick: Margolus has abandoned the traditional view of the cell in which each interacts with all of the cells—and thus all the data—in a given neighborhood. Instead, CAM-8 shifts vast sheets of data around in three or more dimensions and each bit of data only interacts at the place where it lands.

"We move the data so that all the data that we need on one spot land there," said Margolus. "It is like the physical collision of data—data can only interact if they land at the same spot. The scheme is ideally suited to lattice gas simulations of physics where particles move and collide. . . . It now seems obvious but we were dumb enough that it took a decade— and other people never noticed." The result is CAM-8, a computerized environment containing several dimensions, like the real world, and custom made for the activity of seething masses of cells. Toffoli likes to think of its one million million cells as "programmable matter."

The more a computer is like the world, they reason, the better it will model the world's complex features—hence the attempt to design machines able to handle ever greater numbers of automata and reproduce the graininess of the universe. In the words of Margolus, "If you build a computer that has the same structure as the world, then a little piece of the world can correspond to a little piece of the computer, and two parts of the world that are next to each other can correspond to two parts of the computer that are next to each other. Thus hardware can be piled up to make a computer world as big as one likes."[35]

This work has had many applications, modeling crystallization, complex fluid flows, and chemical reactions. There are also practical uses for producing special effects in videos and for the real-time computer generation of holograms. One of the most evocative spin-offs is an eight-year research project that has been underway in the ATR Human Information Processing Research Laboratories in Kyoto, Japan, since 1993. Within the environment of CAM-8, Hugo de Garis hopes to evolve vast networks of processing elements into what he is optimistically calling an artificial brain. The first milestone was passed by the summer of 1994, when over 1,100 rules had been developed to govern how a two-dimensional network wired up in the "CAM-brain" project (see Fig. 4.5).[36] Three-dimensional networks are now being simulated, and by the end of the millennium de Garis hopes to be able to evolve a billion "neuron" network consisting of a trillion cells within nineteen months. This will be used to control an automaton with a limited repertoire of behaviors, say, a "robot kitten."[37] Similar work is now also being conducted by Todd Kaloudis at MIT and Eduardo Sanchez at the Swiss Federal Institute of Technology in Lausanne, Switzerland.

CEMENT: A HARD PROBLEM

Cellular automata may still sound rather highfalutin and esoteric to the casual reader but they are being applied to an increasing number of extremely practical problems. Among their many uses, one is helping to reveal the secrets of an old and traditionally trusted material that has been widely employed since at least Roman times.[38] Cement is ubiquitous and not very glamorous, but remains one of the most complex materials we use—and one of the least understood. Through a lack of understanding of its properties, all manner of bridges, roads, and other concrete edifices built several decades ago are now crum-

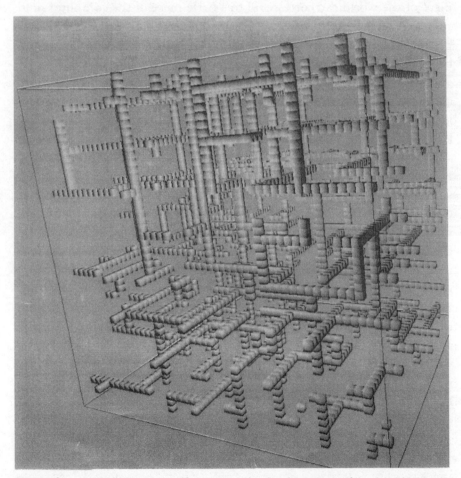

Figure 4.5   A 3D brain. Some of the networks that have evolved in the CAM-Brain project.

bling. It is not difficult, however, to see why the physics and chemistry of cement and concrete are so mysterious.

Cement powder is extremely complex. It consists of multisize, multimineralic, irregularly shaped particles ranging in size from less than 1 micron (one-millionth of a meter) to 100 microns (one-ten-thousandth of a meter). When this powder is mixed with water, a huge diversity of chemical reactions occurs that ultimately convert the initial slurry into a load-bearing, porous material.

Just as a cake mixture sets during baking to hold firmly the berries and nuts in a fruitcake, so cement sets to form the "matrix phase" for concrete, a cement-sand-rock composite. The process of setting is horrendously complicated. Various chemical components within the cement powder react with water at different rates and interact with one another to form various new hydration products. Some of these deposit on the remaining cement particle surfaces; others form as crystals in water-filled pores between the cement grains.

Microscopic interactions affect the microscopic structure; this in turn determines whether the cement is strong enough to hold up a tower or bridge. Some of the hydration products contain pores that are of the order of a nanometer, a billionth of a meter. Some rocks used in the concrete measure a few centimeters, so that important effects extend over a length scale spanning ten million orders of magnitude. Cement is big business, so a major effort has been undertaken to understand the link between its final properties and the way in which it is made.

Cellular automata are very useful for simulating hydrating cements. Here a vice becomes a virtue. We have pointed out that the programmable matter in a cellular automaton occupies length scales somewhere between the grainy molecular picture of the world and macroscopic continuum models analogous to the Navier-Stokes equations used in fluid dynamics. This is similar to the range occupied by the ingredients of cement, from pores to rock aggregates.

Though the precise chemical makeup of a cement is not known in detail, a cellular automaton can at least handle the vast range of irregularly shaped and sized components. The chessboard or honeycomb lattice of the cellular automata can be "colored in" from a real-life example. Digital images obtained from electron microscopy and chemical analysis can be used as starting points for a two-dimensional simulation. Starting states can also be derived from other computer simulations of idealized materials, such as computer-generated images, formed from arrays of "pixels" or "voxels" (abbreviations, respectively, for "picture elements" (2D) and "volume picture elements" (3D) picture elements); (see color plate 2).[39]

Armed with this detailed description of what materials are where, we can then get the cellular automata to model what happens when a builder adds water to the cement. From the starting digital image, and using known chemistry, it is possible for us to update the chemical changes, step by step, as the hydration processes act on the various materials present. Thanks to computer graphics, the operator can watch what is going on in the cement at any level from a grain of sand to a rock. (See color plate 3.)

Such a model can reveal the extent of interconnectivity of the solid grains following the addition of water to the cement since it allows us to watch the evolution of the microscopic structure. Indeed, it enabled scientists to realize that the moment of setting would be a well-defined *percolation* point, the point at which isolated clusters of suspended material connect to form a solid network that spans the full dimensions of the system, whether the cement is plugging a borehole or holding up a skyscraper. Beyond the percolation point, the remaining isolated clusters gradually link up to this backbone. When a cement passes through the percolation point, a sharp transition in its properties takes place. As a result of this work, we have, for the first time, a precise definition of what is meant by the setting of cement. In turn, the concept of a percolation point allowed scientists to pick a measurable change that should occur at the same time: after the percolation point, but not before, ultrasonic shear waves—high-frequency sound, beyond human hearing—should be able to pass through the cement slurry. This proposal, first made on the basis of the cellular automaton model, was subsequently confirmed experimentally.[40] Time is worth a great deal of money in many cementing operations, so that conceivably ultrasonic measurements could catch on as a way of detecting precisely when cement has set, thus enabling other work to continue.[41]

The virtual cement in a computer has other uses. It can help determine the rate of corrosion of steel cables placed under tension in concrete structures for strengthening purposes. Set cement is porous, so that ions, particularly the common chloride ion found in table salt and ground waters, enter the body of the material, where they can then attack, corrode, weaken, and ultimately fracture the iron cables, leading to major failures in the concrete properties. The cellular automaton model enables us to calculate the rate at which such ions diffuse through and bind to the complicated porous cement structure. Indeed, long-term properties of these materials can thus be accurately computed.[42] Here is a particularly important technological issue for construction cements: information of this kind is needed to design improved concretes and to better understand their effective lifetimes when subjected to the elements.

ANNEALING PROBLEMS

Cellular automata were partly inspired by crystal growth and partly by the desire to understand biological self-replication. What blossomed was the ability to mimic a huge range of patterns, using local rules of interaction between neighboring cells to copy the role played by the forces between atoms and molecules that exist on a microscopic level. Within nature, however, other systems exist that are helping scientists to tackle hard problems that, although simple to state, are devious and difficult to solve. These include the NP problems we first met in Chapter 2. What is the most economical layout for the printed circuit board of a computer? What is the shortest route for a parcel delivery service among a specified set of cities where each is visited only once? What composition should a cement powder have to set at a specified moment?

Nature provides us with a powerful metaphor for expressing the quest for an optimum solution to these types of problems: the hunt for the tallest peak or the deepest valley in a rolling landscape. The first task is for us to couch the problem in a mathematical form. We can often do this by writing down a "cost" or "fitness" function that measures how good any particular solution is. In the case of the traveling salesman, we provide the mileage. Then the problem of calculating the best solution translates into finding the optimal value of the function (i.e., its minimal or maximal value, depending on the application). The cost function is best depicted as a landscape of potential solutions where the height of each feature is a measure of its cost. This undulating landscape is called the *search space* (see Fig. 4.6).

In the case of our traveling salesman, the optimal location is found at the base of the deepest valley in the landscape. This, the shortest possible trip, is the *global* optimum, a term used to distinguish it from

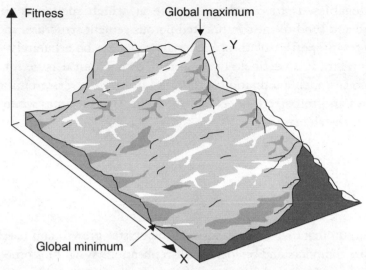

Figure 4.6   The landscape concept. A mountainous terrain showing the locations of the global maximum (highest peak) and global minimum (lowest valley). The height of a feature is a measure of its fitness, shown here to depend on the variables x and y.

local optima at the base of more shallow valleys. In the case of the circuit board problem, we are looking for the most prominent peak, where the greatest manufacturing efficiency resides. Such optimization problems are difficult because the mathematician or computer scientist faces the same problem as a myopic mountaineer searching for Everest in the Himalayas: information is only available about the local area. An exhaustive scrutiny of every peak would be impractical and soon deplete our mountaineer's rations. The next simplest strategy he could adopt would be to pick the direction that seems to lead uphill most quickly and set off that way. The mission of our myopic mountaineer would be to follow the direction of steepest ascent at all times in order to boost his altitude as quickly as possible. This method ensures a speedy ascent to the top of the local peak. However, it is by no means the same as finding Everest. Almost all real problems have local optima of this sort, which makes this simple approach inadequate.

A more effective search strategy would employ the capability of occasionally moving downhill in order to escape from a local peak.[43] This strategy accepts small short-term penalties in exchange for potentially larger long-term rewards. The analogy holds equally, of course, when the search is for a global minimum—the deepest valley in the landscape. This arises in the problem of locating the optimum arrangement of the atoms in a cooling metal. The perfect configuration for a metallic crystal—the one with the least energy[44]—is the one in which all the atoms are neatly aligned in a lattice, a three-dimensional version of soldiers on parade. Annealing, a trick blacksmiths use to persuade atoms in a metal to adopt a reasonably orderly arrangement, not only provides a valuable method for finding the global optimum, but can be applied in a wide variety of circumstances. In fact, on the hunt for optimal solutions to the range of NP problems, Scott Kirkpatrick, Charles Gellatt, and Mario Vecchi at the IBM Thomas J. Watson Research Center used this analogy to propose a clever strategy of searching for the best solutions called simulated annealing.

Annealing is a way to bring a metal close to a perfect atomic configuration, although the perfect configuration of atoms in a metal is never realized in practice. Sure enough, as the metal cools, atoms move less and line up with their neighbors. Often, however, although the atoms in one small region are lined up, the region itself can be misaligned with respect to neighboring areas. Whether the metal achieves a local energy minimum associated with the latter situation, or succeeds in finding the global energy minimum associated with maximal long-range order, depends on how quickly it is cooled.

A sprinkling of heat is enough to provide the balance between shorter-term penalties and longer-term rewards in the search for the optimal atomic arrangement carried out by the cooling metal. Heat adds randomness: at any temperature above absolute zero, individual atoms within the lattice jiggle about. The energy of their dance depends on the temperature of the metal—the greater the temperature, the more frenetic their motions. At high temperatures, it is easier for atoms to jump between sites on the underlying regular lattice, even if in so doing they temporarily increase the total energy of the atomic configuration.

Heat offers a way for the metal to jump out of a local optimum so that, for instance, the atoms in one region can line up with the local arrangement of atoms in an adjacent one. As the temperature drops, less energy is available to permit alterations in the atomic arrangement. As the temperature is decreased toward absolute zero, the location of all the atoms becomes fixed. The more gradually the metal is cooled, the more nearly perfect the crystal it forms is likely to be, and the nearer its energy will be to that of the global optimum. It is important to stress the word "near" here. Perfect crystals are never found because their formation would take an infinite amount of time.

A rapidly quenched metal has many defects and dislocations at the atomic level that make it brittle. However, these defects in its crystal structure can be ironed out to a degree by annealing—heating the metal up and then cooling it down slowly. Blacksmiths use this method to make metals such as copper malleable and less brittle. In reality, the atomic arrangement obtained by annealing will be in some local minimum energy state whose value is likely to be very close to that of the global optimum. That is good enough. Engineers and scientists prefer to settle for good, cheap, and fast solutions than taking forever to find a perfect solution, one that is probably little better. This temperature-dependent behavior is the basis for Kirkpatrick's optimization technique of simulated annealing, which can be applied to a wide range of optimization problems in computer science, neurophysiology, biochemistry, and evolution, as well as the dilemma of the ubiquitous traveling salesman.

SPIN GLASSES

Kirkpatrick's motivation for developing simulated annealing was the need to understand the very complex behavior of a certain class of magnetic alloys called *spin glasses*. Spin glasses are some of the simplest disordered systems found, and their study, beginning in the 1970s, marked a move by physicists toward understanding complex systems rather than ignoring them. The behavior of spin glasses will no doubt appear fairly abstract to the general reader, yet the processes that go on within them embody optimization processes similar to those taking place in annealing. And, in certain cases, we can calculate exactly the properties of spin glasses by mathematical analysis. They are thus very valuable prototypes of complex systems. Indeed, our knowledge of the behavior of spin glasses has generated entirely unexpected but deep insights into complexity: as we will describe in Chapters 5 and 9, major progress in our understanding of artificial neural networks as well as important aspects of brain function have emanated from this work. For these reasons, it is well worth our making an attempt to grasp the key features and properties of these strange magnetic materials.

The chemical composition of a spin glass is unremarkable. It can consist of a few iron atoms scattered in a lattice of copper atoms. The term "glass" arises by analogy with ordinary glassy materials such as windows, which are obtained when molten silicates are cooled too quickly for their component atoms to freeze into the lowest energy state with the greatest long-range order. Instead, the random atomic motions are quenched, producing a rather disorganized solid state. The same thing happens in a spin glass, but now the randomness is caused by the disorderly arrangement of the spinning electrons within their constituent atoms, which give rise to magnetic effects. Although spin glasses usually possess the regular atomic arrangement of a crystalline lattice (unlike ordinary glasses), the tiny atomic magnets (on iron but not copper in this case) are not themselves aligned in an orderly manner. A spin glass is a confection of positive and negative feedback effects, as each magnetic atom attempts to align its associated magnet with those of its neighbors. The resulting magnetic properties are confoundedly complex and sometimes unpredictable.[45]

To understand spin glasses, physicists use the discipline of statistical mechanics, the field that relates the microscopic world of atoms and molecules to the macroscopic world we can touch, feel, and see.[46] Statistical mechanics is concerned with a vast range of issues, including the description of how ice cubes melt to form puddles, how water boils into steam, and how certain materials gain and lose their magnetic properties as a function of temperature. All are examples of *phase transitions* in which the macroscopic properties of matter are abruptly altered

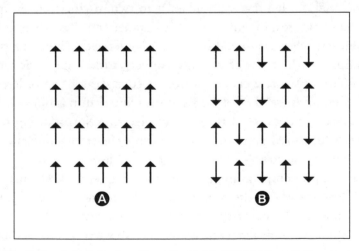

Figure 4.7 Order and magnetism. Two-dimensional illustration of a ferromagnet: (A) at a temperature below the Curie point, all magnetic moments on the atoms—indicated by arrows—point in the same direction, leading to magnetization; (B) at a temperature above the Curie point, where the magnetic moments are randomly oriented and there is no magnetization.

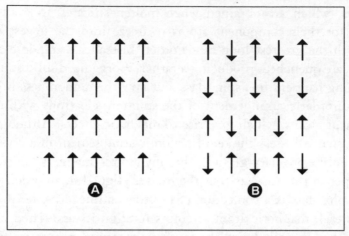

Figure 4.8   Ferromagnets and antiferromagnets. A is a ferromagnet, where all the individual magnetic moments are aligned, giving rise to magnetization. B is an antiferromagnet, where all the individual magnetic moments on neighboring atoms are antiparallel, giving rise to zero magnetization.

at a particular value of the temperature; phase transitions occur because of a macroscopic scale rearrangement of the basic microscopic atomic components of matter.

A common and relevant phase transition is the magnetization of iron. Iron is naturally enough the prototypical example of a ferromagnetic material. Such a material can exist in two distinct phases: one, at higher temperatures, in which it is not magnetized; the other, at lower temperatures, when the metal becomes magnetized. There is a precise temperature at which the phases interconvert, called the Curie temperature. We can picture a crystal of iron as consisting of an ordered lattice of iron atoms (see Fig. 4.7). Each individual iron atom acts like a tiny bar magnet because of the mist of electrons surrounding it. The presence of unpaired spinning and rotating electrons confers a magnetic moment—a north–south magnetic dipole—on each iron atom. Below the Curie temperature, each atomic magnet in a ferromagnet is aligned, and the lower the temperature, the neater the arrangement: the total magnetization of the crystal is then given by the sum of all the tiny aligned atomic magnetic moments. Above that temperature, the atomic jiggling resulting from thermal energy causes chaos in the ranks and the atomic magnets align themselves haphazardly, with a zero net total magnetization.

To investigate the atomic behavior in a ferromagnet—and, by direct

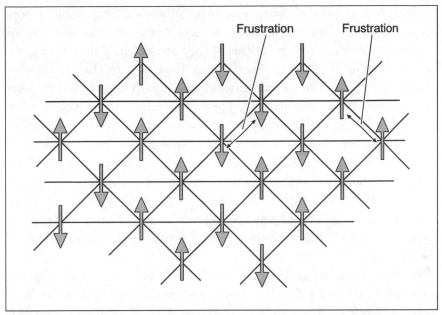

Figure 4.9   A frustrated antiferromagnet. If atoms are arranged in a triangular lattice and interact antiferromagnetically, it is impossible to satisfy the desire of each magnetic moment to be antiparallel to its nearest neighbors (as illustrated in two instances). The underlying symmetry of the lattice leads to frustration. The same effect occurs in spin glasses, but then it arises owing to the random nature of the magnetic interactions, which may be a combination of ferromagnetic and antiferromagnetic.

extension, in spin glasses—we need to construct a model based on known laws of atomic physics. Finding the balance between the randomizing influence of heat and the organizing influence of the microscopic magnetic interactions is the central issue determining the final state of such systems, and is the main aim of equilibrium statistical mechanics. It accomplishes this through the concept of "free" energy, a quantity that represents the balance between thermal randomization and organizing interactions and achieves its lowest value at equilibrium. The minimization of the free energy provides a theoretical explanation for the existence of a magnetic phase transition in ferromagnets: below the Curie temperature, the free energy is lower for the ordered (magnetic) state, whereas above this point it is lower for the disordered state.

Now let's consider a spin glass composed, for example, of a dilute solution of *antiferromagnetic* manganese in nonmagnetic copper metal. An antiferromagnetic material is one in which the lattice of spins prefers to align in a mutually *antiparallel* way; that is, nearest-neighbor spins tend to be aligned pointing in opposite directions (see Fig. 4.8). Problems occur if an odd number of antiferromagnetic atoms are arranged in a loop. For example, it is impossible to satisfy globally the desire of all the spins to align antiparallel to one another if they are arranged on a triangular lattice (see Fig. 4.9). This effect is writ large within the manganese/copper spin glass. Competition between the various interactions that align the spins leads to *frustration*. There may be local patches of stability in which all the atomic magnets are aligned in an antiparallel sense, but regions will always exist in which nearest-neighbor spins conflict and are forced into energetically unfavorable mutual orientations. Herein lies the headache for anyone trying to figure out the lowest free energy state of a spin glass. For in a spin glass are a great many local minima in the free energy, differing from one another by only a small amount because of the ease of realizing energetically similar spin configurations on the lattice.[47]

An important advance in understanding the complexity of these materials came in 1975, when David Sherrington, then on leave from Imperial College, London, and working with Kirkpatrick at IBM, published a paper concerned with the statistical mechanics of an infinite-range spin glass model.[48] Their model allowed each atomic magnet in the spin glass to influence all others through infinite-range magnetic interactions. No real magnetic materials have such infinite-range interactions, but, as Sherrington recalled, "I knew this was the kind of model that had the potential to be exactly soluble."[49] On this basis, Sherrington and Kirkpatrick were able to work out the properties of these peculiar magnetic materials.

As before, it helps to conjure up a search space. Imagine again a rolling landscape, this time corresponding to the free energy. The height of a feature corresponds to its free energy and the location to the particular configuration of spins. In general, we are dealing with a very complicated energy surface (see Fig. 4.6). How easily neighboring valleys (minima) in the energy can "communicate" will depend on the temperature: the height of each peak between neighboring minima is a measure of the amount of heat energy needed to flip the state of the material from one global spin state to another. A grassy knoll will require much less heat than an Everest.

The large number of different local free energy minima leads to many fascinating attributes. For instance, if a few parts of iron are mixed with 100 parts of copper, the iron atoms, which usually interact ferromagnetically, can now also interact antiferromagnetically. But if we ask which state is the lowest and most stable, it is hard to find the answer. As in the case of our itinerant salesman, this is an NP problem. If we attempt to solve it with simple deterministic algorithms such as gradient descent, there is a great risk that the method will lead into the nearest valley—a local minimum—rather than finding the global minimum. The simulated annealing technique of Kirkpatrick and his collaborators overcomes this problem.[50]

Simulated annealing, like the blacksmith's annealing, seasons the method of steepest ascent or descent with a little randomness. Each step taken, whether upward or downward, is assigned a probability that is dependent on the corresponding change in energy,[51] divided by the "temperature" used for annealing. If the temperature is low, only steps that lead to a decrease in energy will occur, and the recipe is the same as the deterministic steepest descent method. The higher the temperature, the more chance we can clamber up the energy surface, thereby avoiding becoming trapped in local minima. Then there is a greater probability of falling into a deeper minimum elsewhere. As time proceeds, this algorithm will with increasing probability drag the state of the system into deeper and deeper minima. To reach the lowest point or global minimum, the "temperature" (which is the source of noise or randomness) must be gradually turned off. It has been found that simulated annealing often substantially outdoes traditional methods.[52]

Originally devised to study spin glasses, simulated annealing has improved solutions to a number of very practical optimization problems, real-world variants of the traveling salesman dilemma, that range far and beyond the apparently esoteric confines of spin glass physics. These range from the best ocean routes for oil-laden supertankers to the design of microchips. Simulated annealing is used to reveal atomic arrangements within crystal structures from the way the atoms scatter X rays: these so-called diffraction patterns do not give enough infor-mation to deduce the crystal structure.[53] Simulated annealing has helped oil companies to reconstruct geological information and hence locate hydrocarbon-bearing strata from the waves propagated by seis-mic detonations. It is also being employed to help find the three-dimensional structures of protein molecules given knowledge of their atomic composition, an extremely challenging problem that, when fully solved, will help to make drug design more systematic.[54]

THE GLASS BEAD GAME

Confronting nature's macroscopic complexity head on has born remarkable fruit. In the previous chapters, we have seen how Turing and von Neumann's interests in the complexity of the brain helped spawn the modern computer. Von Neumann's fascination with the complexity of self-replication gave us cellular automata, a novel virtual laboratory with which to explore emergent phenomena. And Sherrington and Kirkpatrick's work on the complex properties of spin glasses has provided a powerful technique for tackling some of the toughest optimization problems.

In the next chapter, we will see how these ideas, together with others drawn from biology, are helping advance our understanding of inanimate complexity across a very wide range of scientific disciplines. It is fitting that the origins of these studies can be traced to a student of Arthur Burks, who helped keep von Neumann's ideas alive in the wake of his death. The same John Holland, prior to becoming Burks' first computer science Ph.D. student in Michigan, had grown disenchanted with his mathematics studies, dominated at that time by the ultrapurist school of thought propounded by the French "Bourbaki school." Holland had filed away his ideas on emergence, adaptation, learning, and a host of other topics in manila folders labeled *Glasperlenspiel 1*, *Glasperlenspiel 2*, and so on after Herman Hesse's novel, *The Glass Bead Game*.[55] For Holland, the book's futuristic game of interweaving and uniting diverse themes played by an intellectual elite epitomized his own research. Of the eclectic set of ideas Holland encountered at the time, one of the most significant was a book on genetics, *The Genetical Theory of Natural Selection*, written by one of Britain's foremost medical statisticians, Ronald Fisher. Fisher's mathematical approach to evolution—and its inadequacies—would inspire Holland to explore how natural selection can solve complex problems.

Chapter 5

EVOLVING ANSWERS

Nature cannot be ordered about, except by obeying her.
—FRANCIS BACON

The fabulous complexity of life has been fashioned in the process of evolution by natural selection, a powerful concept that has withstood a great deal of abuse since Charles Darwin unveiled it in 1859. As the foundation stone of modern biology, the theory of evolution has emerged unscathed from more than a century of stringent evaluation by scientists and attacks by creationists. Now it has spread its influence to the inanimate world, where evolutionary metaphors are taking root.

Biological evolution provides a powerful paradigm for tackling intractable problems, using what are known as genetic algorithms and genetic programming. These methods capture the process of evolution within a computer, making it possible to "breed" a better turbine blade design for an aircraft or find the most efficient pathway for sending information across a vast electronic communication network. These are fiendishly complex optimization problems, which can be represented by Himalayan fitness landscapes. Digital Darwinism provides a powerful tool for solving them.

The human brain is in a sense one solution to the optimization problem posed by biological evolution. Because of the brain's immense capabilities, it furnishes another source of inspiration in its own right.

One of the ironies of recent attempts to come to grips with intelligence by simulating it is that all the insight and opportunity afforded by our understanding of evolution had been set aside in favor of an approach now derisively called "Good old-fashioned artificial intelligence." We will see how this turned out to be an intellectual dead end and that, by returning to the inspiration of nature, a more fruitful approach has now developed. That approach is already being used in a vast range of applications, from grading pork carcasses to predicting the rise and fall of shares on the stock market.

SURVIVAL OF THE FITTEST

The catch phrase "survival of the fittest" usually springs to mind when people try to summarize Darwin's thinking about evolution. Plants, animals, and insects seem to have evolved in order to refine various features, whether the wings of a bee, the white fur of a polar bear, or the spots on a leopard's coat. Each of these is called an adaptation, a term used to designate any open-ended process by which a structure evolves through interaction with its environment to deliver a better performance. These structures may range from proteins, through brains, to interacting ecologies of organisms such as the wildlife of Tasmania. The kind of adaptation these examples of living art undergo can be amazingly specific.

The bee orchid, *Ophrys apifera*, has evolved to look like a female bumblebee so that it can be fertilized when male insects are lured for sex. These plants have not deliberately optimized their appearance; their "goal" in life is simply reproduction. Among the decaying matter at the bottom of ponds, the thiobacilli bacteria have evolved to cope with high levels of sulfur by means of a complex suite of enzymes enabling them to use sulfur in place of oxygen. In the process of surviving through reproduction, even the lowliest species must be capable of adaptive improvement, for otherwise they would be eliminated in the biological arms race.

Species caught up in this struggle are engaged in an attempt to solve a complex optimization problem. The resulting adaptations, which crucially take into account the creature's environment, are in some sense nearly optimal, leading to such features as the shape of the bee orchid, the hydrodynamic features of a whale's fins, and the eyesight of an owl. Although adaptation produces organizations and interactions that are highly refined, they are invariably still improvable and not truly optimal.[1] Finding effective improvements, not optimization, seems to be the heart of the Darwinian process. Defined in its most general sense, adaptive processes have a critical role in fields as diverse as psychology, economics, control engineering, and computational mathematics. Evolution would, therefore, seem to be a good metaphor to plunder for inspiration. In the past ten years or so, many have done just that, reproducing natural evolutionary structures within computers and using them to model—and find connections between—the recognition of visual information, the setting of oilwell cements, learning, memory, intelligence, and artificial intelligence.

By the standards we have considered so far, biological evolution is an adaptation process of staggering complexity. Imagine, for example, a simple seaweed that is described by 1,000 genes, each of which comes in two varieties so that it is responsible for a characteristic protein that may control a green as opposed to a brown color, wrinkly as opposed to smooth leaves, and a short rather than a long reproductive cycle. If each gene acts independently, there are $2^{1,000}$ combinations of genes that can be on or off—an enormous landscape of opportunities.[2] Like simulated annealing, evolutionary optimization works well because the effective search strategy it employs incorporates the capability to occasionally move downhill—through mutation—in this vast landscape in order to escape from a local peak and so find a combination of high fitness.

Evolution occurs whenever there is reproduction and competition for finite resources. A reproducing system might intend to produce exact copies of itself, but no copying process works perfectly. Usually mistakes will impair the copy's ability to reproduce successfully. The mutant will probably perish, or at least its genes will become less numerous than those of its parent. Sometimes, however, the mutant will be superior, in which case its genes may be more successful than its parent and thus become more common. Because resources are limited, competition usually occurs between different organisms and species. In view of all these contributing factors, it is helpful to think of organisms as having a "reproductive fitness." All sorts of things will contribute to this fitness. For example, in a monkey, such factors as visual acuity, agility, attractiveness to other monkeys, intelligence, and strength play a role in determining how likely a monkey is to survive.

THE GENETIC ALGORITHM

John Holland sought a way to make computers—or at least their programs—mimic living things by evolving adaptively to search for the solution to a problem.[3] As he remarked, "Living organisms are consummate problem solvers. They exhibit a versatility that puts the best computer programs to shame."[4] His fascinating concept of a genetic algorithm as a population of computer programs that evolves by reproduction, random changes, and competition drew on all the ideas and language of biologists, ranging from the chromosomes that contain an organism's genetic blueprint to the crossovers and mutations that can occur to alter it. The first major work describing his computational use of the evolutionary metaphor was *Adaptation in Natural and Artificial Systems*[5] published in 1975, more than a decade after his initial work.

"When this book was originally published I was very optimistic, envisioning extensive reviews and a kind of 'best seller' in the realm of monographs," he recalled in a later edition. "Alas! that did not happen. After five years I did regain some optimism because the book did not 'die', as is usual with monographs, but kept on selling at 100–200 copies a year. Still, research in the area was confined almost entirely to my students and their colleagues and it did not fit into anyone's categories. 'It is certainly not part of artificial intelligence' and 'Why would somebody study learning by imitating a process that takes billions of years?' are typical of comments made by those less inclined to look at the work."

The start of the 1980s saw a surge of interest in his research for several reasons. One was the recognition that genetic algorithms could crack hard-to-solve (NP) problems. Studies began to appear that used genetic algorithms in areas ranging from the design of integrated circuits and communications networks to compiling market portfolios and designing aircraft turbines. Another application was in artificial intelligence research where, after languishing at the periphery for many years, the importance of *learning* from the environment began to be regarded as central in the study of intelligence. More recently still, genetic algorithms began to be used as a general theoretical tool for investigating "complex adaptive systems"—an umbrella term for nonlinear systems specified in terms of the interaction of large numbers of individual agents—including economics, political systems, ecologies, immune systems, developing embryos, and brains.

Holland's work has shown that in expansive and very rugged land-scapes that we can think of as representing the *fitness measures* or cost functions for traveling salesmen, spin glasses, or the reproductive capability of a monkey, genetic algorithms tend to regularly improve on known solutions. They thus appear to provide an effective method for solving "intractable" problems computationally. In Holland's view, "Pragmatic researchers see evolution's remarkable power as something to be emulated, rather than envied. By harnessing the mechanisms of evolution, researchers may be able to 'breed' programs that solve problems, even when no person can fully understand the programs' structure."[6]

In the case of biological evolution, however, we have to modify our mental picture of the landscape furnished by the fitness measure. The hills and valleys not only depend on properties of one species but also on the performance of rival organisms. In Darwinian evolution, each species hunts for beneficial adaptations in a complicated and constantly changing environment. In other words, the landscape moves and mutates, with soaring peaks indicating the success of one species that might simultaneously push another into a deep trough of extinction. The evolutionary fitness of rabbits taken in isolation could theoretically soar for all sorts of reasons, ranging from oversupply of carrots to extra sensitive hearing that enables the early detection of predators. However, a fox population boom could still drive the poor rabbits into a fatal bolt hole.

Initially, it is easier to put aside the notion of such adaptive, coevolving landscapes and instead focus on the ingredients of biological evolution that Holland exploited to locate the peaks and troughs representing optimal solutions to many simpler kinds of intractable problems. Evolution has two essential components: selection and reproduction. Selection is a filter that determines by competition which individuals within a population survive to reproduce. Reproduction permits innovation through the alteration of genetic traits and thus provides a route for the evolution of increasingly successful individuals. Just as breeding and selection have allowed us to "optimize" the appearance of a rose and the quality of the wool on a sheep, Holland realized that it should be possible to use a mixture of innovation and filtering to optimize the solution to a complex problem.

Holland sought to capture computationally the processes that alter this chromosomal makeup: mutations, crossover, and inversion. Mutations, an important source of genetic variation, can be caused by high-energy radiation (typically ultraviolet and X rays) in the environment, or by the incorrect copying of genetic material by cellular machinery. Mutations provide variety through randomness and occasionally introduce beneficial material into a species' chromosomes.

Crossover, which occurs in species that reproduce sexually, exchanges genetic material from two parent chromosomes, allowing beneficial genes from each to be combined in their offspring. Inversion occasionally rearranges genes so that those far apart in the parents may be placed close to one another in the offspring's chromosomes. Sometimes all this gene shuffling and shifting makes little difference, sometimes it may be deadly, but from time to time it may confer real advantages.

Crossover is particularly important to genetic algorithms. It allows useful alterations in the genes of a population—in other words, improvements in the solutions to a problem—to accumulate over time. For an individual to acquire two beneficial and unlikely mutations without crossover, one of the rare mutations must occur by chance in a parent, and the second improbable mutation must occur by chance in one of that parent's offspring.[7] Species reproducing without crossover could have a population containing members with one or the other of the two mutations but lack any members possessing both. When crossover occurs, beneficial mutations on two parents can be combined immediately as they reproduce; if the most successful parents reproduce more often than less successful ones, the probability becomes high that this will occur. Holland's algorithm captures the ability of many individuals in a population to explore the search space (i.e., the fitness landscape) in parallel and combine their best findings through crossover. The combination of reproduction and the information exchange of crossover gives genetic algorithms much of their power.

But how is it that these biologically inspired operations are smart enough to hunt down the solution to a difficult problem? After all, the very notion of a systematic search does not sit comfortably with the element of chance these operations introduce. The answer had already been glimpsed in the mathematician Jacques Hadamard's discussion of inventiveness: "It is obvious that invention or discovery, be it in mathematics or anywhere else, takes place by combining ideas."[8] Crossover allows a very large number of attempted solutions to be tested simultaneously. This inherently parallel property permits charmed combinations of units to be found from a huge number of alternatives by a process that *seems* to be random, but which is, in fact, guided by Darwin's hand.

DIGITAL DARWIN

Conceptually, the construction of Holland's genetic algorithm is clear. However, its actual translation into a computer program is not easy because of the problem of "brittleness": most mutations or random changes made to conventional computer programs cause them to crash. Even a period in the wrong place can make them fall over. To harness his evolutionary techniques, Holland had to develop a robust way of representing solutions to a problem in terms of "chromosomes" that could be modified by the operations of crossover and mutation. The result was his binary classifier system: a solution is expressed as a binary string of information. Each bit corresponds to the presence (1) or absence (0) of a particular property in the same way that a chromosome contains genes corresponding to the presence (or absence) of a trait such as hair color. For example, in finding the solution to the problem of which animal is man's best friend, dogs can be encoded as a string containing 1s for the bits corresponding to "hairy," "slobbers," "barks," "is loyal," and "chases sticks" and 0s for the bits corresponding to "metallic," "speaks Urdu," and "possesses credit cards," and so on. Cast in the form of binary strings, the search for a solution to this problem is the search for the correct string of ones and zeros.

As always, a good solution corresponds to a special combination of features in the search space. We can again think of our landscape metaphor where the location corresponds to the combination of zeros and ones and the height measures how good the solution is. To calculate the latter we need a "fitness" function, which is the same as the cost function we encountered before. Then it is over to Darwin. The algorithm (see Fig. 5.1) starts out with a "population" of bit string chromosomes, each acting as a potential solution to the problem at hand. Usually, each is a guess or a random stab at the problem. Then their respective fitnesses are evaluated. In a loose analogy with molecular biology, the individual bits or the collections of bits on the chromosomes can be referred to as genes. The higher-ranking chromosomal strings are then mated using the crossover and inversion operations, and the resulting offspring will replace the lower-ranking chromosomes. In addition, a small number of genes (perhaps one in every 10,000 symbols) is randomly flipped from 0 to 1 or vice versa, the equivalent of an environmentally induced mutation that adds a little more variety. In this

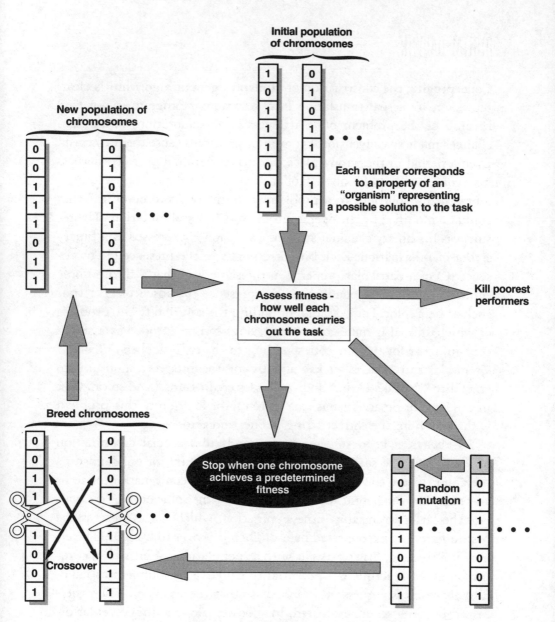

Figure 5.1   Digital Darwinism. The genetic algorithm.

way, a combination of innovation (mutation, crossover, and so on) and selection (retaining only the fitter chromosomes within the population) converges generation by generation on the optimum solution. On a parallel computer, the individual members of a chromosomal population can concurrently test, sample, and explore large portions of the search space independently of one another.

The effectiveness of this binary classifier system can be demonstrated in the search for an optimal design for commercial jet engine turbines, a problem involving more than 100 variables. The search space contains more than 10^{387} points. Evaluating the suitability of a single design can take about thirty seconds on a typical workstation, so that a purely random search for the optimal design would very probably last far longer than the age of the universe. It took an experienced engineer about two months to come up with a reasonable turbine design. However, using a genetic algorithm, it took only a couple of days to "evolve" a design with three times the efficiency.[9]

Work by one of Holland's students, David Goldberg, now at the University of Illinois, has shown how genetic algorithms can control gas flow in pipelines, adjusting that flow to meet daily and seasonal fluctuations in demand.[10] This is an example of what is called an adaptive control problem. The genetic algorithm responds to the demand for gas at different times, and adjusts the gas velocity through the pipelines accordingly. Even Holland was surprised by that application: "My first reaction was that this was too difficult a problem for a dissertation—there are no closed analytic solutions to even simple versions of the problem, and actual operation involves long, craftsman-like apprenticeships. Dave persisted, and in a surprisingly short time produced a dissertation that, in turn, produced for him a 1985 National Science Foundation Presidential Young Investigator Award. So much for my intuition as to what constituted a reasonable dissertation."

An extension of the genetic algorithm is the concept of *genetic programming* pioneered by John Koza of Stanford University. In the basic genetic algorithm, there is only one type of solution algorithm, represented by a fixed-length chromosome. But in genetic programming, programs of varying shapes and sizes themselves undergo adaptation. The task is to search through the space of all possible programs to find the fittest one for a given problem. Thus, one breeds hundreds or thousands of computer programs using survival of the fittest together with reproduction utilizing genetic recombination (i.e., crossover).[11] The offspring are, in general, of different sizes and shapes from their parents. One example is the genetic breeding (evolution) of a computer program that can control the motion of a truck moving along a road, through the application of a force, so as to bring it to rest at a specified location in the minimum amount of time, starting from a knowledge of Newton's laws of motion.

As with genetic algorithms, no mathematical proof is currently available that enables such a technique to be used for solving problems in general. All that is available is "a large amount of empirical evidence to support the counterintuitive and surprising conclusion that genetic programming can be used to solve a large number of seemingly different problems from many different fields."[12]

There are, however, critics of the evolutionary computing approach. The computer codes in question "evolve" in ways that resemble natural selection and so they frequently solve complex problems by means that even their creators do not fully understand. This is both a strength and a weakness. From a fundamental point of view, there is no rigorous mathematical theory of their capabilities, so that we cannot be sure of their reliability. As the pragmatist David Goldberg put it, "We have been concerned only with what works and why. . . . Nature is concerned with that which works. Nature propagates that which survives. She has little time for erudite contemplation, and we have joined her in her expedient pursuit of betterment."[13]

ARTIFICIAL INTELLIGENCE

Genetic algorithms and genetic programming can solve many "intractable" problems, and in certain respects they may be viewed as acting "intelligently." In the drive to solve complex real-world problems, scientists are not only mimicking the evolutionary process, but are also attempting to simulate the highest product of evolution—the human brain—by creating neural networks, a class of computer programs that, like genetic algorithms, apply nature's strategies in a wider context. Like genetic algorithms, neural networks are also capable of solving hard problems and, through the central role they play in learning and memory, may possibly lead to the creation of true artificial intelligence (AI).

We should state at the outset that AI is one of the most controversial of all contemporary academic disciplines, mainly because no one can agree on a definition of intelligence.[14] The possibility of neural computing was at first ignored by most members of the artificial intelligence community in favor of good old-fashioned AI (GOFAI).[15] This "top-down" approach tries to compartmentalize intelligence into discrete "modules" that deal with specific types of knowledge and information, such as perception, planning, and executing actions. Each module is equipped with explicit models of the external world, and they are supposed to interact by means of logical rules contained in an "inference engine" to produce intelligent behavior in a robot, such as grasping or avoiding objects. In its heyday in the 1960s, there were plenty of bold claims about the power of GOFAI, and the possibility of simulating intelligence within a computer. One of the most ardent advocates of GOFAI, Marvin Minsky of MIT, said in 1970 that "in from three to eight years we will have a [GOFAI] machine with the general intelligence of a human being"—a claim that ignominiously failed to materialize.[16] The mathematical physicist Sir Roger Penrose has achieved prominence for his attack on this position, also known as "strong artificial intelligence," and his arguments "to show the untenability of the viewpoint that our thinking is essentially the action of some very complicated computer."[17]

Allen Newell and Herbert Simon of the Rand Corporation helped GOFAI effectively get off the ground when they explicitly showed toward the end of the 1950s that the bit strings within a computer could be used to represent not only numbers but also more general symbols.[18] These symbols could be taken to signify aspects of the world in general, and their logical manipulation could be effected using rules provided by suitably structured programs. Since human brains also appeared to manipulate symbols, whether musical notes, pictures, or text, and carry out thoughts of infinite variety and ingenuity, it seemed to the early artificial intelligence community only a matter of time before GOFAI could embrace anything a human can do.

Evidence mounted that successful performance depended on rapid searches of large repositories of knowledge, cast into symbolic form. For example, AI used for medical diagnosis would have to emulate a doctor, who recognizes patterns corresponding to disease symptoms, and thereby gains access to his or her knowledge about the disease, treatment, and further tests. Experiments suggested that a human expert in any given domain is capable of recognizing 50,000 or more such familiar patterns, and uses these to key into relevant information stored in long-term memory.[19] Thus, GOFAI places faith in digital cramming: it is argued that a machine can be made clever by stuffing it with as many rules and facts as we can extract from such designated human "experts." The GOFAI community proceeded to design computational simulations of human intelligence along these lines. Such *expert systems* are comprised of three parts: a user interface, by which the user interacts with the computer and supplies relevant information; a knowledge base, the backbone of the system that stores information relevant to the problem supplied by experts; and an inference engine, which provides answers using systematic *logical operations* carried out from the knowledge base and user-supplied information. In the case of a medical expert system, the inference engine should be able to infer a case of hepatitis from flu-like symptoms, loss of appetite, malaise, and a yellowing of the patient's skin.

Early progress was made at Stanford University in 1956 by Edward Feigenbaum, who worked with Nobel laureate Joshua Lederberg to develop DENDRAL. This expert system could infer the structure of a molecule from the readout of a mass spectrograph, which itself analyzes chemicals by breaking them down and "weighing" the resulting fragments.[20]

The success of expert systems in "knowledge engineering"[21] rests on their restriction to a narrow domain. Indeed, expert systems have been useful in dealing with a wide range of highly specific yet complicated tasks, including financial and manufacturing problems, airline flight and passenger scheduling, medical diagnosis, and drug design. Though they are found in many fields of human endeavor,[22] their capabilities are tightly circumscribed by the nature, quality, and consistency of the human rules they are based on; there is always the danger that, if insufficient attention is paid to the needs of the users, they will become "data tombs"—vast repositories of useless information.

Expert systems suffer from brittleness: they have no "common sense" to speak of and are unable to cope with imperfect or ambiguous information. Take, for example, the task of persuading computers to understand speech or typewritten input in a language such as English or French. In the 1960s, a program called ELIZA was developed for reproducing the conversational skills of a psychotherapist.[23] Superficially, dialogue between it and a human interrogator looked impressive, but after a while it became apparent that any skills in its repartee were purely illusory: it would turn replies into questions and was programmed to respond aggressively to any mention of mother, father, or dreams—no matter what the context. A statement such as "Nelson Mandela is the father of the new South Africa" would be met with: "Tell me about your father."

The grandiose claims of the GOFAI approach began to go awry in the early 1970s when it was shown to be incapable of coping with children's stories. The hurdle was how to program a computer with enough general knowledge to attain an understanding that could match that of a four-year-old.[24] Natural languages are far more complicated than was originally thought: they encode a vast amount of context-dependent knowledge that enables us to imbue verbal utterances with meaning. The statement: "Peter saw the computer in the store window and wanted it" would leave a GOFAI computer program baffled as to what he wanted: the computer, the shop, or the window. There were attempts to overcome this puzzle by simplifying the problem. One small success story was the SHRDLU program developed by Minsky's student Terry Winograd that could use language unambiguously—so long as all you wanted to discuss was an idealized "microworld" consisting only of movable colored blocks.[25]

Today, most people accept that the grand ambitions of this style of AI will never be realized. The long-standing American AI critic Hubert Dreyfus believes that for a "mind" to have common sense, it needs a body to investigate the world: "Anybody who has children must be struck by the number of years they can spend playing around with sand, even just playing around with water, just splashing it around, sopping it up, pouring it, splashing it. It seems endlessly fascinating to children. And one might wonder what are they doing? Why aren't they getting bored? How does this have any value? Well, I would say, they're acquiring the 50,000 water sloshing cases that they need for pouring and drinking and spilling and carrying water. And they've got their 50,000 cases of how solids bump in, scrape, stack on, fall off. Common-sense knowledge would, in this story, I believe, consist in this huge number of special cases which aren't remembered as a bunch of cases, but which have tuned the neurons, so that, when something similar to one of those cases comes in, an appropriate action or expectation comes out. And that's what underlies common-sense knowledge."[26]

The real bane of GOFAI are the tasks we perform totally effortlessly, such as recognizing a face in a crowd, speaking and interpreting language, and avoiding flying rocks, arrows, and spears. A primary reason for the ease with which we handle such tasks—which have been vital for human evolution and survival—is the structure and function of our brains. Unless we understand how the brain has been adapted to these tasks, we will not be able to emulate it effectively, much less turn it to advantage for dealing with complex systems in general. As we will repeatedly stress, one of the most important features of the brain is its ability to learn by interaction with its environment. Another is its inherent parallelism. While you are reading this sentence, many other bodily and thought processes are taking place simultaneously, each controlled by the brain's nervous system. Some you may be conscious of, others not. Today, there is new hope for artificial intelligence, thanks to a return to the original inspiration that fired Alan Turing and John von Neumann.

NEURAL NETWORKS

One of the first attempts to design a computer based on the structure of the brain came from von Neumann's EDVAC report, which reveals how he was struck by the mathematical description of brain cells put forward in 1943, when Warren McCulloch and Walter Pitts of the University of Illinois published their paper entitled "A Logical Calculus of the Ideas Immanent in Nervous Activity."[27] Their pioneering work suggested that the brain cells—neurons—used for thinking can be considered to act as logical switches operating on the basis of Boolean (binary) arithmetic. McCulloch and Pitts argued that, by combining collections of such elementary units into "neural networks," they could perform any operation within the propositional calculus (the laws of logic).[28]

McCulloch and Pitts proposed a bottom-up approach to artificial intelligence modeled on the massively parallel interconnections of brain neurons. In the brain, a neuron receives signals from other neurons through a web of gossamer connections called dendrites. The neuron sends out spikes of electrical activity through a thin strand called an axon. In turn, each axon sprouts into many fine branches; each branch ends in a structure called a synapse that can pass on a signal to a neighboring neuron. Learning occurs when connections between brain cells sprout, wither, or alter in strength. To construct a computer-based model, we need an abstract logical representation of the structure of each neuron. This representation is shown in Figure 5.2. McCulloch

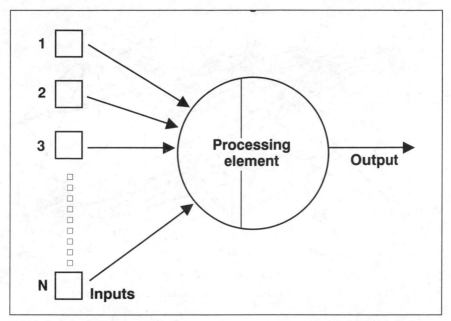

Figure 5.2   Bio-logic. The logical structure of a biological neuron according to McCulloch and Pitts.

and Pitts only considered two-layer networks, in which the first layer of neurons takes information from the outside world; the connections of these to the neurons in the second layer lead to a response that is output to the external world.

Artificial neural networks are highly simplified logical representations of large collections of such simple and identical McCulloch-Pitts neurons, simulated in software or actually built into silicon chips. They share several features with the cellular automata discussed in Chapter 4. Each neuron (also referred to by computer scientists as a processing element) in such a network is connected to a great many others that can be arbitrarily far away (Fig. 5.3), whereas the sites in a cellular automaton only communicate with their closest neighbors.

Another stimulus for the interest in artificial neural networks came from the American mathematician Norbert Wiener, whose 1947 book *Cybernetics* proposed a new science dedicated to the study of control mechanisms in man and machine.[29] He saw theoretical parallels between devices that control machines and the biological mechanisms that allowed the brain to rule the body.[30] A little later, the psychologist Frank Rosenblatt of Cornell University produced a particular neural

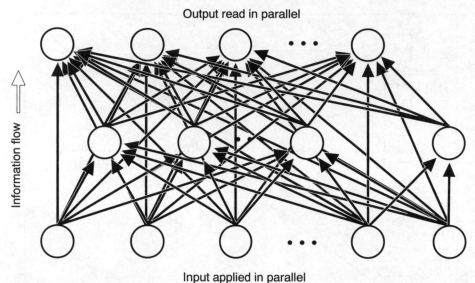

Figure 5.3   Typical architecture of an artificial neural network.

net structure, which he called the perceptron.[31] His Mark 1 Perceptron, unveiled in 1960, consisted of 400 photocells, which crudely represented the light-sensitive neurons in the retina, and a set of association units. Each unit merged signals from several photocells and sent that signal to a bank of response units. The network could be trained to recognize letters of the alphabet by "punishing" it if it guessed wrong and altering its connections until it achieved the desired result.

Although Rosenblatt's work appeared promising, the field unfortunately suffered a setback shortly afterward. In 1969, a book entitled *Perceptrons*, coauthored by Marvin Minsky and Seymour Papert, delivered a highly damaging critique of Rosenblatt's neural networks. These authors favored the top-down GOFAI approach. "Why bother, after all, to design and build new, complex and very probably unworkable machines when the digital computer was already tried, tested, and apparently on the very brink of becoming intelligent?" wrote Igor Aleksander of Imperial College, one of the researchers who struggled on in the wake of this attack. "With the benefit of hindsight, we can see that this dismissal of the bottom-up approach was both foolish and unjustified."[32]

At the heart of Minsky and Papert's book was a theoretical analysis of the performance properties of the two-layer perceptron. In fact, they did demonstrate that the perceptron, in its most elementary form, was devoid of any interesting properties.[33] Almost two decades later, a new

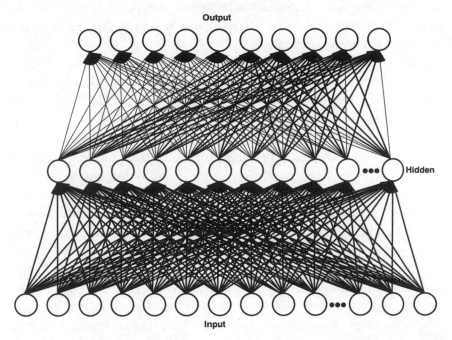

Output

Hidden

Input

Figure 5.4 An example of a multilayer perception with one hidden layer. Note that the connections are only between neurons in adjacent layers in the network.

edition appeared that was unrepentant: its prologue declared that "little of significance had changed since 1969, when the book was first published."[34] The authors conceded that "some readers may be shocked at this" and that "discoveries have been made that may, in time, turn out to be of fundamental importance" but argued "there has been little clear cut change in the conceptual basis of the field."

This view is not shared by those in the "connectionist" community, and symbolists such as Minsky and Papert cheerfully admit to being at war with them.[35] The simplest modifications to two-layer perceptrons lead to all sorts of exciting new possibilities. The "fundamental" shortcomings pointed out by Minsky and Papert were overcome in the mid-1980s by David Rumelhart and James McClelland of the University of California at San Diego, who extended the simple perceptron's capabilities by making two modifications. First, they added more neurons in a well-defined third layer—the hidden layer—sandwiched between the input and output neurons. There is full connectivity between neurons in adjacent layers, but none within any given layer, or across more than one layer. Figure 5.4 shows an example of a three-layer network, with just a single hidden layer, although it is quite possible (though un-usual) to introduce as many layers as one wishes.[36] The result is the *multilayer perceptron*, or MLP.

The second modification was the implementation of a novel algo-rithm, called "the backpropagation of errors" or, more simply, "back-propagation," which enabled the network to learn effectively.[37] This method was first proposed by Paul Werbos in 1974 while he was working for a doctorate at Harvard University.[38] The network is liter-ally trained on examples of problems for which both input and output data are available. For instance, it may be trying to recognize alphabet-ical letters in handwriting. The network initially predicts letters when given the scrawl. Its guesses are then matched up with the true letters, and the errors in the predictions are recorded. At this point the back-propagation algorithm comes into play; it *minimizes* the network's er-rors by propagating these errors back through the network, adjusting the connection weights—the strengths of connections between indi-vidual neurons—until ideally the global error minimum is located. When the error minimum has been found, the network is said to have been trained and in a condition suitable for "generalization"; in other words, it is ready to be asked to decipher a range of previously unseen handwriting.

We can envision what is going on in more detail by depicting an er-ror landscape, which is conceptually similar to the many other fitness landscapes we have described for other complex problems. The multi-layer perceptron's learning algorithm searches for the lowest feature (the lowest error) on that landscape. For a given piece of handwriting under the network's scrutiny, the landscape is a very complicated func-tion of the many connection weights in the net. Like the previously encountered examples, the search for the global error minimum is a generically "hard" optimization problem, since the landscape lies within a high-dimensional space: for instance, if there are twenty connections of interest, the error surface would be in a twenty-dimensional space (see Fig. 5.5). The backpropagation algorithm is designed to locate the global minimum. A number of criticisms can be leveled at the algo-rithm, however. The algorithm has a tendency to get trapped in local minima rather than the global minimum; it can also be very slow to converge if the landscape is complicated. But there are ways of over-coming these problems.[39]

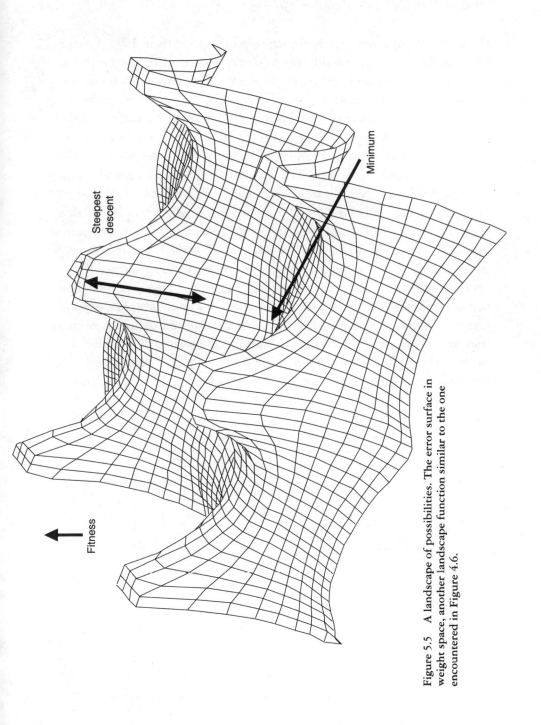

Figure 5.5 A landscape of possibilities. The error surface in weight space, another landscape function similar to the one encountered in Figure 4.6.

FROM SPIN GLASSES TO NEURAL NETWORKS

Little did Sherrington and Kirkpatrick realize that their 1975 infinite range spin-glass model would help to clarify aspects of the way artificial neural nets function and, indeed, in certain respects, how the brain works. The relevance of spin glasses to neural networks does not come from any physical similarity (there is none) but rather from a more abstract conceptual analogy. It is important, however, for it shows that a two-way exchange can occur between ideas on complex systems developed in the inanimate and animate worlds. The link between spin glasses and brain-like assemblies of nerve cells stems from the fact that in each model, simple units (spins or neurons) can influence each other over very large distances. Because the connection strengths between neurons in a network can be either excitatory or inhibitory—that is, they can either respectively increase or decrease the propensity of a neuron to pass on a signal—many conflicts of interest can arise between the neurons. These conflicts in a neural network—the equivalent of frustration in a spin glass, described in Chapter 4—make it hard to determine what the neural network's overall state will be when it is subjected to some kind of external stimulus. Indeed, working out what such stimuli can do to such a large and heavily interconnected network turns out to be very similar to setting off on the spin-glass free energy landscape, in search of the deepest valley.

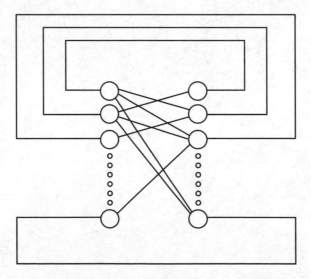

Figure 5.6   The Hopfield network—a fully connected artificial neural network.

The profound analogy between the problem of understanding neural networks and the magnetic behavior of a spin glass was first realized by John Hopfield of Caltech. He showed that there is a mathematical mapping of the Sherrington-Kirkpatrick spin-glass model onto a simple type of fully connected neural network model (Fig. 5.6) now called the Hopfield network.[40] A fully connected network is one in which each neuron is connected to every other neuron; the Sherrington-Kirkpatrick spin glass is one in which each spin feels the presence of every other spin. Note that Hopfield's net differs significantly from the MLP we described before. All the machinery developed in statistical physics for the analysis of spin glasses could therefore be applied wholesale to understanding various problems in neural network research. In two respects, the physics of spin glasses has helped clarify how neural nets work. First, spin glasses account for the global behavior of networks of given architecture and connection strengths. Second, they provide a recipe for choosing these connection strengths in order to get the optimal performance from a network, whether recognizing a signature or the license-plate of an automobile.

The rekindling of interest in the study of neural networks internationally (and particularly in the United States) following the blow meted out by Minsky and Papert was due in considerable measure to Hopfield's paper published in 1982. Hopfield believed that his neural network provided the basis for a practical implementation of a form of memory ideal for large databases, called content-addressable memory. Also known as *associative memory*, it captures, in essence at least, the way we as humans store and access our memories.

In conventional computers, data is filed in a structured way, usually in labeled (numbered) files. The labels can be thought of as addresses. To get access to any piece of data, users must specify the appropriate address and the data will be retrieved. This method is brittle in the sense that any error in the address will lead to failure—either the system will crash or the information will be retrieved from the wrong address. Content addressability is intended to remove this brittleness. The user specifies some keyword or words from the file, which may even be incomplete or incorrect, and the system should then be able to search through the database and retrieve the full information record desired. From "Pluck from the memory a rooted sorrow, Raze out the written troubles of the brain," a literary database would associate Shakespeare's *Macbeth*. This idea was not new: it was in circulation among engineers and computer scientists

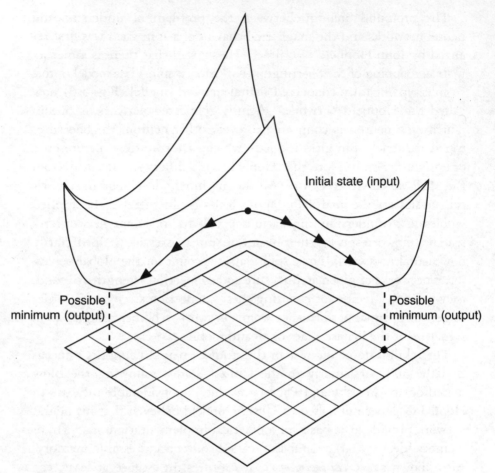

Initial state (input)

Possible
minimum (output)

Possible
minimum (output)

Figure 5.7   Attractive landscape. The stable patterns of activity within an attractor neural network. The particular pattern that is output by the network depends on the "stimulus" input to the network, and is one of the network's so-called "attractors."

and had already been implemented.[41] The novelty of neural nets resides in their tolerance for only partial or erroneous data.[42]

It might be thought that the list of which neurons are firing at any given moment in time in a Hopfield net would be haphazard, but in general the patterns that appear while the net is active evolve into organized forms, in which specific neurons fire constantly and others remain quiescent. The pattern that emerges depends on the initial input "stimulus" to the network—that is, which neurons fire initially (see Fig. 5.7). Because many feedback processes couple the neurons to one another, leading to significant nonlinear effects, the network constitutes a nonlinear dynamical system of the type we will discuss in more detail in

Chapter 6. For a well-defined set of initial neural activation levels (the input or initial conditions defined by an external stimulus), the network will eventually settle down into a specific activity pattern, which is an *attractor* for this input set. Just as a suitably trained handwriting recognition MLP network can deduce the letter "a" when the word "aardvark" is written by several hands, so several inputs can produce the same attractor in a Hopfield network. The set of input states (stimuli) that tend to roll into a given attractor state is called the *basin of attraction*. Hence, feedback neural networks such as Hopfield's are also sometimes known as *attractor neural networks*.[43]

Our memories work in a similar way: all input stimuli within a single basin of attraction lead a network of neurons in the brain to converge into a single firing activity pattern. This pattern can be thought of as encoding the underlying information content associated with those stimuli. For example, the inputs might be different pictures of the same face, in different shades of light, with short and long hair, with and without a beard, and so on, and the attractor state could be "Joe Bloggs." Thus, the network can essentially store as many different pieces of information as there are attractors in the network. These attractors can be regarded as the positions of *local* minima in an error surface defined in a similar way to that in the multilayer perceptron.[44]

For Rumelhart and McLelland's MLP, learning was a central issue, yet it did not play any explicit role in Hopfield's initial ideas. The information that could be memorized was just a consequence of the initial description of his net. Hopfield did, however, go on to recognize that in neurobiology the basis for learning is enshrined in a rule described by the Canadian psychologist Donald Hebb in his book *Organisation of Behaviour*, published in 1949. Hebb proposed that, during learning, neural connection strengths would be increased if a neuron fires at the same time as one or more of its synaptic inputs are firing.[45] Hopfield showed that, if a Hebbian learning rule is used to alter the neural connection strengths during and as a consequence of their firing activity, it has the effect of altering the shape of the net's error landscape by deepening some of the wells, making others shallower, and so on. Thus, repeated exposure to external stimuli—learning or training—*does* alter the neural activity patterns in a Hopfield network that uses Hebbian rules.

Working with David Tank, Hopfield went on to show how his type of network could deal with "hard," that is, NP-type, problems, including that of the traveling salesman.[46] They found that, although the search for solutions to such problems could become trapped in local minima on the error surface, the network in practice often finds solutions close to the required optima. Nevertheless, the danger posed by having the search become trapped in local minima casts some doubt on its general capabilities. In 1984, Geoffrey Hinton, who was then at Carnegie Mellon University, Pittsburgh, and Terry Sejnowski, then at Johns Hopkins University in Baltimore, proposed the use of simulated annealing in a so-called "Boltzmann machine," an extension of the Hopfield network named after the great Austrian physicist, Ludwig Boltzmann, one of the pioneers of statistical mechanics.

Just as the random thermal motions of atoms that occur during annealing help a metal crystal to settle into its most organized atomic arrangement, so simulated random noise generated in a computer can be made to shake the neural network out of local minima and guide it toward the deepest valley on the error landscape.[47] As before, the volume of noise is gradually reduced so that the network evolves toward progressively more stable—and more information-rich—minima.[48] The Boltzmann machine thus guides the neural net to converge into states with a probability dependent on their individual errors ("energies")—that is, on the depth of the landscape feature.[49] In addition to improved solutions to NP problems, Boltzmann networks have enormous potential for storing large amounts of information and recalling it speedily. Given one input, they can retrieve a vast amount of associated information without the need for formal, rule-driven search procedures. Their drawback is that they are among the most sluggish of all networks.[50]

THE SHOCK OF THE NEW

Your brain is adapted to deal with novel or unfamiliar situations—say, how to behave when a nearby apple cart is overturned. Yet simple artificial neural networks, such as the multilayer perceptron, can be thrown by novelty. One nugget of new information can often upset the connection weights of the net. As a result, the landscape (or, equivalently, the internal network representation) sprouts new hills and valleys. This means that the network must be trained all over again, usually a lengthy process. Often one would prefer a network to be capable of adaptive learning, by which is meant "learning on the job." New information added to old, one feels, should not upset the apple cart. Clearly, an optimally trained net must be at the same time stable and receptive to new information, a trait called *plasticity* in the trade: this is the stability-plasticity dilemma. Many networks are too unstable.

A neural network that continually learns has been developed by the American computer scientist and mathematician Stephen Grossberg, working with Gail Carpenter. It is a kind of self-organizing neural network, that is, it is composed of nets that do not need to be trained or supervised.[51] Grossberg and Carpenter call them adaptive nets because they are based on certain biological models of behavior and cognition. Using so-called adaptive resonance theory (ART), and including only two layers, these networks can detect and learn generalizations very rapidly.[52] The name derives from an analogy with the phenomenon of resonance, wherein a small oscillation of the same frequency as the natural vibrations of a mechanical or electrical system can set the system oscillating with a large amplitude—with shattering results, as when an opera singer hits the resonant frequency of a glass. Similarly, information that propagates through an ART network oscillates between the neural layers; during this resonant period adaptive learning occurs. ART exploits to the fullest one of the inherent advantages of neural nets we have already discussed—massive parallelism. One variant (called ART-3) employs equations that model the dynamics of neurotransmitters, the chemicals that carry messages in the brain.[53]

Testimonial to their benefits can be found in the marketplace. Like other nets, ART nets are being used in a steadily increasing number of industrial applications. At Boeing in Seattle, for example, ART nets are employed to group engineering designs so that when a new design is presented to a database any similar preexisting designs can be found from the classifications the network has created. Significant savings in terms of process planning and manufacturing may arise if an earlier design is reused.[54]

DO-IT-YOURSELF NEURAL NETWORKS

In laboratories across the world, computer scientists are building neural networks for a huge range of applications. The most common variety has

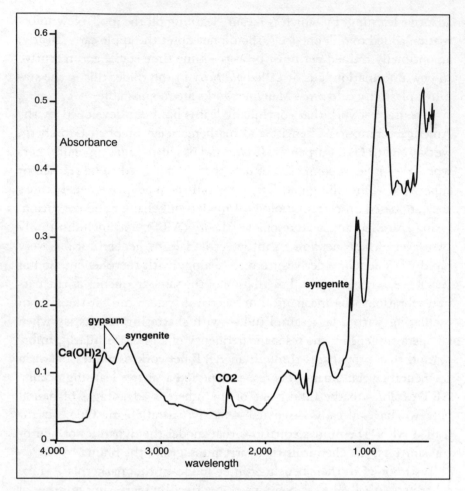

Figure 5.8a   Cement analysis. The infrared spectrum of a typical oilwell cement.

been the multilayer perceptron with the backpropagation learning algorithm. Their inherent flexibility means that these nets are able to model essentially any arbitrary relationship, whether that between pen strokes and letters of the alphabet or stock movements and currency rates. Instead of talking in abstract terms, let us examine a specific application: the design of neural networks able to predict how a cement slurry will set.[55] This problem is of considerable importance both within the oil and construction industries, since one needs to know how much time is available for manipulating the slurry before it solidifies.

Precisely because we know so little about cement, and in particular the relationship between the properties of the powder before mixing with water and the way it hardens after mixing, neural networks offer

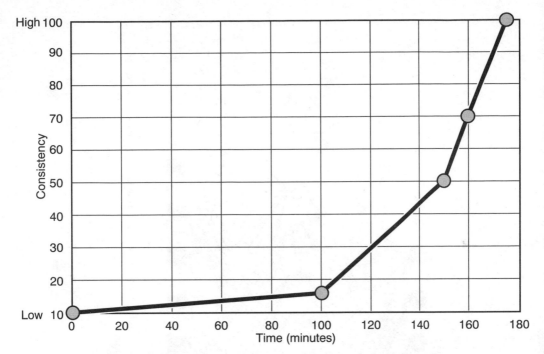

Figure 5.8b   Cement hardening. The profile shows how the cement thickens.

an excellent technique for gaining control of this material. A group at the Schlumberger Cambridge Research Laboratory has been able to link a measurable property of the powder—the signature of infrared light it absorbs—to how long it needs to harden. In essence, one takes the infrared spectrum of the powder as the input to a neural network. That is, the amount of radiation absorbed at every frequency is fed into the input layer of the network. Figure 5.8a shows an example of a typical cement infrared spectrum. The output returned by the net is a curve depicting how the cement slurry hardens after mixing with water (see Fig. 5.8b).

The network has a fixed number of input and output neurons, but nothing is said about how many neurons should be in the hidden layer. Selecting the overall number within this layer is frequently something of a black art. Only a few rules of thumb provide guidance, so often all we can do is try out a range of different numbers of hidden layer neurons. Figure 5.9 shows the architecture of one network used. More sophisticated ways of dealing with this are emerging, however. A technique has already been developed that allows the optimum number of neurons to be selected with a genetic algorithm. Using Holland's

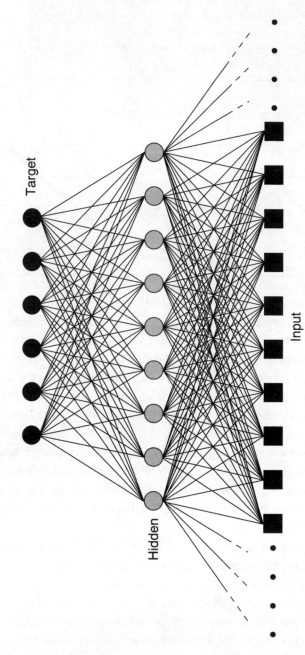

Figure 5.9 The neural network used to predict the setting times of oilwell cements. The input layer receives infrared data; the output layer provides estimates of the performance of the cement.

methods of evolutionary programming, the algorithm chooses the best network for the job.

Once we have connected up the network, it has to be taught or "trained" to deal with the problem at hand. In the present case, it must be given a large number of examples of cement powder infrared spectra. As each one is shown to the net, the correct "answer"—the associated setting curve—is also shown to the output nodes. At the outset, an untrained network takes each spectrum and propagates it through the many connections to the output neurons, modified by the strength of the connection or weight. These weights are initially set at random values. However, as the process is repeated, connection strengths are varied according to the backpropagation algorithm so that it learns the link between spectra and setting.

The inner workings of the net are as follows. Each neuron in the hidden layer receives a signal from every neuron in the input layer. Neurons add up the signals they receive, and, depending on whether that sum reaches a threshold, decide to "fire" and send on a signal themselves—like flood gates that open when all the streams leading to the dam reach a certain height. In practice, each neuron is rarely either fully on or off. Instead, as the signal received at a neuron increases in intensity, there is a smooth but rapidly varying level of neural firing that passes on information from the input—the spectrum—to the hidden and then to the output layers.[56]

At first, during teaching, the output based on the infrared spectrum of the cement bears no resemblance to the true cement setting curve (see Fig. 5.8b). However, as time goes by, these errors are altered through the adjustment of the myriad connection strengths between neurons in the net. This is done in such a way so as to minimize the difference between calculated and actual values. This process, which continues until the error on prediction for these known cases drops below some acceptably small value, is the first stage of the training or learning exercise. In the present case, training times for the MLP based on around 120 cement spectra lasted between two and three hours.[57]

Now the network may be put to its first real test. We can present it with a new spectrum, this time denying it the right answer. The network has no "teacher" to assist it, and is being asked to do its best to "generalize" from the cases which it has encountered and been trained on. Provided that the training was done properly, and the new spectra look a bit like those used for training, then it is reasonable to expect that the network's prediction will be good. Initially, its prediction can be compared with the known values. If its performance turns out to be good, then the network can be turned loose with a vengeance on cements for which no performance information is available. This is precisely what has been done in the present case: a neural network is now in operation on a small personal computer that can take infrared spectra of cement powders and predict their setting time in about fifteen minutes rather than waiting six hours to see what will happen to the slurry.[58] Almost all the fifteen minutes is taken up with making the measurement—the net's prediction is essentially instantaneous. This represents an enormous saving of labor, and greatly speeds up the process of analyzing a wide range of samples.

As we saw earlier, the multilayer perceptron is somewhat prone to stumbling into local minima. For many problems a slightly different type of neural network, called a *radial basis function* network, is more effective.[59] These nets have the same layered structure as the multilayer perceptron, but by automatically preclustering the initial data into groups with common attributes (such as particular infrared spectral features in the case of cements), these nets have a more appropriately selected hidden layer[60] and as a result the supervised learning taking place can now all be done using linear mathematics. This makes the task very fast and guarantees that the trained network will hunt down the global minimum.

As this example demonstrates, the fascination of neural nets is that they require no explicit programming to get results. In the case of the cement powder, the network does not have to be told anything about the details of the physics and chemistry: it is smart enough to work out the link between composition—as expressed by the spectra—and the setting properties without any other information. This is probably beyond the capabilities of any human being. Despite this awesome power, the neural net approach is inimical to some people, who are worried that understanding is being sacrificed for black art and a black box. In fact, it is often possible to interrogate a trained network to unravel its findings and interpret them in ways that we can comprehend. Nonetheless, the concern expressed is one aspect of a dichotomy that occurs throughout this book, between those who seek clean, comprehensible mathematical answers and those who do not: if the former criteria were applied for choosing problems to work on, then the vast majority of real-world problems might remain forever unsolved.

NEURAL NETWORK

Neural nets are today being applied to an impressive range of problems. One of the first nets to reach the market was Wisard (Wilkie, Stonham, and Aleksander's Recognition Device), which could be trained to recognize objects or, more whimsically, whether its British creators were smiling or frowning. There are a host of other applications in pattern recognition; for instance, identifying a submarine target from a sonar trace, screening cervical smears for abnormal or cancerous cells, and recognizing handwritten postal code characters on mail.[61] The "Airline Marketing Tactician" has a multilayer perceptron to predict demand for airline seats, while other MLPs can spot abnormal heartbeats, predict stock market fluctuations, and perform "loan scoring" for banks (i.e., deciding to whom it is worth lending money). One of the most successful applications can be found in Europe, where a neural network has been designed by Danish computer scientists to grade pork. This method, developed by Hans Thodberg of the Danish Meat Research Institute, is now used to grade the amount of meat and fat in up to 20 million carcasses each year.[62] Each carcass passes inside a robot where nine optical probes are inserted that respond to the difference between the white fat and red meat. From these sensors, the neural network can assess the meat content of about 360 pigs every hour to within 1.5 percent, a greater accuracy than that achieved by human inspectors. Yet its 200 or so "neurons" give it brain power closer to that of a worm. "It is a simple brain but it is tailored exactly to that problem," according to Thodberg.

An important consequence of the highly distributed, parallel processing nature of neural networks is that they are fault tolerant: if one neuron stops working (if it dies, biologically speaking), the whole system can continue to perform without significant error. As more and more neural processing elements fail, the overall performance undergoes a graceful degradation. This robustness has obvious advantages, both in nature and in our technologically oriented society: it is often preferable to use a fault-tolerant machine with many processors than having to rely on the perfection of a single one, as with contemporary serial computers. If that processor fails, everything else fails: this may lead to major disasters, such as the crashing of an aircraft or the misdirection of a billion-dollar satellite.

It is also possible to mix ideas from neural networks with those from Darwin. Shara Amin and José-Luis Fernández of British Telecommunications Laboratories in Martlesham Heath, England, have put evolution into action—in a self-organizing network of nodes—to produce an efficient solution to the traveling salesman problem for 35,000 cities.[63] It is the latest in the long succession of large-scale solutions that started in 1954, when mathematicians at the Rand Corporation in California tackled a tour of forty-nine cities. By 1992, a team from AT&T Bell Laboratories in New Jersey, Rice University, Rutgers University, and Bellcore managed a tour of 3,038 cities, using eighteen months of computer time.

The BT team started out with a random solution to the problem, consisting of a network of virtual nodes corresponding to stopover points, and then bred new solutions. The program carries out two steps: first, it works out how close each node used in the initial attempt is to a city by, in effect, calculating the gravitational tug between them. The program then moves the nodes in the direction of the tugs, and duplicates the ones that are pulled by two or more different points. The process of reproduction follows rules taken from biology: a newly bred stopover point cannot reproduce until the next generation, but rather, like a real child, has to wait until it reaches puberty. "If I allow points to breed immediately, the simulation is 20 percent worse," Amin observed.

To complete the Darwinian picture, the program wipes out those stopover points that are hardly pulled at all. This process of reproduction and death is then repeated in an algorithm that always gives a reasonable answer, however many cities are specified, and within a reasonable amount of time. "This is a good compromise between the time and accuracy of the solution," said Amin. Indeed, the method is so successful that British Telecom is already using it for solving routing problems.

NATURE'S LESSON

The class of problems that are computable in theory but intractable in practice have succumbed to the techniques outlined in this chapter. It is fitting that many of these modern computational methods have been directly inspired by nature. They provide a powerful means of understanding complexity in general.

In the next two chapters, we will see how complex processes akin to those described here and in the previous chapter can emerge from the simplest circumstances. For many physicists, this would mean attempting to express complexity in terms of the most basic units of matter and the most fundamental forces—a futile reductionist quest. A more appropriate language for depicting the complexity of the real world is chemistry. The full gamut of complexity can emerge from the simplest chemical circumstances, providing a powerful nonlinear paradigm for the patterns found in nature.

Indeed, we will discover how chemistry can even solve mathematical problems. To find the quickest route through a maze, or advise a robot on how to negotiate around the rows of shelves in a warehouse, follow the path of a "chemical wave." This is a self-propagating chemical reaction front that can explore a labyrinth in a manner akin to the way a ripple of chemical activity can probe the networks of the brain. Again, there is a symbiosis between computing and complexity. Just as chemistry can compute, so computers can unravel the secrets of chemical complexity, ranging from the highly organized to the utterly chaotic. In turn, these chemical processes may be honed in the environment by Darwinian evolution and tempered by a series of contingent and fortuitous events, to generate the seething diversity of the living world, from wriggling bacteria to charging elephants.

Chapter 6

NATURE'S ARTISTRY

Anon out of the earth a fabric huge
Rose like an exhalation
—JOHN MILTON

Complexity arises at many levels in nature, fashioning pattern within pattern and endless tiers of design. The result is a variety of styles that span extremes as diverse as cave drawings, Renaissance art, and Abstract Expressionism. In this chapter and the next, we will take a stroll down the gallery of natural art, examining how the computer is helping to provide insights into the "self-organizing" chemistry that paved the way to the emergence of living things, the processes that turn within the cells of our bodies, and the patterns formed by societies of creatures.

There have been many false dawns in the quest to understand the genesis of natural complexity. The celebrated English theoretical physicist Paul Dirac claimed in 1929 that with the rise of quantum mechanics "the underlying physical laws necessary for the mathematical theory of a large part of physics and the whole of chemistry are thus completely known."[1] Given that the human body is a glorified chemical reaction, does that mean the myriad processes turning in our cells, guts, and brains could be reduced to a quantum description? It is an intriguing thought. Unfortunately, after struggling for over half a century with quantum mechanics, chemists have only managed to apply it to a meager assortment of small molecules, a far cry in complexity from proteins in the human body. The fundamental problem with this view is not that one simply needs large computers to handle large problems but rather that it fails to recognize the possibility of fundamentally new phenomena arising from the collective interactions of large numbers of particles.[2] Our quest to understand how chemical complexity can emerge is off to a poor start, let alone any bid to understand "self-organizing" chemistry. For instance, how do such processes produce regular color changes in a chemical clock, or create stable spatially extended patterns? And how did these same processes lead to the emergence of life in a primeval puddle many billions of years ago?

It is not that we have run up against Gödel; using quantum mechanics to describe chemical reactions is rather like chartering an airplane to cross a road. This would be a sledgehammer approach in which a computer puts the machinery of quantum mechanics through its paces. When combined with fancy mathematical footwork typically involving the use of approximations, this can go some way toward simulating a range of very simple chemical reactions, yet the results fall far short of dealing with "real-world" chemistry.[3] The method is still generally intractable, with the time taken to perform a calculation depending on a very high power of the number of electrons in the problem being studied. A radically different, yet far more powerful approach is to combine computers with a mathematical description of chemistry that works at a "higher" level.

By higher level, we mean a description couched in terms of macroscopic properties, such as the regular changes in the color (i.e., concentrations) of a mixture of chemical species that occur in a so-called chemical clock reaction, rather than a microscopic one based on trillions and trillions of component molecules. If we are to describe these emergent properties, we need concepts to handle the *global* properties of large collections of molecules, whether in terms of concentration or temperature changes, to avoid becoming overwhelmed by redundant detail. This is the business of the science of complexity. Equipped with tools such as nonlinear dynamics and nonequilibrium thermodynamics, together with computers, we can understand the tick of a chemical clock and many other complex phenomena.

THE GLOBAL VIEW

In physics and chemistry, the conceptual palette used to paint the big picture is conventionally thermodynamics, the science of heat and work born during the nineteenth century, when steam powered the Industrial Revolution. Instead of attempting to describe a steam engine in terms of its inner workings (incorporating an account of the many pistons, valves, and levers), thermodynamics builds a description of its behavior through overall macroscopic properties such as the operating temperature, the mechanical work done on or by the engine, and the heat exchanged with the surroundings.[4]

In a similar way, when a chemical reaction takes place, the thermodynamic description ignores the trillions of individual molecules and discusses instead their global properties, such as temperature, pressure, and so on. In traditional thermodynamics, attention is focused on the state of *equilibrium*, the full stop at the end of a chemical reaction. In many respects, it is the least interesting state of matter: for any living organism it corresponds to death. But for well over a century—since its origins in the middle of the nineteenth century—most physical scientists have concentrated their efforts on understanding matter in this dreary state.

The guiding principle of thermodynamics is the second law of thermodynamics, which expresses the fact that a macroscopic system, left to its own devices, will evolve to a state that maximizes the *entropy* of the universe. One can think of the entropy in thermodynamics as a quality that determines a system's capacity to evolve irreversibly in time. Loosely speaking, we may also think of entropy as measuring the degree of randomness or disorder in a system. As everyone knows, things left to their own devices have a tendency to wind up in a more disordered state, whether it be a back garden, an office desk or one's financial affairs. The second law of thermodynamics can be thought of as merely enshrining that observation as a scientific principle.

The landscape metaphor can help us visualize this guiding principle at work. Now the hills and valleys represent variations in such dependent thermodynamic parameters as the concentrations of the reacting chemicals, as a function of independent variables, such as temperature and pressure, that control the system. Equilibrium thermodynamics is only concerned with a single point on that landscape—the end state of the chemical reaction predicted by the second law. It has nothing to say about how the reaction ended up on that spot, which, for obvious reasons, is called a fixed point attractor.

However, the real world has little to do with the uninspiring state of equilibrium. Life consists of many highly organized biochemical processes, from those that take place in cell division and heart beat to digestion and thinking, all of which can only occur because they are maintained out of equilibrium. To describe these dynamical systems we must turn to *nonequilibrium* thermodynamics, which allows a herd of more complex attractors.

REAL-WORLD THERMODYNAMICS

One force drives the living economy of our planet away from equilibrium to ensure that the Earth does not collapse into a barren wasteland: unimaginable quantities of light particles (photons) raining down from our local star, the Sun. Photosynthesis, the harnessing of sunbeams by plants to convert carbon dioxide into simple sugars, is the primary source of living energy. So long as sunlight beats down onto the Earth, it will not wind down to dead, icy equilibrium.

Nonequilibrium thermodynamics divides naturally into two parts: the "linear" version describes behavior close to equilibrium, while the "nonlinear" version is concerned with behavior far from the equilibrium state. Even linear nonequilibrium thermodynamics is devoid of interest if we relate it to the study of complexity. As the Belgian Nobel laureate Ilya Prigogine showed in 1944, for a very wide range of situations thermodynamic behavior is universal, being reminiscent of equilibrium.[5] Far from equilibrium, however, no simple laws state how thermodynamic evolution will occur, other than that it must involve the dissipation of energy, where dissipation can be thought of as synonymous with increasing entropy. Dissipation occurs in every process of energy conversion, whether an athlete turning muscle fat into marathon miles or an incandescent bulb converting electrical energy into light. As a result of this dissipation taking place far from equilibrium, genuine complexity can emerge, whether a chemical clock or a convection pattern churning in a kettle. As we come to grips with complex real-world phenomena, we find that we need a richer language than thermodynamics, one supplied by nonlinear dynamics.

NONLINEAR DYNAMICS, COMPLEXITY, AND COMPUTERS

Nonlinear dynamics, like thermodynamics, provides a macroscopic description of processes that take place far from equilibrium.[6] However, these mathematical descriptions of the instantaneous change of behavior with the passage of time are much more quantitative. They can deal just as readily with the rate of change in the population numbers of a North American gypsy moth as with the varying concentrations of ingredients in a chemical reaction. Precisely because these equations are nonlinear, they permit a vast range of different possible behaviors, from organization to "chaos," and can account for much of the richness of the world we inhabit.

We have only begun to exploit the full scope and power of dissipative nonlinear dynamics within the past twenty-five years, thanks to the exponential increase in computer power. Programs for studying the nonlinear jungle are notoriously voracious consumers of computer memory.[7] One must be able to handle and manipulate very large amounts of data, either gathered from theoretical model equations or experiments, or from their combination. Whenever we simulate such behavior on a computer, the maximum possible amount of the computer's random access memory (RAM) needs to be used. Previously scientists had to resort to large "mainframe" computers; today desktop high-performance workstations have effectively replaced these larger machines for such applications, and even personal computers can help. With such tools at our disposal, we can begin to model nonlinear phenomena.

COOKING UP COMPLEXITY

Even something as simple as a puddle of oil can demonstrate one of the most intriguing nonlinear effects of complexity—*self-organization*, when patterns in space and time spontaneously emerge from randomness. If a thin layer of silicone oil is heated carefully from below, the initial

Figure 6.1   Order out of chaos, Bénard cells in a thin layer of heated silicone oil.

featureless uniformity of the liquid suddenly gives way to an array of hexagonal convection cells, forming a honeycomb pattern.[8] This pattern of so-called Rayleigh-Bénard cells, caused by convection, can be easily seen if some aluminum dust is sprinkled into the oil.

Concepts from thermodynamics and nonlinear dynamics can describe this behavior, up to a point. Thermodynamics signals that interesting things may happen, yet without describing what will take place; it focuses on the temperature difference between upper and lower fluid surfaces at which the hexagonal cells appear. This is a *bifurcation* point, which in effect says "expect novel behavior here." The word "bifurcation" means that the liquid can do one of two things as the temperature exceeds the critical value. A close look at the honeycomb pattern reveals this choice: neighboring cells have fluid convection currents turning in opposite directions. But the direction of rotation of convection currents within neighboring cells can be clockwise/anticlockwise or anticlockwise/clockwise. The actual outcome is a consequence of a remarkable cooperation between chance and determinism.[9]

The direction in which individual cells rotate depends on ever-present tiny and random fluctuations caused by molecular motions within the fluid; they are vital to seed self-organization. Near equilibrium the convection currents in the fluid are small and controlled and exert a negligible effect. Like a dying whisper, the fluctuations quickly fade away. But far from equilibrium, positive feedback can amplify these fluctuations to build up a microscopic convection current into an organized state that invades the entire dish and turns it into a liquid honeycomb.

Scientists versed in equilibrium thermodynamics would suggest that the more heat applied, the more the countless molecules should swarm willy-nilly around the dish. But oil in the honeycomb state is clearly more organized than it was before we applied heat. Vast collections of molecules move in unison over great distances to make the hexagonal "convection cells." To underline how unexpected this is, imagine a box filled with a uniform mixture of white and black marbles. The analogue of heating the fluid would be to shake the box violently. Imagine regular patterns of balls forming in this seething activity, in which, for example, all the black marbles are found at one end and all the white ones at the other.

The honeycomb pattern is no less remarkable: it implies that there is cooperative behavior of vast numbers of individual molecules in both time and space. If anything, it is more extraordinary, since the organization of the molecules stretches over a much greater distance than that of the marbles, relatively speaking. The forces that allow one molecule to influence another usually only extend over distances as little as one-hundredth of a millionth of a meter. Yet the self-organization that arises in a dissipative system stretches over comparatively enormous distances as great as centimeters. The system takes on a life of its own: it can no longer be viewed as a haphazard mix of randomly moving molecules. The molecules in the fluid have spontaneously self-organized. This organization changes as the fluid is pushed further from equilibrium: the honeycomb can give way to rotating spiral patterns[10] or targets[11] and then form parallel rolls of convection. Self-organization is no mystery, however; this complex behavior can be modeled in the heart of a computer, using the nonlinear equations of fluid dynamics. It is the inevitable consequence of far-from-equilibrium physical laws.[12]

SELF-ORGANIZATION IN CHEMISTRY

Our bodies contain vast numbers of complicated chemical processes that are far from equilibrium, just like the heated layer of oil. To maintain our bodies in such a state we rely on the activities of feeding and excreting. It is no coincidence that even simple chemical reactions can spontaneously form patterns in time and space when pushed into this nonlinear regime. Whether fleeting or static, they are created by coupling sequences of chemical reactions—interlocked by feedback processes where one reaction affects another—and diffusion of the chemical species involved. The dimensions of the resulting chemical patterns depend only on the recipe of the chemical cocktail and how far it is from equilibrium. Quantum theory could give few if any insights into such processes. Thermodynamics fares better by labeling bifurcation points where interesting chemical capers emerge. Armed with nonlinear dynamics, however, we can use a computer to explore the rich possibilities of complex behavior.

It is simple to maintain a mixture of reacting chemicals far from equilibrium—they need only be fed through a stirred chamber, like a flow reactor of the kind found in a chemical manufacturing plant. Chemicals enter the reactor, are stirred to ensure mixing and reaction, and the products drawn off downstream. Nonlinearities will occur if a product of a reaction (say, X) catalyzes its own production, a feedback process known as *autocatalysis*. The amount of X formed at any instant depends on how much X was there in the first place—the signature of a nonlinearity that is analogous to the feedback found in the heated layer of oil or between a microphone and a loudspeaker. Thus, without too much trouble, we have all the ingredients necessary for the production of complexity via self-organization.

Alan Turing was the first to envision these nonlinear patterns. His ideas were laid down in a remarkable paper published in the *Philosophical Transactions of the Royal Society* (Part B) in 1952, when he was forty years old and working at the University of Manchester. Turing was interested in furnishing a chemical basis for the means by which shape, structure, and function arise in living things, a process known in biology as morphogenesis.[13] He sought to understand how an organism transforms a chemical soup into a biological structure, and, following fertilization, turns a spherical bundle of identical cells into a fully fledged organism. It is one of the greatest puzzles of life. Take the process of gastrulation ("stomach formation"), in which an early embryo—a sphere of cells—loses its symmetry as some cells begin to develop into the head and others the tail. Starting off with a ball of cells, one might expect that the constant irreversible diffusion of the biochemical reactions controlling development would retain this symmetry and we should all be spherical blobs. But in his paper, Turing showed that the breakdown of symmetry necessary for a fertilized egg to develop into the complicated form of a creature can indeed arise. It is the same concept as one we have already encountered: at or near equilibrium, the homogeneous state of maximum symmetry is stable, but it can become unstable at greater distances from equilibrium so that a fluctuation can smash its symmetry. Turing illustrated this with a mechanical analogy: "If a rod is hanging from a point a little above its center of gravity it will be in stable equilibrium. If, however, a mouse climbs up the rod the equilibrium eventually becomes unstable and the rod starts to swing."[14]

In reality, few eggs are precisely spherically symmetric and in any case various factors, such as gravity, break this symmetry. However, Turing's ideas have indeed led to some impressive descriptions of pattern formation in both the animate and inanimate worlds, and form the mainstay of a large part of modern theoretical biology.[15] He made the important discovery of how nonlinear effects in a chemical soup can lead to the emergence of spatial patterns if several colored substances with different diffusion rates react with each other in a liquid. The theory is counterintuitive, because one would expect any type of irreversible mixing process to wash out preexisting patterns or structures just as the patterns made by milk added to black coffee eventually disappear.

Again way ahead of his time, Turing formulated a mathematical recipe for oscillatory patterns, which can be seen as wave-like ripples of color. He also suggested that if the reactants diffused at different rates, and the faster components inhibited reaction while the slower components promoted it, stationary patterns might be set up by these antagonistic processes. The bifurcation point at which the system is driven far enough from equilibrium for stationary patterns to emerge is now known as the Turing instability.

For the following fifteen to twenty years, Turing's work went largely unnoticed by chemists and biologists. There were several reasons for this hiatus. To make the mathematics tractable, Turing had to tame his nonlinear equations. He took the reasonable approach of assuming that the mathematics behaved in a linear, predictable fashion for a limited distance beyond the bifurcation point that breaks the spatial symmetry. Turing was thus able to work out when a pattern might be formed but not any subsequent changes in that pattern if the system were pushed still further from equilibrium. Turing realized that what would be needed to make further progress was a fast computer, yet none was then in existence. And in the decades that followed his pioneering work, no one could find a chemical reaction that laid down his stationary patterns.

THE BRUSSELATOR

Two developments in particular led to a surge of interest in self-organization in the wake of Turing's pioneering work. One was a symposium held in Prague, where Western scientists first learned about the intriguing Belousov-Zhabotinski chemical reaction (which we will describe shortly) and began to make a comparison with oscillations that occur in biology to help organisms harness energy, such as glycolysis and photosynthesis.[16] The second also happened in 1968 with work published by René Lefever and Ilya Prigogine,[17] who had become familiar with Turing's work after visiting Manchester the day Turing completed his morphogenesis paper.[18] Citing this seminal paper, they formulated a nonlinear dynamical model of a chemically reacting system compatible with their own nonequilibrium thermodynamic theory and possessed some of the necessary ingredients for spatial self-organization. This idealized model was later dubbed the "Brusselator" by John Tyson of Virginia Polytechnic University in 1973, due to its geographical origin.

The Brusselator lays bare how feedback and nonlinearity are essential for self-organization, where enormous numbers of molecules seem to "communicate" with each other in creating complexity. The Brusselator involves two chemical substances, A and B, which are converted into two products, C and D. To generate interesting nonlinear features four simple steps are included, involving two chemicals X and Y as intermediates. The details are straightforward: a molecule of A is first converted into one of X, which reacts in the second step with a molecule of B to form Y and product C; in the third reaction, two molecules of X combine with one of Y to produce *three* molecules of X. The final reaction is the direct conversion of X into product D. The springboard to self-organization—nonlinear feedback—is held in this third step of the Brusselator, where *three* molecules of X are formed from *two* by a reaction with intermediate Y. The feedback occurs because one molecule—X—is involved in its own production, an "autocatalytic" step. The nonlinear nature arises because for every two molecules of X reacting, another one is produced, making three in all.

The Brusselator can be held far from equilibrium by replenishing the reacting chemicals before they are used up. All we have to do is to carry out the reaction in a well-stirred open-flow reactor. There we can maintain the concentrations of A and B at the appropriate values by controlling their flow rates into the reactor, and similarly for the products C and D that are removed; only the concentrations of X and Y can vary with time. To learn how, we must write down and solve the mathematical equations describing the Brusselator—a pair of coupled differential equations describing X and Y. If the Brusselator is put through its paces, one gets a feeling for the elaborate behavior locked up in the mathematics. The model is simple enough so that we don't need a computer and the results are rather colorful if we assume that the chemical X is red and Y blue.

Let us begin somewhat perversely at the end of the reaction. The ingredients are mixed together, they react, and the reaction achieves equilibrium, when all chemical change ceases. At equilibrium, we have an uninteresting purple soup—a mixture of blue and red molecules. Little changes if the concentrations of A and B are maintained close to these equilibrium values. The interesting behavior begins when the flow of chemicals A and B into the reactor is cranked up to levels beyond a certain threshold from the equilibrium concentrations. For the chemical reaction, no matter what the starting concentrations of X and Y, oscillations occur beyond the critical point. The reaction mixture can turn red, then blue, and so on at regular intervals. This is known as the Hopf instability, after the mathematician who discovered it.

We can describe these color changes by appealing to the landscape metaphor, this time located within what is called "phase space," a term referring to the set of chemical concentration variables that change with time. The course of the chemical reaction can be represented as entrapment within a circular valley or moat, or less metaphorically in a closed loop or cycle within the phase space, called a *limit-cycle*. All one has to keep in mind is that, as the chemical reaction changes from blue to red, it moves around this valley like a ball bearing in the rim of a hat. This is another type of attractor. At one compass point, the

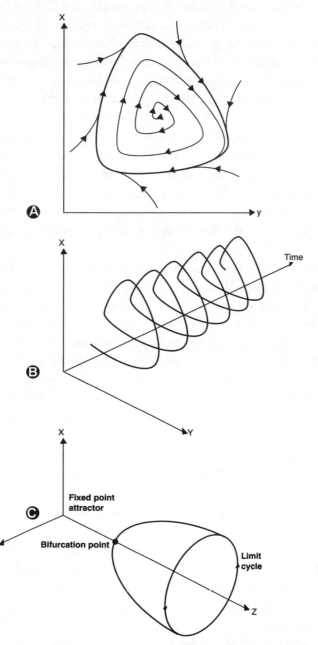

Figure 6.2 The chemical clock. (a) Two-dimensional projection of the attractor: X and Y are concentrations of chemical intermediates (colored blue and red, say). (b) Three-dimensional view, showing time going on as the color changes. (c) Limit cycle attractor emerging from a point attractor. The parameter z measures the distance of the system from chemical equilibrium. At the bifurcation point, the limit cycle unfolds and the chemical clock begins to tick.

chemical mixture corresponds to blue, at another, red. When no more chemicals are added, the chemical reaction has run its course, with the phase space trajectory dropping into a funnel in the landscape that represents the fixed-point attractor of thermodynamic equilibrium.

The Brusselator highlights how order may be created from the unlikely wellspring of mayhem through the process of self-organization. By holding the system sufficiently far from equilibrium—through constantly adding chemical ingredients—the chemical soup in the reactor turns periodically from blue to red to blue and so on indefinitely rather than forever remaining a featureless violet mixture. Such an oscillating reaction is often called a chemical clock because of the regularity of its chemical cycles.[19] Only if there is a continuous flow of reagents into and out of the Brusselator will the color changes continue indefinitely.

For anyone who has experienced nothing more than mundane classroom chemistry, such behavior is startling. Remarkably, all the molecules in the Brusselator seem able to "communicate" with each other over vast distances, all knowing when to turn blue, and when to turn red. The "tick" of this chemical clock as it rolls around the limit-cycle moat is a function only of certain physical properties of the Brusselator. It is completely independent of the initial concentrations used. Like the Bénard cells, this is an example of a *dissipative structure*, a term introduced by Prigogine to emphasize the origin of such self-organization in a far-from-equilibrium thermodynamic process. The term draws attention to the fact that dissipation, frequently associated with the winding down of isolated systems into randomness, can in fact play the very opposite role in underpinning the evolution of complexity. Other dissipative structures include human beings, ecosystems, and the organized light generated by lasers.

The concept of dissipative structures struck a chord in many diverse areas. It helped to create scientific, as opposed to purely mathematical, interest in nonlinear differential equations. The chemistry of oscillating reactions was taken up because they were a controllable and relatively easy-to-model manifestation of complexity. In turn, this effort paved the way for attempts to develop mathematical models of biological processes in single cells and multicellular ensembles, suggesting that biochemical cousins of oscillating reactions, which can also be described by limit-cycles, have major significance for life. Life, after all, is the most dazzling example of an ordered chemical phenomenon in nature. All this gave great impetus to the study of complexity.

CHEMICAL PATTERNS AND WAVES

Chemical clocks hold many of the secrets of complexity. We have dealt with patterns in time. What about Turing's spatial patterns? So far, we have ignored the possible role played by another nonequilibrium process—diffusion. We assumed that the chemical cocktail in the reactor was strongly stirred so that all the species (A, B, C, D, X, and Y) were uniformly distributed. This is unrealistic: the various chemicals milling around in the reactor take some time to meet. Certainly when the reactor is not stirred we cannot suppose that intermediates X and Y and products C and D are automatically formed in equal amounts in all regions of the vessel. Thus, we should think about what happens when we have to take into account how little reservoirs of reacting chemicals may form in one part of the reactor vessel and then have to migrate to other regions to participate in the feedback reaction.

To model the way reacting molecules mill around in the reactor before they bump into each other, we must consider the effect of diffusion. This is readily incorporated into the analysis by stirring in a term from "Fick's law," which relates the concentration of a substance at a given point in space with how it changes in time, for each of the different chemicals in the broth. The conversion factor in Fick's law that links the two is the diffusion coefficient, a property of a given chemical species: it reflects the fact that fat molecules diffuse slowly, slim ones diffuse quickly. It also takes into account the viscosity of the solution they have to move through.

With the help of Fick's law, we can intimately link the behavior of the chemical mixture in time with the emergence of patterns in space. The latter features place much heavier demands on the computational methods used to describe and simulate the development of complexity. Correspondingly, the chemistry becomes far richer. Now the limit-cycle resulting from a Hopf instability can turn in space as well as time. One familiar object that changes in space and time is a wave— think of one out at sea. And indeed, in an oscillating reaction where the Hopf instability rules, we should expect to see a ripple of red or blue pass through the reactor rather than the entire solution in the reactor simultaneously changing color to red or blue instantaneously. There is also the possibility of evolution to a fixed spatial pattern that does not change with time. In the case of the Brusselator, spots or stripes could appear in a hypothetical test tube. This is the Turing instability we mentioned earlier.

By putting the Brusselator through its paces we have been able to describe various permutations of complex, self-organizing behavior. The message is that the simple nonequilibrium Brusselator can become self-organizing in time, in space, or in space and time. There are limitations to the behavior of this mathematical model, however: the attractors are all two-dimensional limit-cycles. More complex behavior would demand a more complex shape of attractor. Instead of a neat circular orbit between red and blue, we could choose a squiggly orbit that spends more time producing the blue color than the red. Yet there is an important restriction: the trajectories on the landscape must not intersect, for that would lead to mathematical absurdities. In two dimensions, this severely limits the range of dynamical behaviors. But by mixing in a third chemical concentration variable beyond X and Y to the reaction scheme, these restrictions disappear and an almost unlimited range of complex exotica can emerge, each described by an attractor.

THE CROSSCATALATOR

Although there is nothing quite like it known in nature, the crosscatalator is a paradigm that illustrates how simplicity is the mother of complexity.[20] To create the crosscatalator, all that is effectively required is to add one more reaction step to the Brusselator—say, a chemical reaction in which the blue chemical is turned into a green one. What results is a mind-boggling range of possible behaviors. Studied in detail by one of the authors, working with colleagues in Schlumberger and Cambridge University, its antics can only be fully explored with a computer.

Arguably the simplest chemical scheme capable of producing any significant degree of complexity at all, the crossscatalator is a remarkable antidote to the sterile debate about complexity. It provides a simple recipe for just about the entire repertoire of possible complex behaviors found in chemistry. What is even more striking, perhaps, is that if *more* steps are added to the five that make up the crosscatalator, some of the complexity can be washed away: the moral of the crosscatalator is that merely adding more components does not necessarily mean more complexity.[21]

To put the crosscatalator through its paces, one must first hunt down the bifurcation points, the "signposts" that appear as the reaction is pushed away from equilibrium and indicate qualitative changes in behavior. We then need a powerful workstation to explore in detail the landscape around those regions, altering one parameter, then another, and so on. In this way, the range of behaviors we encountered with the Brusselator can be unearthed again: point attractors that represent a steady-state chemical reaction, when the mix of red, blue, and green molecules remains constant; and limit-cycles, when the color of the reacting mixture of chemicals undergoes rhythmic color changes.

A two-dimensional limit-cycle has a much broader repertoire of behaviors than its one-dimensional cousin in the Brusselator. One variant, associated with what are called "mixed-mode" oscillations, shows a large surge in the concentration of one component, a rapid fall-off, followed by a sequence of smaller peaks in concentration, before the whole process repeats (see Fig. 6.3). By altering the relative concentrations of the reacting chemicals, it is possible to observe any number of small peaks, from one on upwards. Intriguingly, these oscillations have

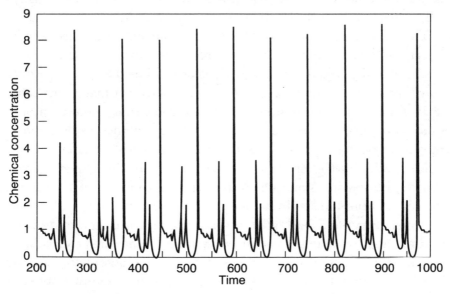

Figure 6.3   Mixed-mode oscillations of the crosscatalator.

a real-life counterpart in nature: within your head, the firing of individual brain cells during conscious and unconscious nervous activity produces similar kinds of patterns. The electrochemical behavior of each neuron within a neural network (as we discussed in Chapter 5) is reminiscent of the crosscatalator.[22]

BASINS OF ATTRACTION: MEMORIES ARE MADE OF THESE

The crosscatalator contains a peculiar feature that is worth a short diversion to highlight another similarity with the behavior of large collections of neurons, notably with the artificial neural networks encountered in the previous chapter. In particular, these models can help us understand *association*, one of the most poignant features of our memories—the way, for instance, a scent of perfume can evoke the memory of a kiss.

A potent metaphor for this association of smell with memory can be found in the rugged landscape of the crosscatalator. In some regions of its parameter space, more than one possible pattern of long-term behavior exists for the hypothetical chemical mixture. In the language of landscape metaphors, this corresponds to the situation where a mountaineer could fall into more than just one valley; *which* one he actually falls into depends on his location on the mountainside.

The simultaneous existence of more than one possible attractor is a direct result of the nonlinear equations we use in the crosscatalator. Mathematically speaking, the equations describing the system permit *more than one* solution for a given set of conditions. For many such conditions, only one state is stable. However, sometimes more than one solution becomes stable and the crosscatalator can then explore more than one attractor; in which case the system is said to become "multistable." Which attractor the crosscatalator ends up in is then dependent on its starting state. One can build up a picture of the set of initial conditions that always lead to a given attractor, called the "basin of attraction." Each attractor has its own basin—meaning that as time passes some initial conditions will lead to one attractor, others to another attractor. The attracting states of cellular automata (classes I to IV, cf Chapter 4) also possess basins of attraction; examples of these are shown in color plate 4.

The same phenomenon can be found in feedback (recurrent or Hopfield) neural networks, which are also governed by nonlinear differential equations. As we have already noted in Chapter 5, the storage of information in attractor neural networks (such as Hopfield's) is associated with the presence of a set of local minima in the energy (more correctly, the error) landscape. Different input patterns are associated with different basins of attraction and so give rise to different memories.

We often *associate* memories—for instance a once-popular love song with an old flame. Using the language of attractors, such associative memories occur when the basin of attraction for a piece of music is shared with one linked with a lover, perhaps caused by recollections of passionate embraces enjoyed on the dance floor. The attracting memory state contains a representation of both song and lover. The dissipative structures set up by neural networks are as much an example of self-organization as the myriad virtual color changes of the crosscatalator.

CHAOS AND THE CROSSCATALATOR

While wandering across the crosscatalator landscape, it is possible to fall into another attractor, in which the reaction displays *chaotic* behavior. Then the color changes occur seemingly quite irregularly. However, what is apparently random arises, paradoxically, due to an overdose of order. To illustrate this intriguing aspect of chaos, we need a way to display the menagerie of possible behaviors of the crosscatalator as it is pushed further and further from equilibrium. The generic behavior can be summarized concisely using the bifurcation points we have already encountered. Figure 6.4 captures each of these milestones in behavior in a bifurcation diagram

The distance down the vertical axis corresponds to the distance from equilibrium. At the first bifurcation point away from equilibrium, the reaction branches and two possible oscillatory clock behaviors are available. In turn these branches sprout more and more branches, twigs, and so on, indicating that more and more different states become stabilized. Eventually, however, these bifurcations occur so densely that they merge at the bottom of the diagram into a dense clot of possibilities.

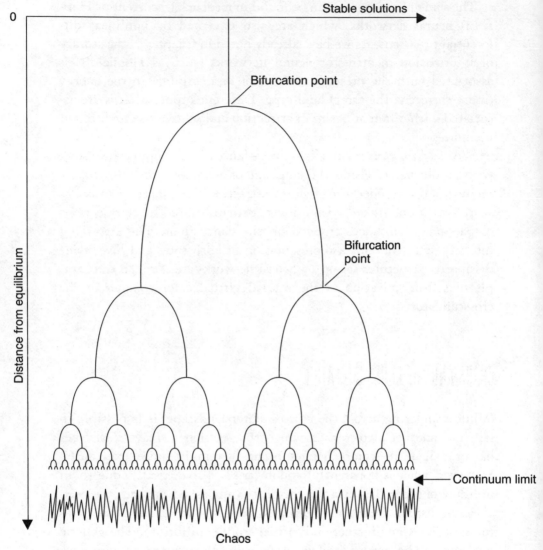

Figure 6.4   **The bifurcation cascade to chaos in a simple nonlinear system, also known as period doubling.**

A dazzling range of self-organized behaviors is possible in this clot, because of the enormous number of potential states densely packed together. The chemical clock is no longer confined to a limited number of branches. Its behavior *appears* to be chaotic because it is free to sample an infinitude of ordered possibilities. The measurements recorded in one experiment can never be repeated. The next time around, another apparently random set of time intervals between color changes will appear. This is the same kind of behavior found in Class III cellular automata. Note, however, that the meaning of the word "chaos" here is quite distinct from its everyday usage: it is used to describe processes that are *not* random but *look* random.

One might think that the further one goes from equilibrium, the more chaos would occur in this bifurcation tree. But the situation is more like that of a cedar tree, where there are gaps between layers of foliage. One can encounter "islands" or "windows" of regularity between regimes of chaos, and vice versa; inside them are windows embedded within windows. This kind of behavior is not limited to the crosscatalator but is a signature of the complexity that abounds in nature.

COMPUTERS AND CHAOS

Although it was glimpsed around the end of the nineteenth century by Poincaré, the first clear inkling of the existence of chaos was uncovered by means of a computer in studies carried out in 1963 by Edward Lorenz, a professor of meteorology at MIT. Lorenz was trying to make sense of the all-too-frequent discrepancies between what weather forecasters say and what actually happens.

With his primitive machine and an unusually strong mathematical background for researchers in that field, Lorenz sought to make as simple a mathematical model of atmospheric weather flow as possible, while retaining the essential physics. Lorenz's equations give an approximate description of a horizontal fluid layer heated from below, similar to the Rayleigh-Bénard cell we encountered at the start of the chapter. The warmer fluid is less dense, tending to rise and stir convection currents. If the heat is sufficiently intense, the ensuing convection takes place in an irregular, turbulent manner.

While he was studying the complicated Navier-Stokes equations of fluid dynamics, it gradually dawned on Lorenz that tiny variations in the initial weather conditions fed into the computer could have a dramatic effect at later times. When the computer had pushed the numbers through the equations, the resulting solutions—the weather forecast—often changed totally in a very short time, no matter how slight the variation in starting conditions. It would have been tempting to blame this sensitivity on some problem with his computer, as others must have done, but it was undoubtedly Lorenz's meteorological training that led him to be so receptive to this result—and in that respect he was years ahead of his time. He realized that his work was doomed to failure unless he could construct an easily understood system of equations whose solutions behaved chaotically. He achieved this with a drastically simplified representation of the Navier-Stokes equations, consisting of just three coupled nonlinear differential equations. As he later wrote, "Chaos suddenly became something to be welcomed."[23]

Lorenz had discovered that the quest by meteorologists across the planet to produce ever more accurate weather forecasts using fluid mechanical principles and with ever greater computational power faces one huge drawback: the sensitivity of the fluid dynamical equations to the initial input data means that tiny fluctuations could wreak an enormous change in weather patterns. This fact led Lorenz to coin the phrase "the butterfly effect," a vivid image capturing the idea that through chaos the smallest of events can lead to the most massive of consequences. With some poetic license, the beating of a butterfly's wings in Brazil could spark off a tornado in Texas.

The butterfly effect is easy to visualize with an attractor. A far cry from simple point attractors or limit-cycles, the Lorenz attractor consists of a spaghetti of trajectories that form what looks like a pair of owl's eyes or, more appropriately, a butterfly (see Fig. 6.5). The slightest shift in the position on the attractor leads to a different trajectory and thus a different weather pattern. In its entirety, the attractor represents the climate: though Britain's weather varies a great deal, it is always within a certain repertoire of possibilities, excluding monsoons or months without rain.

STRANGE ATTRACTORS, CHAOS, AND FRACTALS

The butterfly shape of the Lorenz attractor in Figure 6.5(c) is an example of what is called a *strange attractor*, to highlight its complex character when compared with its simple cousins such as point attractors and limit-cycles. The beast was first recognized and named in 1971 by David Ruelle, a Belgian mathematical physicist at the Institut des Hautes Etudes Scientifiques at Bures-sur-Yvettes near Paris, and Floris Takens, of the University of Groningen in the Netherlands.[24] They

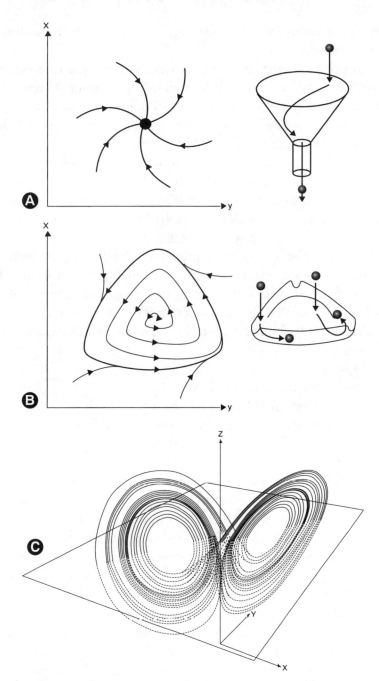

Figure 6.5　Degrees of attractions. (a) A fixed-point attractor and its mechanical analog, a ball bearing rolling into a funnel; (b) a limit-cycle attractor and its mechanical analog, a ball bearing rolling around the rim of a tricorne hat; (c) the Lorenz strange, or chaotic, attractor.

were studying turbulence in fluids but the scope of their conclusions has proved far wider.[25]

Two distinct ingredients of the strange attractor produce complex behavior. First, unlike a limit-cycle, it displays an immense sensitivity to the starting or initial conditions. The long-term behavior of a system trapped on a strange attractor depends on the minutest details of how it was launched, as highlighted by Lorenz's work.

Second, unlike the limit-cycle, it is a *fractal* object. The word "fractal" has been in existence since 1975. It was coined by Benoît Mandelbrot, then at IBM, to describe the peculiar geometry of irregular shapes that look the same on all scales of length. In the same way, regardless of how much a fractal object is magnified, it contains essentially the entire structure of an object. This property of endlessly manifesting a motif within a motif within a motif is known as *self-similarity*. The motif is mirrored at every scale of length: the edges of a clover leaf will be bristling with smaller clover shapes that will bristle with still smaller clover shapes, and so on indefinitely. Jonathan Swift's fleas fall into this category:[26]

> So, naturalists observe, a flea
> Hath smaller fleas that on him prey
> And these have smaller fleas to bite 'em
> And so proceed *ad infinitum.*

A fractal's form is the same no matter what length scale we use to view it. Figure 6.6 shows one example produced by Mario Markus of the Max Planck Institute of Molecular Physiology in Dortmund, with Javier Tamames of Madrid's Universidad Complutense.[27]

Mandelbrot's work has shaken the way we think of dimensions. We are all familiar with the idea that a line has a dimension of one, while the area enclosed by a rectangle is two dimensional. But in fact these are almost always idealizations: objects can be one-and-a-bit dimensional. The "and a bit" means it is a fractional or fractal dimension.[28] Mandelbrot illustrated this in a paper in which he posed the question: "How long is the coast of Britain?" A little thought shows that the answer depends on the length-scale chosen to measure the coastline. Fractal shapes abound in nature, from cauliflowers to clouds; they can even be found in a dripping tap, when an infinite cascade of structure forms between tap and falling drop.[29]

Figure 6.6   Patterns within patterns. A fractal repeats at every length scale.

A chaotic chemical reaction, represented by a ball moving over a strange attractor, can explore a never-ending series of states because of the attractor's fractal properties. Indeed, whereas fixed-point and limit-cycle attractors have a dimension of zero, one, two, three, and so on, a strange attractor can be defined as one of fractal dimension. The fact that the dimensionality of a strange attractor is fractal should prepare us for chaos. The attractor contains an infinity of possibilities, albeit confined to a finite region: the system samples an infinity of different configurations as time passes, never repeating itself. One can imagine the system endlessly tracing out patterns within patterns within patterns. At first sight this is difficult to grasp. Yet armed with the notion of fractal forms, it becomes easier to see how the crosscatalator, though restricted to a finite region—its own strange attractor—can nevertheless discover unlimited opportunities for exploration. Once the crosscatalator chemistry is sucked onto a strange attractor, it is totally impossible to predict in any detail its long-term chemical behavior.

We can highlight this complexity by comparing it with that of a chemical clock trapped on a limit-cycle. No matter how the ball representing a chemical clock was tossed into that landscape of possibilities, it would end up rolling around the bottom of a circular valley. But suppose that the ball had rolled into a strange attractor and you wished to repeat the complicated path it followed. You would find that any neighboring starting point you subsequently chose, which would always be different no matter how close it was to the first, has a trajectory on the landscape that rapidly diverges from the original, leading to a completely different motion on the attractor—a different path through the fractal's infinite patterns within patterns.

Deterministic chaos arises from the infinitely complex fractal structure of the strange attractor. Only if an observer knew with infinite accuracy what the starting conditions were in an experimental study of such a chaotic system would he or she be able to make a cast-iron prediction. But the slightest uncertainty—always the case in the real world—denies this, since no matter how small the imprecision, it will be amplified exponentially as time passes. Yet there is deep order present. That is why we call it *deterministic* chaos—it results from deterministic dynamical equations. Conceptually, therefore, chaos is intrinsic to the system and clearly distinguished from the effects of random or "stochastic" fluctuations in the external environment. A typical source of these fluctuations is heat, which can be detected by tiny random temperature changes in the surroundings. However, such stochastic processes can generate random, chaotic-looking behavior in a system not trapped in a strange attractor. Telling deterministic and stochastic chaos apart is one of the principal hurdles that confronts "chaologists"—scientists working with potentially chaotic systems.[30]

Deterministic chaos also blurs our intuitive ideas of order and disorder. Confused popularizations of the idea have put forward chaos as an explanation for everything complex, not only things that are unpredictable or unstable, but even when self-organization would be more appropriate. However, no one should be blinded by this buzzword. Order *and* deterministic chaos spring from the same source—dissipative dynamical systems described by nonlinear equations.

Chemical chaos was first proposed by Ruelle as long ago as 1973, although his idea, like so many other really good ones, was initially ignored.[31] In a chemical reaction, the presence of a strange attractor is betrayed by the absence of any kind of clock-like regularity in the color changes from red to blue. In 1971 Ruelle asked a chemist who specialized in oscillating reactions if they ever had a chaotic time dependence. "He answered that if an experimentalist obtained a chaotic record in the study of a chemical reaction, he would throw away the record, saying that the experiment was unsuccessful. Things fortunately have changed and we now have several examples of nonperiodic chemical reactions." The most famous real-life example of chemical complexity, including chaos but also much more, is the Belousov-Zhabotinski reaction.

BELOUSOV AND ZHABOTINSKI'S MAGIC

Half a century ago Boris Pavlovitch Belousov stumbled on an amazing chemical reaction while he was head of a biophysics laboratory attached to the Soviet Ministry of Health. During his research he devised a cocktail of chemicals intended to resemble and so throw further light on aspects of the Krebs cycle, a crucial metabolic pathway by which living cells break down organic foodstuffs into energy and carbon dioxide gas. To his surprise, the mixture oscillated, with clockwork regularity, between being colorless and having a yellow hue. There is some evidence that, in subsequent investigations, he also observed the formation of spatial patterns. Thus, Belousov provided the first example of a real chemical reaction that supported the notion of self-organization through the far-from-equilibrium processes of reaction and diffusion, as predicted by Turing at the same time.

Unfortunately for Belousov, the reaction was so peculiar that he had great trouble in convincing the scientific establishment it was real. In 1951 a manuscript of his work was rejected. The editor told him that his "supposedly discovered discovery" was quite impossible. Belousov submitted another analysis six years later but the editor would only offer to publish a savagely cut version in the form of a brief communication. Belousov's work eventually appeared as an obscure contribution in the proceedings of a symposium on radiation medicine. It consisted of two pages before another of his papers. He was neither the first nor the last to meet with overzealous scientific skepticism.[32]

It was only when Anatoly Zhabotinski studied Belousov's oscillating recipe that interest in the reaction really began to take off. During the 1960s, as a graduate student of biochemistry at Moscow State University, he played around with Belousov's ingredients in a number of ways—for instance, by producing a more distinctive color change from red to blue. Eventually, he managed to capture the imagination of his conservative peers. Others began to take up the study of this amazing system and since then research on self-organizing chemical reactions has grown into a trendy field of investigation. Testimonials to the importance of the work were sought from scientists worldwide in 1979, and in 1980 both Belousov and Zhabotinski were awarded the Lenin Prize, along with Valentin Israelovitch Krinsky, Genrik Ivanitsky, and Albert Zaikin. Unfortunately, Belousov died in 1970, before the belated international recognition of his seminal contribution.

The reaction originally discovered by Belousov and the many variants subsequently developed have come to be known as the Belousov-Zhabotinski (BZ) reaction. The chemistry of this remarkable and complex reaction has been analyzed in depth: some thirty distinct chemical species are thought to participate, including shortlived intermediates that act as stepping-stones in the various interlocking cycles of chemical reactions turning in the BZ reaction. These were summarized by a group at the University of Oregon in a simplified yet important model scheme comprising just five separate steps and six species that has since been dubbed the "Oregonator."[33] The model is capable of describing in many respects the behavior of real BZ mixtures, such as the limit-cycle that generates chemical oscillations. The general message from this work, as elsewhere, is clear: we can often explain complexity on the basis of simple nonlinear mathematics.

The subtleties of these complex reactions are still being pursued. Spiral waves are well known in the BZ reaction (see Fig. 6.7). Strong evidence for the existence of a strange attractor underlying the chaotic regimes in the BZ reaction has been found by one of the most active groups in the field, led by Harry Swinney at the University of Texas at Austin.[34] From a detailed study of the reaction dynamics, this chaotic behavior has been modeled using a set of differential equations containing three variables, the minimum necessary.[35] Others have shown how to control chemical chaos, locking on to one of the usual unstable periodic orbits embedded in a strange attractor by a series of small adjustments to the flows of chemicals fed into the reactor.[36] In this way, scientists have grasped some of the ordered yet apparently randomly changing patterns of complexity.

Figure 6.7   Spiral waves formed in the BZ chemical reaction.

Scientists classify any such changes in pattern or color as changes of state. An excitable medium—like the soup of chemicals in the BZ reaction—is one that changes its state when subjected to a stimulus that exceeds a certain threshold level. After excitation, such a medium becomes refractory (i.e., unresponsive), only returning to the initial receptive state through a series of states that may themselves be excited. Everyday examples can be found in the spiral waves generated by heart muscles, the feeding activities of primitive organisms called slime molds, and the electrical activity within our brains, where neurons demonstrate similar kinds of behavior.[37] Indeed, the analogy between the BZ reaction and the behavior of heart muscle is now well established.[38]

Chemical cocktails that display Turing's 1952 prediction of *static* patterns have proved much trickier to find, however. Experimentalists had so much difficulty creating these patterns that there was growing skepticism about their reality in the physical world. They were in danger of being written off as yet another mathematical idealization. Recently, however, a group at the University of Bordeaux have managed to produce complex, but genuinely static, Turing patterns.[39] For example, they generated a line of spots, a pattern similar to one obtained when putting the Brusselator through its paces.[40]

Conditions for Turing patterns are stringent and need some kind of positive feedback, such as autocatalysis, and two chemical species called an activator and an inhibitor. The French group developed an appropriate chemical concoction and used a flat piece of transparent gel for their reactor, wedged between reservoirs of chemical reactants. A video camera studied the yellow and blue color changes, revealing stationary lines of spots, the first experimental evidence of a genuine Turing structure.

This work has been extended by Harry Swinney and his colleagues to create static but irregular patterns with a very "organic" look to them.[41] These are reminiscent of what Turing had achieved by slogging through his equations with a desk calculator. Turing's mother once described how "he showed me some of these [patterns] and asked whether they resembled the blotches of color on cows, which indeed they did to such an extent that the sight of cows always calls to mind his mathematical patterns."[42]

Again using gel as the reactor, Swinney and his colleagues studied how the patterns made from a palette of blue and yellow reacting chemicals depend on the conditions. When the temperature was lowered, a hexagonal pattern formed. By changing the concentration of reactants, they found that striped patterns were laid down that could be maintained for days, without much change. The patterns were familiar to Swinney, however, who had already created something similar in computer simulations. In further work, his team created a reacting system wherein blue spots undergo a continuous life-like process of birth by replication and death by overcrowding.[43] By at least one account, the behavior bore an uncanny resemblance to that of living things.[44]

THE CHEMICAL COMPUTER

The BZ reaction contains other surprising properties. In an elegant experiment by Oliver Steinbock, Agota Tóth, and Kenneth Showalter at West Virginia University, the BZ reaction was harnessed to solve a tricky mathematical problem.[45] To find the quickest route through a maze, a serial computer typically works out every possible path in turn and then chooses the fastest. Instead of resorting to such brute force, the West Virginian team followed in the wake of a chemical wave to find the shortest route. In effect, they used a parallel approach to what normally would be tackled in an iterative manner.

The wave—a self-propagating reaction front—has various useful properties for solving the problem: it moves at a steady speed; it negotiates barriers without breaking up; and it vanishes at dead ends or when it collides with another wave. To create a wave within a maze, the team took a polymer membrane and soaked it in the ingredients of the reaction, then cut out the appropriate pieces to create the "walls" of the maze. The excitable reaction was set off at one point by touching it with a silver wire and then a video image was taken of the wave front at fixed time intervals to produce a composite image showing its progress

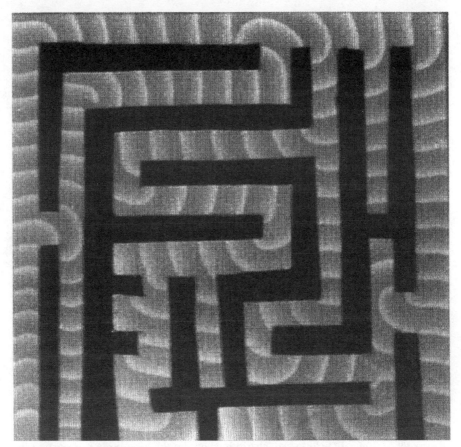

Figure 6.8a   Chemical computer. A chemical wave propagating through a membrane labyrinth. A sequence of fifty images obtained at 50-second intervals was superimposed to form this image.

as it explored (see Fig. 6.8a). From the image, they generated maps that determine the path length from any location in the maze to a given target point. With the help of a computer, they could also derive a grid of vectors showing the direction of the local flows (see Fig. 6.8b). By picking any point at random, it is possible to follow this vector flow to find the optimal—fastest—way back to the starting point.

It is fascinating to note that the operations of the brain must depend in part on a similar process that governs the way signals ripple and branch through all the neuronal pathways. "Could path optimizations in networks of neurons rely on the properties of excitable media, as suggested in our optimal path determinations?" asked Showalter. His team has also successfully simulated the way chemical waves ripple through

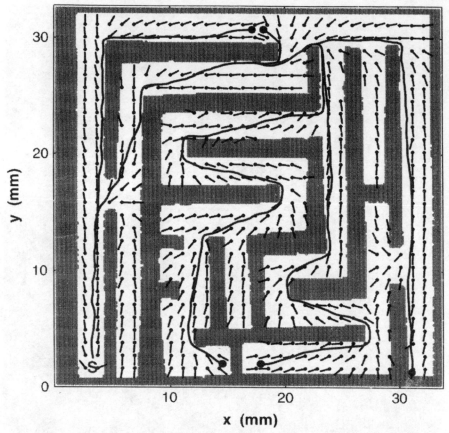

Figure 6.8b   Velocity field in the labyrinth, showing the local wave propagation direction. Small black dots represent the origins of vectors. The quickest paths between five points and the target point S are shown by dark solid lines.

mazes within a computer.[46] The resulting algorithm could have useful applications: for example, it could help a robot find the quickest route to a specific item among the miles of shelves in a warehouse.

The regular "tick" of a BZ reaction could one day play the role of the clock in a chemical computer, according to John Ross of Stanford University. Working with Allen Hjelmfelt, Ed Weinberg, and Adam Arkin, he has identified the basic logical functions from which virtually any computer can be constructed—including analog and fuzzy logic—in the biochemistry of glycolysis, a process found in living things that we will examine in more detail in the next chapter.

His team has also shown that it is possible to simulate neural networks using chemical reactions that behave analogously to nerve cells—

there is a threshold below which an input signal dies away and above which it is amplified—and has already built a prototype that can carry out pattern recognition.[47] "What can be learned by studying the computational functions of simple chemical networks may lead to a better understanding of biologically relevant networks," remarked Ross.

CELLULAR AUTOMATON CHEMISTRY

The complexity of the Belousov-Zhabotinski reaction can of course be re-created using "programmable matter" within a computer—the cellular automata (CA) we encountered earlier in models of turbulent fluid flows in Chapter 4. These computer simulations do not rely on a detailed picture of molecular behavior but on a higher level description: simple rules describing excitability, an essential feature of the BZ reaction that enables it to be pushed into action, so that a pattern forms or a chemical clock ticks in what would otherwise be a quiescent medium.[48]

To imagine how a CA can simulate chemical pattern formation, think of an extended two-dimensional checkerboard with the squares representing individual reaction sites. A given square responds to squares in its neighborhood; that is, the state into which a given square evolves depends on the states of the squares in its immediate vicinity. Instead of becoming bogged down in the chemistry, the model describes the state of a square—receptive, refractory, or excited—by a whole number. If the number is zero, the site is receptive, meaning that it can be excited from a quiescent state; otherwise it is either excited (an upper limit) or refractory (any of the intermediate integers). Refractory states may themselves be excited to the same excited state.

To play this cellular automaton "game" and create complexity, we must specify a set of transition rules that determine how the state of each square changes from one moment in time to the next as a result of the states of its neighbors. If no excitation occurs at a site, the associated integer decreases by one. Beautiful patterns can emerge in this way. If we ascribe a color to the value of the number at each site, we can translate the CA's digital representation into a colorful image on a computer screen.

Such models were devised by Mario Markus and Benno Hess at the Max Planck Institute in Dortmund.[49] Before they could achieve realistic

Figure 6.9   Chemical wave simulations. Random cellular automata were used to simulate the spiral waves shown in Figure 6.7.

simulations, however, they had to overcome a departure from reality inherent in previous cellular automata: molecules can jiggle about in any direction while the directions on a CA checkerboard are limited to up, down, and sideways. The classic patterns produced by a BZ reaction—and other excitable media—are spiral waves. However, the computer patterns assume a square appearance if a checkerboard CA lattice is used, or become hexagonal if the lattice is hexagonal: the underlying symmetry of the lattice shines through the simulation, just as surely as houses made of bricks tend to be rectangular or square.

The solution Markus and Hess hit on was to randomize the lattice so that it became effectively isotropic; that is, there was no preferred set of directions for molecular motion.[50] Then they could simulate a wide range of spiral wave phenomena, including deterministic chaos, which were in good agreement with experimental observations. Compare Figure 6.9 with Figure 6.7. Indeed, in subsequent work, square spiral waves *have* been created in the laboratory by carrying the reaction out on a rhodium crystal: just as the underlying CA lattice produced square waves, so the rhodium produced square-shaped chemical concentration waves in a reaction of nitric oxide (NO) and hydrogen.[51] Cellular automata are not only capable of simulating known macroscopic complexity in the nonlinear world, they can even anticipate it.

This approach has also succeeded in simulating irregular patterns such as stripes and spots, with spots being reminiscent of the patterns found on the cheetah, and stripes looking rather like the different domains of magnetization that can be seen in a thin film of a ferromagnetic material (see Fig. 6.10).[52] Working with Ingo Kusch, Markus has also used cellular automata to model the patterns found on shells; these patterns can be explained on the basis of the Class IV type of cellular automata we encountered in Chapter 4, which provides the mix of chaos and order required to mark the mollusk shells (see Fig. 6.11).[53]

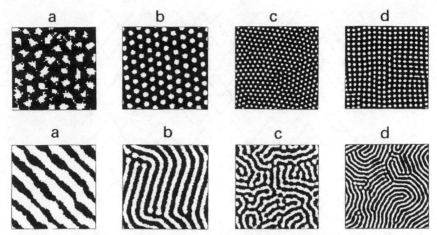

Figure 6.10   Stationary Turing patterns revealing spots (top) and stripes (bottom), generated by M. Markus and H. Schepers in Dortmund.

TO SEE A WORLD IN A GRAIN OF SAND[54]

The computer can give us other insights into the diverse patterns of nature by modeling something apparently even simpler than the mixture of chemicals in a BZ reaction. An avalanche on a pile of sand in cyberspace provides a paradigm for how complexity can emerge, honed by an apparently inordinate mix of many separate small- and large-scale events. The process may be as much at work in the financial markets as within earthquakes. A huge tremor can be the culmination of many sizes of subsidiary strikes and slips as two plates in the Earth's crust grind past each other. Similarly, a financial crash costing billions can result from a myriad contributing factors of varying sizes, ranging from a pensioner withdrawing her savings in Salt Lake City to a war in Europe.

Analyzing such "cascade" events in detail is awkward because an apparently small event can be as important for the outcome as a large one, rather like how a camel's back can be snapped by an insignificant straw or a hard-to-ignore stine block from the pyramid of Cheops. Earthquakes, avalanches, and financial crashes do have a common fingerprint: the distribution of events follows a simple power law. Take avalanches, for example. If in a given time interval there is one event of size 10,000 measured in some arbitrary units, there will be approximately 10 events of size 1,000, roughly 100 events of size 100, about 1,000 events of

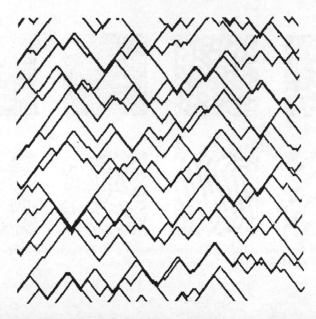

a

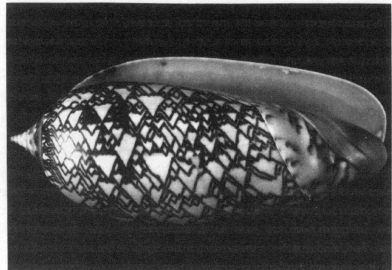

b

Figure 6.11   Cellular automata simulations (a) of patterns on shells (b).

size 10, and so on. This power law means that the physics of small avalanches is the same as that of large ones. There is no characteristic time scale separating large-scale from small-scale behavior.

The power law fingerprint of such cascade events can be detected through two distinct but related "signatures": from a fractal spatial arrangement of the locations where the events happen; and from the appearance of so-called "flicker noise" (or $1/f$ noise, where f is the frequency of occurrence), a telltale, ostensibly random, pattern found by charting the sequence of events over time. The ubiquity of flicker noise is a classic puzzle of physics. It has been found in systems as diverse as resistors in electrical circuits, the trickle of sand through an hourglass, the flow of the river Nile, and the luminosity of stars, including the Sun.[55] Similar fluctuations have been seen in economic indicators such as the Dow Jones index. Equally odd is the observation that objects such as the cosmic strings of matter strung out in space, mountain landscapes, and coastal lines appear to be self-similar fractal structures. Both spatial and temporal signatures can be found in some phenomena: turbulence is associated with self-similar structures in time and space.

Flicker noise cannot be explained by a conventional reductionist approach that focuses on the individual behavior of all the millions of elements of a complex system. For Per Bak of Brookhaven National Laboratory, one of the advocates of a new approach to analyzing such cascades, "The global features such as the relative number of large and small events, do not depend on the microscopic mechanisms. Consequently, global features of the system cannot be understood by analyzing the parts separately."[56]

Computer simulations by Bak and his colleagues at the Brookhaven National Laboratory suggest a generic "domino effect" of all sizes and durations that can unite these kinds of events.[57] To explain the occurrence of flicker and fractal signatures in such events, Bak's group played with sand. Rather than spend time in a sand pit, they constructed computer simulations of one-, two-, and three-dimensional sandpiles, using cellular automaton models. The sandpiles were idealized and built from uniform cubic grains added one at a time. The locations for the added grains were chosen at random so that the pile grew with an uneven surface. However, when the height difference between neighboring sites reached a threshold—in other words, when the slope was too large and the pile far from mechanical equilibrium—the surface grains could flow from higher to lower regions. This in turn could trigger further trickles in a nonlinear fashion.

Armed with their cellular automata, Bak and his colleagues found self-organization occurred during their computer simulations. The steepening of the slope of the sandpile increased until one particular value of the slope—the so-called critical slope—was reached.[58] Further grains added to this pile merely caused surface grains to slide off, leav-

Figure 6.12   Critical avalanche. The dark grains in the center represent the path of an avalanche that was initiated by the addition of a single grain to this self-organizing critical sandpile.

ing the slope unchanged. In the computer simulations, at the self-organized critical state, sand flowed from the pile as single grains, or as landslides, or as all possible cases in between. Regardless of whether the pile starts out too steep or too shallow, it always ends up at this critical state.

The most striking feature of this self-organized state is its lack of length and time scales. As in the case of the fractal and flicker phenomena, no particular event size or frequency stands out from any others. Remarkably, in the critical self-organized state, two events are equally likely to act together, whether or not they occur close to each other in space and irrespective of how much time has elapsed between their individual occurrence. Physicists are familiar with this insensitivity to time and distance in what are called second-order equilibrium phase transitions. In a vessel of carbon dioxide at its critical point, for example, there is no way for us to distinguish between gaseous and liquid regions. Only a single state of matter is present. The motion of every molecule throughout the vessel of fluid influences in equal measure that of *all* others, no matter how remote.[59]

Bak's model sought to combine the equilibrium concept of criticality with the nonequilibrium concept of self-organization. Called the theory of *self-organized criticality*, its final outcome is the "self-organized critical state."[60] Once the final sandpile slope has attained this critical state, it is relatively difficult to disturb. Elegant experiments in which grains of sand were dropped onto a high precision balance have found similar features in small sandpiles.[61]

It is helpful to think of the thermodynamics of the sandpile. The long-term equilibrium state (fixed point attractor) would be attained when the sand has been scattered across the floor. But the growing sandpile is far from equilibrium. The grains of sand in the computer simulation behave in a "cooperative" manner: a single grain may affect all the others. As sand is continuously added, the system evolves into a critical state characterized by large periods of static behavior, or stasis, interrupted by intermittent bursts of activity. The heap gradually becomes steeper, and there are bigger and bigger avalanches, until the heap builds up a critical slope that produces avalanches of all sizes. If the computer model is adapted to simulate clammy instead of dry sand, the pile grows steeper for a short time, until it reaches a *new* critical state with avalanches of all sizes.[62]

The behaviors of the "creatures" in Conway's *Game of Life* show signs of self-organized criticality. Per Bak, Kan Chen, and Michael Creutz of Brookhaven National Laboratory studied whether the number of "live" sites in the game would fluctuate over time, in the same way as the size of the avalanches in the sandpile model. Once the game had settled down, they added a single organism at a random position, waited until the system settled, and then repeated the procedure. Next, they measured the total number of births and deaths in the "avalanche" after each additional perturbation. Lo and behold, the distribution was found to follow a power law, indicating that the system had organized itself to a critical state.[63]

The fingerprint of self-organized criticality has been found in earthquakes, helping to explain their evolution and the distribution of their epicenters. The epicenters of smaller earthquakes—the sites of slips that trigger off other slips that ultimately make a quake—are not spread evenly over the earth's surface. A map of the earth reveals that earthquakes occur in string-like sequences that look similar on all length scales. In other words, they are fractal. The theory of self-organized criticality predicts that the energy released during a tremor varies with the inverse of the frequency of that event. At this level, earthquakes fit the pattern over time consistent with self-organized criticality. This pattern is also known as the Gutenberg-Richter law, the power law put forward by geologists Beno Gutenberg and Charles Richter in 1956.[64]

The connectedness of phenomena based on the activity of many agents, whether simulated grains of sand or real molecules, is striking. Self-organizing criticality has been a powerful spur in efforts to find underlying similarities between diverse and complex phenomena. Since Bak and his colleagues announced their sandpile model, an avalanche of activity has taken place hunting down examples of self-organized criticality. This bandwagon effect among theoretical physicists is reminiscent of the frenzied hunt for strange attractors in the wake of Ruelle and Taken's discovery. There are now many claimed examples, ranging from volcanic activity and forest fires to friction, spiral protein chains, the change of airway resistance during the inflation of lungs, and conductivity of an isotope of helium.[65] The frequencies of pulsars and light emitted from quasars follow a similar fractal/flicker pattern, with bursts of all sizes.[66] Economies may also show this effect.[67] Extrapolating from all this, Bak feels that the idea may be relevant to aspects of brain activity: "Throughout history, wars and peaceful interactions might have left the world in a critical state in which conflicts and social unrest spread like avalanches. Self-organized criticality might even explain how information propagates through the brain. It is no surprise that brainstorms can be triggered by small events."[68]

There is also skepticism that ranges from "so what?" to polemics about whether the effect is even real. For example, people have failed to find the power law distribution of avalanches in real sandpiles.[69] "While the theoretical appeal of self-organized criticality is beyond dispute, its relevance to the dynamics of real sand is questionable," commented British sandpile experts Anita Mehta and Gary Barker.[70] Though the debate on its relevance to the real world continues, its motif has been spotted in many places on nature's canvas.

THE ARTIST'S MOTIF

The Rayleigh-Bénard cell demonstrated in an elegant way a defining feature of complexity—that self-organization is a natural consequence of the time evolution of vast aggregates of simple agents, namely, molecules in a liquid. By making these agents interact in a more complex way—for example, through the interlocking chemical steps of the Belousov-Zhabotinski reaction—we could create an even greater variety of behaviors, such as spiral structures reminiscent of galaxies, hurricanes, and living things. It is gratifying to see that the "programmable matter" within cellular automata can mimic this process to create natural-looking patterns, emphasizing the connectedness of complexity research.

A long-held dream, dating from the original proposals of Alan Turing, has been to connect the ideas of pattern formation developed for relatively easily understood physical or chemical systems to the creation of structure and form in living creatures.[71] The range of patterns that the BZ reaction forms can be found in living organisms so long as the underlying units—whether molecules or organisms—interact nonlinearly, and operate under far-from-equilibrium conditions. As the theoretical biologist Art Winfree has written, the BZ reaction "shares many of the features that make living systems interesting: chemical metabolism, self-organizing structure, rhythmic activity, dynamic stability within limits, irreversible dissolution beyond those limits, and a natural lifespan."[72] It is now time for us to study life itself.

Chapter 7

LIFE AS WE KNOW IT

If among all possible worlds, none had been better than the rest, then God
would never have created one.

—GOTTFRIED WILHEM VON LEIBNIZ

Biologists spent much of the nineteenth century looking for the spark
of life in the tissues of living things. They never found it. Chemists
put forward the notion that some molecules—"organic" ones based on
strings of carbon atoms—were unique to plants and animals.[1] However, they quickly learned how to make these supposedly natural components of life from "inorganic" substances in test tubes.[2] Physicists,
blindly conditioned by the second law of thermodynamics and its
gloomy suggestion that everything tends toward randomness and disorder, argued (wrongly) that it was odd that life existed at all.[3]

Today we have built up an astonishingly detailed picture of the
complexity of life by pooling the enormous collective knowledge of
biologists, chemists, and physicists. The fusion of these disciplines has
produced molecular biology, a field concerned with the *molecular* basis
of life. There are countless examples of the awesome power of this
reductionist science. Consider Marfan's syndrome, a potentially fatal disorder linked to a wide range of symptoms: abnormal height, a deformed
chest, eye problems, and a dangerous dilation of the blood vessels leading to the heart.[4] In 1991 it was discovered that sufferers were making

an unusual form of a single protein called fibrillin, one found in the connective tissue that holds together flesh, muscle, and organs.[5] Depending on the precise molecular defect in the fibrillin gene, a wide range of complaints results, from those of the eye to those of the heart.

The beautiful molecular structure of the foot and mouth virus (Fig. 7.1), the bane of livestock breeders, has recently been discovered.[6] The structure of these tiny agents—one million of them put together would measure just an inch across—was found by studying how crystals of the virus scatter radiation in the technique of X-ray crystallography. Using a modern ultrahigh resolution scanning force microscope capable of imaging atoms themselves, we are able to witness the birth of a virus as it escapes from a living cell.[7] Nobel laureate Gerd Binnig and colleagues glimpsed its exit by using the electronic tip of the microscope's probe to scan a cell, itself measuring only one-hundredth of a millimeter across, with one ten-millionth of the pressure of the needle of a recordplayer.

The picture provided by molecular biology is extraordinarily compelling, offering a detailed understanding of many aspects of life. Through our knowledge of so many of its molecular processes, we are

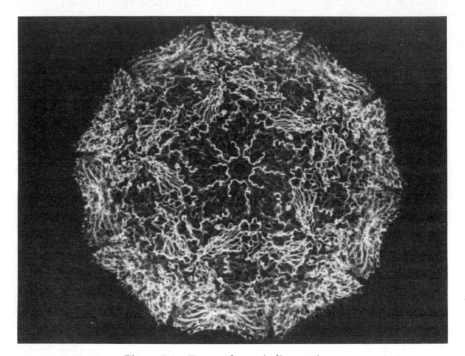

Figure 7.1   Foot and mouth disease virus.

today in an unrivaled position to treat disease, avoid illness, and genetically engineer crops. We have detailed descriptions of many of the
molecules of life, whether they are individual protein molecules floating within a cell, the structure of itinerant viruses, or excerpts of the
genetic blueprint, DNA. And we know that by manipulating these
molecules, we can alter the course of life.

It is largely because of the remarkable success of molecular biology
that the reductionist doctrine has intoxicated the minds of so many
scientists. As the Harvard geneticist Richard Lewontin has written,
"the ideology of modern science, including modern biology, makes the
atom or individual the causal source of all the properties of larger collections. It prescribes a way of studying the world, which is to cut it
up into the individual bits that cause it and study the properties of
these isolated bits."[8] Thus, the popular view has evolved that sees us as
being totally controlled by our genetic complement, itself comprised
of self-replicating molecules of DNA, in the way advocated by Richard
Dawkins in his book *The Selfish Gene*.[9]

But while molecular biology is powerful, it is far from omnipotent.
We have already shown in the last chapter that a purely molecular description is not sufficient to account for the complexity of many chemical systems, such as the Belousov-Zhabotinski reaction, which are
prototypes for biological ones. Such complexity demands analysis at
a macroscopic, rather than a microscopic, scale because it is a consequence of interactions between many units whose properties in isolation tell us virtually nothing about important global behavior. Indeed,
life itself can only be understood at this macroscopic level. While it is
difficult to formulate a simple, watertight definition of life, *self-replication*
is one of its central properties. As we have already discussed in Chapter
4, in his studies of automata von Neumann laid down the logical rules
of self-reproduction. He emphasized the importance of a minimum
level of complexity before a machine could replicate successfully, and
stressed the need for redundancy and evolvability so that self-replication
could be maintained in a changing environment.

To find out more about the inherent complexity of life, we need to
explore how it thrives on both self-organization and evolution. We
need to explore how the myriad components of living processes mesh
together, beginning with the search for the self-replicating chemistry
that seeded life itself. At that critical moment when the watershed be-

tween inanimate and biochemical reactions was crossed, there is little doubt that self-organization played a crucial role. Life was no accident.

We need to understand how the secret art of complexity led from interlocked "societies" of chemical reactions to complex societies of creatures. Patterns can be discerned in the chemistry of individual living cells, the cooperative behavior of millions of cells in a beating heart, the evolution of species and their rich behaviors, to the vast cauldron of interlaced and interleaved living and inanimate processes that make up our planet. By revealing these patterns, a plethora of connections between diverse fields will become apparent. The patterns of Turing and the BZ reaction will appear in the biochemistry of cells, in hearts, in strange creatures called slime molds and in a little freshwater polyp called the hydra. The patterns of mass extinction and other natural catastrophes may have a similar basis as avalanches of sand. And by using computer models of behavior, we will find a powerful means of explaining in Darwinian terms the evolution of cooperation, that most pleasing pattern of behavior and one that is all the more striking given that nature is supposedly red in tooth and claw.

WHAT IS LIFE?

Like many other terms discussed in this book, such as order, disorder, emergence, consciousness, and intelligence, it is hard to define what we mean by *life*. One dictionary definition describes life as "the property shared by living things that differentiates them from non-living ones," another calls it the "state of being alive." Tautologies such as these are often used in desperation. However, biologists can list a whole set of features possessed by nearly all living things. As well as the ability to reproduce, these include the existence of genetic information, complexity, organization, and so on. But exceptions can always be found. For example, the ability to reproduce is not possessed by every object we might expect to call "living": sterile men, postmenopausal women, mules, and viruses are all incapable of self-reproduction. A sneeze demonstrates how a common cold virus can persuade cells in the human respiratory tract to reproduce and spread it, even though it is little more than a stretch of genetic programming wrapped in protein. Nonliving things also show some "vital" signs. Crystals, for instance, are capable of self-reproduction during growth.

The eminent Austrian physicist, Erwin Schrödinger, fared somewhat better when he suggested in his book *What Is Life?* that a fundamental property of life was its tendency to produce increased order and seemingly unlikely arrangements of things, whereas the second law of thermodynamics suggested that, left to themselves, things tend to end up in the most probable state, a disordered mess.[10] By using external sources of energy—the sunlight streaming down onto the surface of the planet—life creates apparently highly improbable, low-entropy, orderly states of affairs. Organized complexity exists at every level in living systems, right down to the molecular level at which we find carbon, hydrogen, nitrogen, oxygen, and phosphorous atoms arranged in the beautiful double helixes of DNA. In order to balance the books, an organism is constantly exporting entropy by respiration and excretion or, as Schrödinger put it, "continually sucking orderliness from its environment."[11] Here again, however, problems of nonuniqueness arise since this definition could apply as easily to a self-organizing oscillatory chemical reaction as to a human being.

The most intriguing idea—and a key omission in many textbook lists of life's characteristics—is the proposal that life is intricately associated with evolution.[12] This idea lies behind one of the most general *working* definitions adopted by the Exobiology Program within the National Aeronautics and Space Agency (NASA): life is a self-sustained chemical system capable of undergoing Darwinian evolution.[13] This recipe encompasses the processes of self-reproduction, material continuity over a historical lineage, genetic variation, and natural selection.

To define life, we should shift the emphasis away from surviving individuals and selfish genes toward evolving systems. Evolution is a property that belongs not to a single individual or gene but to a whole system, and does us the great service of pointing away from isolated units toward *interactions* between individuals *and* with their environment. Some, like the British scientist James Lovelock, have suggested that life and its environment act together to form a single self-regulating planetary superorganism, called Gaia.

THE MISSING LINK

Early thinking about how life arose did not distinguish between the origins of life and the origins of the huge diversity of forms that walk and grow on the planet. God, by creating a world full of living things, provided both simultaneously. In the nineteenth century, though, the latter problem was solved by Charles Darwin in a way that had no recourse to God, although it shed no light on the origin of life. Darwin came to the conclusion that all contemporary species have a common ancestry dating back over many eons. The origin of species arose because of variations from random mutation and selection through the competition for limited resources.

Molecular biology has added powerful support to Darwin's ideas by revealing how all living things use the same type of genetic "programming." The diversity of living forms owes itself simply to variations in a particular class of giant nucleic acid molecules, DNA and RNA. Both DNA and RNA store genetic information in the form of nucleotides, chemical units that line up along the backbone of these macromolecules. There are only four different nucleotides, but like letters in words and sentences of the English language, they can form an essentially infinite variety of coded "messages" depending on their order or sequence along the spine of the DNA molecule. The messages are *genes*, which give specific instructions for building the proteins necessary for life on our planet. This chemical form of information technology may sound restrictive, but remember that modern computers use the two-letter language of binary arithmetic (0 and 1). The DNA-bearing chromosomes within the nucleus of a single human cell contain enough information-storing capacity to hold the thirty volumes of the *Encyclopedia Britannica* three or four times over.[14]

Molecular etymology can throw light on the organization of an individual's genetic complement, the *genome*. Quantitative estimates and experiments indicate that the adaptation of individual genes to carry out a given task actually occurred prior to their integration into the giant genome molecule presumably present in our putative earliest common ancestor.[15] The reason is found in the genome's huge length: the present-day error rate during genetic transcription has become adapted to a genome three billion genetic "letters" long, and has been substantially reduced as a result to avoid propagating too many mistakes. Had such a small error rate during replication been standard for the isolated individual genes, their evolution would have drastically slowed down, and they would not have been able to carry out their specific functions as effectively as they do today.

It is possible to use such changes in DNA and RNA to measure evolutionary time. By comparing the genetic material of living and extinct species, molecular biologists have found that DNA and RNA mutate at a fairly steady rate over long periods. These mutations ultimately gave rise to species as diverse as leeches and lichen. An evolutionary "clock" can be defined by these processes, where the tick corresponds to the rate of mutation. We can calibrate the clock with precisely dated fossils and then put it to work estimating when species diverged from one another.[16]

Darwin himself did not know when the clock struck zero. However, a letter he wrote in 1871 reveals that he was wrestling with the idea: "But if (and oh what a big if) we could conceive in some warm little pond with all sorts of ammonia and phosphoric salts, light, heat, electricity and etc., present, that a protein compound was chemically formed, ready to undergo still more complex changes."[17] Scientists continuing Darwin's quest today have good evidence that bacteria are the earliest extant link in this great chain of being. The oldest known fossils are in Australian rocks formed about 3.5 billion years ago, trapping what seem to be a wide variety of bacteria in the process.

In 1988, James Lake of the University of California at Los Angeles announced the results of a sophisticated genetic genealogical exercise suggesting that the most recent common ancestor of all living things was probably a sulphur-eating single-cell bacterium fond of basking in boiling hot springs. His attempt to work out the genealogical relationships at the bottom of the evolutionary tree was one of the preliminary stabs at establishing the relationships between every type of life known on Earth.[18] It clashes with a rival tree of life drawn up by Carl Woese at the University of Illinois, Urbana, Otto Kandler of the University of Munich, and Mark Wheelis of the University of California.[19] Their camp, which has more widespread support, advocates a scheme with three main groupings of living things, formed by dividing the bacteria in two and unifying all other organisms. Either way, bacteria do *not* represent the first link; rather, they simply form the oldest surviving one in the chain of life.

It seems highly probable, therefore, that once upon a time there were even simpler forms of life in existence. The challenge now is to reforge a few of the still earlier, but invisible links, ones that were evolutionarily successful at the time they occurred, but were subsequently superseded. These links were forged between the formation of the earth some 4.6 billion years ago and a billion years later, when it was teeming with organisms that resemble what we today call blue-green algae.[20] With eons to separate us from our earliest biological origins, a great deal of speculation and stormy debate center on the origin of life. Much of what we discuss in this chapter is almost certainly wrong in detail. But the conceptual framework it provides is sufficiently strong to enable life itself to be studied within computers, a subject we consider in the next chapter.

ORIGIN OF ORIGINS STUDIES

In the 1920s, the idea that the origin of life could be understood in terms of plausible chemical and physical processes occurring on the primitive earth was born. In an epoch-making treatise on the origin of life published in 1924, Alexander Ivanovich Oparin discussed the emergence of living organization, using the example of how frost on a windowpane looks like "ice flowers" even though it only consists of water particles.[21] "These particles, complying with the eternal laws of nature which are the same for both the living and dead, arranged themselves in a definite order and, on the simple windowpane, they produce pictures of fabulous gardens, glistening in the sunshine with all the colors of the rainbow."

In scientifically defensible terms, Oparin went on to apply the same laws to explain how "organic" molecules (i.e., molecules containing carbon) could lead to the patterns of living things. By the 1930s, Oparin and J.B.S. Haldane had pointed out that the organic compounds needed for life could not have formed if the atmosphere was as rich in oxygen as it is today. Instead, they proposed that the atmosphere of the young earth was reducing—that is, rich in hydrogen, methane, and ammonia.[22] There was a hiatus until 1953, when the first renaissance in origins studies occurred. Stanley Miller, then a student in the laboratory of Harold Urey at the University of Chicago, wondered what chemistry could take place in this reducing atmosphere.[23]

He created a primeval world using two connected flasks: the one at the bottom contained an "ocean" of water; the upper flask held a stew of simple substances, thought at that time to be similar to the primeval atmosphere. When Miller passed a bolt of simulated lightning through the atmosphere, he found that "the water in the flask became noticeably pink after the first day, and by the end of the week the solution was deep red and turbid." It contained certain amino acids, the basic building blocks of proteins essential to life as we know it.[24]

Since then, a large body of evidence has been accumulated, by Cyril Ponnamperuma, Leslie Orgel of the Salk Institute in San Diego, Sidney Fox of the University of Miami, and others, showing that a range of biologically important molecules can be made in a similar way, including the nucleic acids and energy-storing biomolecules such as adenosine triphosphate (ATP) and sugars. Others can be created with the help of light or catalysis by clays.[25] And even if a few key ingredients were missing on the early earth, astronomers have shown that they could have rained into this stew: carbonaceous chondrites—meteorites thought to be clumps of the dust from which the solar system was made—have been found to contain many of these simple organic chemicals.[26] Thanks to materials made on earth, or imported from space, we have a wide range of starting materials for life.

Drawing on the principles of chemical self-organization, we can speculate about the recipe for life to emerge. If a suitable feedback mechanism was operating in the prebiotic soup on the early earth that could account for nonlinearities, then the general conditions were ripe for self-organization. For instance, if a molecule in that soup catalyzed its own replication, then nonlinear feedback, the hallmark of self-organization, would very likely have emerged. This might have led to a breaking of the spatial uniformity of the medium, triggering patterns and rhythms (along lines broadly similar to those Turing proposed in 1952) just as a chemical clock can show patterns in space and time.

A key ingredient of the prebiotic soup then becomes a self-replicating molecule (or molecules) that catalyzes its own production.[27] Such molecules sound familiar. DNA is nature's best-known self-assembling, self-replicating superstructure. Its famous double helix can be seen with the help of a scanning tunneling microscope, as demonstrated by a team from Caltech.[28] (See color plate 5.) It seems reasonable to speculate that the first primitive life forms were "naked" genes—nucleotide strands with the power to proliferate unaided. The enormous diversity of living matter in nature is made possible by the powers of self-assembly, self-replication, and self-organization implicit within these molecules.

Thus, the central problem in many origin-of-life studies is understanding the first self-replicating, information-carrying material, "the first gene," which provides an evolutionary link between lifeless molecules and living cells. Lacking any fossil evidence, the search for this naked gene is more a quest for the possible than for the probable. Perhaps the earliest functional proteins replicated directly; they "invented" nucleic acids and were ultimately enslaved by them. Another possibility is that early nucleic acids or related molecules replicated directly and then "invented" protein synthesis. Both could have occurred; that is, nucleic acid replication and genetic coding of proteins may have coevolved. Or the earliest form of life might have been based on some inorganic or organic system unrelated to proteins or nucleic acids.[29]

THE NAKED GENE

Let us assume, for a moment, that it is possible to string together molecules of DNA or its sibling RNA from suitable smaller nucleotide precursors. For some years, it was thought that these molecules could not be candidate naked genes. Though equipped to store genetic information, neither DNA nor RNA seemed chemically versatile to orchestrate their own replication. The problem went even deeper because molecular biologists had been misled by studying the way these molecules are used by living things: glibly put, "DNA makes RNA makes protein."[30] Genetic information flowed from DNA into proteins via RNA, a mere gofer molecule that carried the information to ribosomes, where proteins are manufactured. Proteins featured in another dogma of the day: the enzymes that speed up biochemical reactions and the internal building blocks of cells were always proteins. For those seeking a naked gene that directly catalyzed its own production, this would seem to rule out nucleic acids: the naked gene had to be a protein.

But then it was discovered that some viruses contained an RNA genetic blueprint.[31] The 1980s witnessed a dramatic expansion of our knowledge of RNA's chemical repertoire that placed it center stage in early evolution. Molecular biologists discovered that RNA can indulge in the kind of sophisticated chemistry previously associated exclusively with proteins. It suddenly became apparent that, while DNA's forte may be storing genetic information, and proteins may make good enzymes, only RNA can do both. The idea emerged that "in the beginning" both the genetic material and the catalysts were RNA. Those engaged in the quest for life's origins became obsessed by "the RNA world."[32] They argued that the DNA world of today would have grown out of the RNA world as a written culture grows out of an oral tradition, with the DNA acting as a permanent record of what had gone before.

The RNA world-view rested on two earlier hints of a fundamental role for RNA. The first is that all cells make the nucleotide subunits of DNA not from scratch but by modifying RNA subunits. This suggested that primitive cells were making RNA well before they were making DNA. Second, many conventional protein-based enzymes can work only alongside smaller molecules called coenzymes, which are for the most part either RNA nucleotides or close relatives. This suggested that coenzymes are remnants of ancient RNA enzymes. If protein-based enzymes had evolved first, one would expect coenzymes to be made from amino acids, the subunits of proteins, rather than from the nucleotide subunits of RNA. These were only hints, however. Scientists lacked understanding of what led to the demise of the RNA world and the rise of the DNA world. Most of all, they lacked good evidence that RNA could have been central to life.

All this changed in 1982, the year that witnessed a second renaissance in origins studies. Thomas Cech of the University of Colorado surprised biologists by finding RNA molecules with catalytic powers in a single-celled protozoan called *Tetrahymena*. A year later, Sidney Altman and his colleagues at Yale University discovered an RNA enzyme, or ribozyme, lurking inside the well-known gut bacterium called *Escherichia coli*.[33] At the time, this was seen as a kind of living molecular fossil—a relic from the RNA world when RNA played the dual part of gene and enzyme. But it came as a surprise, as Leslie Orgel remarked: "When Francis Crick, Carl Woese and myself suggested that RNA came first in the late 1960s, it never occurred to us that there were actually RNA enzymes still out there to be found in living organisms."[34] Additional pieces of the puzzle then clicked into place. Altman and Cech shared the 1989 Nobel Prize in chemistry for establishing the RNA world-view.[35] "The RNA world provides us with a base camp," said Orgel. "Instead of having the enormous problem of where organisms come from, it can be subdivided into two problems:

(a) where did the RNA world come from, and (b) how did that develop into a world with proteins and DNA? That is easier than trying to do the whole lot in one go."[36]

TEST-TUBE EVOLUTION

Chemists are using other approaches to provide insights into what happened three billion years ago, when primeval chemical stirrings within a muddy pool led to the first self-replicating molecular precursors of life.[37] These approaches are united by the quest of finding a chemistry rich enough to bring an interdependent system of nucleic acids and proteins into being. In seminal work published in 1986, Gunter von Kiedrowski built short chain molecules (oligonucleotides) made of the same nucleotide bases as found in DNA and showed that they were capable of self-replicating in water.[38] In a laboratory at the Massachusetts Institute of Technology, Julius Rebek's group has created other chemical reactions with certain biological features that represent another intriguing step forward in a field Rebek calls "extrabiology," a bottom-up chemical approach to concocting life-like behavior: "Our aim is to express biochemical phenomena such as replication, regulation, transport and assembly with synthetic molecules. If some behavior emerges that has no naturally occurring counterparts, for which vocabulary does not yet exist, well so much the better."[39]

Rebek's team started its search around a decade ago by attempting to design from scratch molecules with the ability to self-replicate. For self-replication to occur, molecular recognition must take place: essentially, the chemical ingredients have to interact with each other before anything can happen. Using computer-based modeling methods, Rebek's team studied the basic principles that govern whether one molecule will dock with another, ranging from the distribution of electrical charge to overall shape. On screen, they could manipulate graphical representations of the molecules and test whether they would fit together in a host-guest relationship, just as one would design the pieces of a jigsaw puzzle.

Two years later, his group achieved its first success.[40] The self-replicating recipe relied on molecules that could gently stick together and then go on to form longer-lived chemical bonds. This paved the way toward a simple self-replicating setup consisting of three molecular

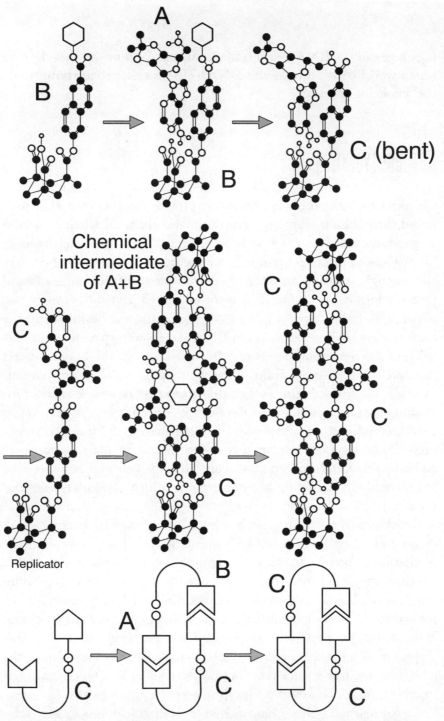

Figure 7.2 Rebek's chemical replicator. Molecules A and B combine to produce the replicator, C, via a reactive intermediate. This replicator has a structure that enables it to bind one further molecule of A and one of B, facilitating their conversion to a new replicator and so on. The final three steps are also shown schematically.

participants: molecules A and B had structures ensuring that they would dock with a parent or template molecule C. Their adjacent positions when connected to C then led A and B to combine to form a copy of C. This copy of C could encourage other A molecules to pair with B molecules and so on, acting like an autocatalyst. Replication continued until the supply of A and B was exhausted.

Such simple behavior helps to make life seem less accidental and more probable if we can introduce the potential to evolve.[41] In another small step in this direction, Rebek's group was also able to introduce a "mutant" into their molecular games. The group developed a new system in which more effective molecular replicators took over in the presence of ultraviolet light, just as a strain of light-tolerant bacteria could overtake a light-sensitive variety in a sunny environment. This time two distinct molecules B and B' could, in turn, lead to slightly different parent molecules, C and C'. Now rival populations of C molecules existed to compete for resources of A, B, and B'. To create the mutant species, they synthesized a C molecule that was sensitive to ultraviolet light. Bathed in this radiation, it fell apart to make A together with B', which in recombination on template C made a new parent molecule, C'. In turn, C' reproduced more rapidly than C.

Having a system with just one mutant enabled Rebek's group to study this severely limited form of evolution with great clarity. The molecules could all reproduce and interbreed. However, C' not only replicated, it outcompeted the unmutated molecule and quickly came to dominate the mixture.[42] Rebek's group is now attempting to encapsulate self-replicating molecules to create something akin to a virus.[43]

Overall, the process modeled by Rebek's group, as with von Kiedrowski's before it, are analogous to the way DNA makes copies of itself as living things propagate, although significant differences exist. Unlike von Kiedrowski's earlier studies, the molecules Rebek considered are not like nucleic acids, and their reactions take place in organic solvents rather than in water. Furthermore, in both groups, the manner of molecular replication proceeds in a "self-complementary" way—that is, a template molecule makes an exact replica of itself. Replication in biomolecules actually occurs in a complementary, instead of a self-complementary manner, so that any given molecule acts as a template for the production not of itself but of a complementary structure.[44] Von Kiedrowski has recently extended his earlier findings by demonstrating that it is possible for short chain oligonucleotides to replicate in the same complementary manner as

RNA or DNA, bringing us one step closer to the formulation of specific prebiotic scenarios whereby replicating reactions based on nucleic acids could have arisen on the earth.[45]

RNA EVOLUTION IN VITRO

Scientists are not only attempting to synthesize self-replicating molecules, they are also persuading existing biological molecules to evolve artificially in the laboratory. The initial groundbreaking work in this area began in the 1960s and 1970s, with work by Sol Spiegelman's group on RNA evolution in vitro.[46] Over the following decade their research was extended by Manfred Eigen, Christof Biebricher, and others in Göttingen.[47] These scientists coaxed RNA enzymes to evolve artificially, doing in a few weeks by "unnatural selection" what nature would have taken millennia to accomplish. This kind of research is likely to shed direct light on the mechanisms of molecular evolution, notably by extending the repertoire of RNA chemistry to, for example, key steps in protein synthesis. It also offers biotechnologists a potentially powerful route to the synthesis of artificial molecules with novel properties that might act as drugs or diagnostic reagents.

The idea behind these "evolution in vitro" experiments is simple: it is the "wetware" equivalent of the genetic algorithm described in Chapter 4. It enables scientists to optimize the design of a molecule, a task beyond the power of traditional analytical mathematics. While the earliest work tinkered with the suitability of an RNA molecule to be a target for a protein enzyme under given conditions, the latest work has been able to evolve new catalytic functions for RNA by a method of "unnatural" selection that offers a powerful alternative to the rational design of a catalyst, which is totally dependent on reductionist understanding.

In recent work along these lines, Gerald Joyce and Amber Beaudry of the Scripps Research Institute in La Jolla, California, started with a population comprising trillions of molecular variants of a single strand of an RNA enzyme, a so-called *ribozyme*.[48] The ribozyme they used for their experiments came from a single-celled protozoan called *Tetrahymena thermophila*. The *Tetrahymena* ribozyme consists of 413 nucleotide units. Naturally occurring ribozymes are strands of RNA that have evolved the ability to snip either themselves or other strands of RNA. This ability, in turn, enables cells to "edit" the information stored in their DNA after it has been copied into RNA—a copying process that is the first step in protein synthesis.

In nature, strands of RNA have evolved through mutation of the genetic code: mutations are favored if the resulting ribozymes are better suited to their cellular environment. To carry out evolution in the laboratory, Joyce and Beaudry used a device—called a DNA synthesizer—that is able to automate the production of DNA and RNA. First they made ten trillion versions of the original ribozyme. To generate diversity, they programmed a synthesizer to insert variable stretches of genetic material, or "oligos" (oligonucleotides), into the ribozyme's genetic code. Each oligo introduced mutations into one of four specific portions of the ribozyme crucial to its catalytic ability.

The resulting population of mutant RNAs was then challenged with a catalytic task: its ability to cleave DNA. If a particular ribozyme mutant was able to cut the DNA thrown into the pool, it would incorporate within itself a portion of the DNA target. This "tag" enabled the researchers to identify "fit" ribozymes and amplify them for the next stage of the cycle. The researchers amplified the ribozyme pool again, this time exploiting the polymerase chain reaction, PCR, a widely used method for multiplying scraps of genetic material into usable quantities. During this second amplification step the successful ribozymes were allowed to mutate by ensuring that the PCR made mistakes, with no bias toward any particular mistake. After two days, Joyce and Beaudry had created an entirely new population of ribozymes. After twenty working days, a tenth generation emerged containing enzymes that snipped DNA one hundred times more efficiently than the starting molecule. Although this could be construed as a modest improvement, it actually represents a singular achievement: the resulting enzyme was the very first to have been adapted in vitro by an evolutionary technique.

The study crowns years of intense interest in the RNA world and in generating novel biological molecules by random alterations instead of specific and deliberate alterations. It is clear that such a Darwinian or "genetic" approach is necessary to effect improvement in the ribozyme's activity. To calculate in a rational way which sites in the molecule should be mutated is an NP problem, well beyond the reach of pen-and-paper mathematics. Natural selection may seem a cheat to scientists who want to make discoveries through sheer deductive brilliance. But there is no effective way of arriving at such molecular structures by a simple algorithm. This was underlined when Beaudry and Joyce analyzed the outcome of their molecular gambling. The molecules that performed best had mutations in quite unexpected sites. And some that initially rose in importance petered out in the long term as they failed to compete with other molecules.[49] (See color plate 6.)

To develop new drugs, pharmaceutical companies have traditionally exploited understanding premised on a reductionist methodology. They use genetic engineering to mass produce a desired protein from a bacterium—so long as they have uncovered the protein's genetic code. An alternative is so-called "rational design," where computers are used to design a drug molecule to interfere with an active site usually within a specific enzyme—so long as the structure of the site is known. Unnatural selection or "irrational design" offers one huge advantage in the quest for new drugs. Scientists may now be able to develop a drug while remaining blind to many of the dizzying details of how it works on a molecular level. All that is needed are the three main ingredients of Darwinian evolution: a mechanism for introducing random mutations in a given population of molecules; a selection pressure that favors some individuals over others; and an "amplification" mechanism that encourages the favored individuals to multiply. In short, the principle is to throw trillions of molecules at a given problem and let the "fittest" pick themselves through evolution. Many other groups are now working along these lines, for instance, to design DNA to inhibit a natural blood-clotting agent, making this one of the most fertile areas in contemporary biotechnology.[50]

Unfortunately, it is not yet obvious how to accomplish with proteins what Joyce has achieved with RNA molecules. In essence, no one has ever managed to reverse-translate a protein to RNA. Sydney Brenner, of the Medical Research Council's Laboratory for Molecular Biology in Cambridge, and Richard Lerner, of the Scripps Research Institute have recently begun to investigate one possible way of overcoming this problem.[51] They resorted to making "hybrid" molecules, each consisting of two parts, one a fragment of a protein (called a peptide), the other a "gene." This gave Brenner and Lerner the best of two worlds. They were able to select molecules according to the chemical "fitness" of their peptide arm and then amplify them using the attached "gene."

THEORY, COMPUTERS, AND LIFE'S ORIGIN

The rise, evolution, and fall of the RNA world may have taken place between 4.2 billion and 3.6 billion years ago, according to one scientific version of the book of Genesis.[52] However, the hunt for the precise molecular fossils of the RNA world is ultimately fruitless. Even if a plausible scenario is found for the emergence of the first naked gene, no one will ever be able to say for certain what happened in a muddy pool or a crack in the ocean floor over three billion years ago. While the work of von Kiedrowski, Rebek, and Joyce has been suggestive, it has not addressed the fundamental issue of how a self-replicating mixture of genetic material and proteins can spontaneously pop out of an incubating collection of subunits. Because self-replicating systems are highly nonlinear, they are very complex and so, once again, computer simulations are providing striking insights into possible answers.

Life is driven by a set of chemical processes far from equilibrium. To maintain these processes, all organisms are open systems; their complexity is founded on feedback involving catalytic molecules that assist reactions without being destroyed in the process. One metabolic or regulatory pathway may produce a molecule that accelerates other pathways, which, through a huge amount of interlinked, intertwined, and interdependent chemistry, may end up supporting the original pathway.

At the Santa Fe Institute in New Mexico, Stuart Kauffman has created idealized theoretical models of such chemical "networks" in which long molecules made of simple units (polymeric macromolecules) react together under the control of catalysts, generating products that may themselves act as raw materials and/or catalysts for further reactions. He maintains that sufficiently complex mixtures of such polymers can replicate as a group, even if no single member of that group can replicate by itself. In Kauffman's opinion, if you start out with a "complicated enough" soup of chemicals, the self-reproducing chemistry necessary for life is bound to emerge. Kauffman maintains that this would have enabled small proteins (peptides) to have become self-reproducing even though they do not have the explicit autocatalytic properties of DNA and RNA. This would have established an integrated protein-based "metabolism" that might only later on have been taken over by RNA.[53]

Some influential figures think Kauffman's chemical networks are distant from the real chemical world.[54] As Leslie Orgel commented, "I don't think that chemicals know about, or behave in a way described by, his equations."[55] The work of Jack Szostak at the Massachusetts General Hospital in Boston is more convincing. Szostak discovered catalytic activity of an RNA molecule selected from a pool of random oligonucleotides of the kind that might have been around at the prebiotic dawn. The RNA catalyst could string together other RNA molecules in a reaction driven by a molecule called adenosine triphosphate. This is the selfsame species that now fuels most of biochemistry. "Such a resemblance supports the idea that an RNA molecule could have behaved like, and preceded, the protein catalysts that today carry out the replication of genetic material in living organisms," said Orgel.[56]

The Nobel laureate Manfred Eigen has shown how proteins can get into the RNA act. Together with Peter Schuster in Vienna and other collaborators, Eigen has sought to elaborate a theoretical framework for such molecular evolution from a prebiotic mixture in terms of so-called *hypercycles*—template-instructed replicating cycles involving catalytic feedback loops where molecule A begets molecule B which begets molecule C which begets molecule A and so on. This concept was introduced nearly twenty-five years ago when Eigen realized that small molecules alone could not encode the machinery complex enough for reproduction without cooperation.

At a later stage, the hypercycle was met with criticism: the evolutionary biologist John Maynard Smith at the University of Sussex showed that in the presence of "parasite" molecules that subvert the mechanism for their own selfish ends, such hypercycles are unstable. Parasites exploit the catalytic properties of hypercycles to reinforce their own numbers, but give nothing in return. Therefore, Maynard Smith argued that the hypercycle would die out over any substantial period of time.[57] His analysis was based on an idealized scenario, in which the hypercycle's ingredients are distributed uniformly in space. Actually, although often cited, Maynard Smith's criticism is something of a straw man, as Eigen and Schuster's original paper written eight years earlier concluded that some form of compartmentalization (i.e., spatial nonuniformity) was required to keep a hypercycle in business.[58]

A more complex and arguably more realistic scenario was analyzed in recent work by Martin Boerlijst and Pauline Hogeweg: they carried out a simulation of hypercycles in a nonuniform medium, using a cellular automaton model to represent the prebiotic chemistry.[59] This behaved in a markedly different way from Maynard Smith's well-stirred soup. Now chemical concentration gradients can emerge during the reaction processes in the manner Turing proposed. Then, the autocatalytic reactions in the hypercycling system lead to multiarmed spiral waves of activity—much like those of the BZ reaction. It turns out that the hypercycles within these spiral waves resist invasion by parasites, which get pushed out to the very edges of the spiral waves.

Evidence that hypercycles turn in the real world has come from work by one of Eigen's students, Michael Gebinoga at Göttingen, on a bacteriophage, a virus that infects bacteria. Viruses are often regarded as being at the borderline of life, consisting simply of nucleic acid and a protective protein coat. Gebinoga found evidence of two feedback cycles, one based on the enzyme replicase, which promotes viral replication, and the other on the viral coat protein, which limits it.[60] The virus would appear to be the self-organized structure formed by the interaction of these components in a hypercycle. "Viruses are useful models for studying how molecules may have organized themselves into self-perpetuating units at the dawn of life," according to Eigen.[61]

This does not mean to say that hypercycles of protein and nucleic acid were spinning at the biochemical big bang. The ever-extending range of RNA chemistry has shown that this giant molecule can hold information *and* carry out functions like a protein. Indeed, Eigen has found that hypercycles actually work better in the RNA world.[62] It does not seem too much of a leap of faith to believe that in an unrelated stew of millions of possible polymer species, a reaction network could self-organize to form an ecosystem of molecules, a kind of metabolism.

EARLY CELLS

If the complexity of life *did* begin with this RNA metabolism, then it must have evolved in very special circumstances, not least because RNA breaks down rapidly in water. RNA-based life would have needed a protective environment, analogous to that found today within a cell membrane. Separated by permeable boundaries from the surrounding aqueous medium, a primitive cell would have allowed the evolution of more complex molecules and processes in a controlled manner. Various ideas have been suggested for how this might have been achieved, including a process whereby amino acids are persuaded to self-organize into microscopic spheres, put forward by Sidney Fox;[63] and the "lipid bilayer model" of Richard Goldacre where molecules of fat join forces to produce simple membrane-like structures.[64]

Pier Luigi Luisi at ETH, Zürich, has for some years been involved in an effort to realize a chemical system that would fulfill the conditions for "minimal life" proposed by Humberto Maturana and Francisco Varela in their definition of life as an *autopoietic* process—one in which self-replication occurs within well-defined boundaries.[65] It is important to recognize that this notion differs from that advocated by Joyce, Szostak, von Kiedrowski, and Rebek, for whom the basic unit of minimal life would be a single molecule with the ability to self-replicate and mutate. In 1990, Luisi's group announced a process that they claimed had just such autopoietic properties and was subsequently connected to possible prebiotic chemistry.[66] His protocell involved a self-replicating cluster of micelles—assemblies of fatty detergent molecules, each with sharp boundaries—that may have been the forerunners of biological membranes. We can describe the formation of these clusters using the techniques discussed in the previous chapter.[67] Moreover, Luisi's group is also studying more realistic core-and-shell reproduction, in which both the fatty cell membrane and RNA within the cell are simultaneously replicated.[68]

THE LIVING BZ REACTION

Once efficient self-replicating molecules have evolved, we may create conditions akin to those found in the Belousov-Zhabotinski reaction, where self-organization appears as the result of millions of molecules adopting coherent large-scale structures in time and space. Within a single cell, concentrations of substances can rise and fall over periods of a few seconds to minutes. The mass of cells constituting a human heart beats around seventy times per minute. And the packages of cells making up plants and animals have built-in cycles of development and reproduction that can last for years. Indeed, the whole human body can be regarded as a complex unit, self-organized in time and space. Ultimately, all these rhythms are controlled by nonlinear processes that we can understand along the same lines as those in the BZ reaction. The only important difference is that the systems we are looking at now are alive.

It is not surprising that biological systems have similar internal feedback to the BZ reaction. When an enzyme is manufactured in the body, it in turn participates in subsequent processes that affect its manufacture. For example, the enzyme may accelerate or suppress parts of the cell's machinery. Such nonlinear processes are tricky to predict, for as the quantities of the enzyme change so do the rules governing its manufacture. Feedback processes are widespread in biology, spanning a vast range of processes from the energy metabolism of cells to the organization of complex societies. They can spawn self-organization in three qualitatively different ways, just as we saw in chemistry with the BZ reaction and the Brusselator. There is temporal organization, corresponding to oscillations, spatial organization, corresponding to patterns, and a combination of the two, when waves of activity propagate through space. Taken together, we will see that these three brands of organization have the power to provide much insight into what makes life tick.

The very genes containing the blueprint for all these feedback processes are responsible for regulating the way they themselves are read and interpreted in our bodies. Thus, the processes that take place within any living being are inherently highly nonlinear. Because of the ensuing complexity, attempts to model or simulate living systems must make use of computers.

SELF-ORGANIZATION WITHIN CELLS

Feedback processes are at work within the cells of plants and animals for generating the necessary physiological order to live. Cells need energy for digestion, the synthesis of biochemicals, to generate muscular contraction, body heat, and so on. The immediate fuel for the body's economy is the energy-rich biomolecule, adenosine triphosphate. Better known by its initials ATP, it carries its cargo of energy in the form of a high-energy chemical bond, like a compressed spring, involving a chemical entity called the phosphate group, a cluster of four oxygen atoms around a phosphorus atom. When the phosphate group is lost, ATP is converted to a depleted form known as ADP. In turn, the ADP can be reactivated to ATP by a chemical reaction called phosphorylation.

One way to make ATP is a fermentation process called *glycolysis*, in which the sugar molecule glucose is snipped in two. Primitive unicellular organisms, such as yeasts, which appear in yogurt and food poisons, draw on glycolysis to survive even when deprived of air. So do creatures such as oysters and green sea turtles, who spend much of their time under water. Even within the human body, glycolysis has a role to play, particularly in areas of limited blood supply—for example, in muscles engaged in frenetic activity.

By about 1940, the entire metabolic pathway of gylcolysis had been mapped out. In 1957, Duysens and Amesz noted for the first time that energy is not always produced in steady quantities during glycolysis: it occasionally oscillates in a regular rhythm, as do the concentrations of various biochemicals involved in the process, among which of paramount importance is the energy-rich molecule ATP. Whether the ATP concentration fluctuates with time critically depends on how much sugar and ADP are around. Feedback is the key to the role these oscillations play in physiological regulation. When there is only a little ATP in the cell (and therefore more ADP), glycolysis switches on to generate the needed ATP molecules, with the cell perhaps drawing on its reserves of starch or glycogen; if there is an abundance of ATP molecules, possibly because the respiratory chain has been working well, the glycolytic pathway cuts off. This regulatory process, known as the Pasteur effect, is essentially controlled by a single enzyme.

The enzyme in question is phosphofructokinase (PFK for short). Specifically tailored for its job over millions of years of evolution, PFK is switched on by high concentrations of ADP, and turned off by high concentrations of ATP. But PFK is a phosphorylating agent: it uses ATP to hitch a phosphate group to a sugar molecule, thereby converting the ATP into ADP. And it is, of course, the presence of ADP that activates the enzyme to operate more quickly. This feedback is precisely the kind of autocatalytic, nonlinear step necessary for self-organization.

One successful theoretical model was set out in 1972 by Albert Goldbeter and René Lefever of the Free University of Brussels. They stripped down the process of glycolysis to its bare essentials. It started out being described with twelve coupled, nonlinear differential equations and ended up with just two, focusing only on the concentrations of the enzyme PFK and the energy-rich molecule ATP.

This massive simplification is repaid by the fact that the glycolytic rhythms are then described by equations similar to those used for the Brusselator. As we saw, the Brusselator's self-organizing properties are described by a limit-cycle. Thus, for appropriate concentrations, the amounts of ATP and ADP in the sugar clock waltz around a repetitive cycle; they vary over a period of about a minute, in good agreement with the experimental values. Glycolytic rhythms therefore became the first confirmed biological example of a temporal self-organizing pattern or dissipative structure. Subsequently, a more detailed mathematical model was formulated by Mario Markus and Benno Hess in Germany, which gave excellent quantitative agreement with experimental observations and also predicted the possibility of chaotic oscillations; indeed, such chaos has now been observed in gycolysis.[69]

Many other examples of such rhythmic biochemistry exist. It has been found in a number of processes involving a single enzyme, such as the autocatalysts horseradish peroxidase and lactoperoxidase, and biochemical processes employing a range of enzymes.[70] These oscillators play a role in the transmission of information, both within and outside cells, that enables cellular and subcellular processes to be properly coordinated, and in the process of cell differentiation, by which cells in a developing embryo turn into, say, brain and liver cells.

There are also examples of spatial patterns within cells where we have found self-organization. In studies of frog eggs, striking corkscrew calcium concentration profiles were observed under a microscope. These patterns resemble those of other excitable media, such as the spiral waves produced in a BZ reaction or in a cellular automaton simulation.[71] Frog cells are large enough, at around one millimeter in diameter, to allow the formation of waves with a crest-to-crest distance of ten microns (ten-millionths of a meter). Size is important: the formation of patterns in cells is not a straightforward scale-down of the BZ reaction, owing to the much greater ease with which molecules diffuse from one side of a cell to another during a single round of the catalyzed chemical reaction that underpins self-organization.[72]

Spatial patterns have also been discovered during the formation of microtubules, believed to be the principal organizers of the cell interior, by Jim Tabony of the Centre d'Etude Nucléaire de Grenoble.[73] These tubular molecular assemblies, composed of the protein tubulin, can be formed in vitro by warming a solution containing tubulin and the nucleotide guanosine triphosphate. It is possible to form stationary horizontal stripes, containing bands up to a millimeter thick in which the microtubules are highly oriented. This nonlinear process behaves in the way expected for the dissipative structures Turing described.

"The results show that complex biological phenomena occur as a result of nonlinear mechanisms," said Tabony. He points out that in developing eggs of the fruit fly, band-like structures form that correspond to the body segments such as the head and thorax. Let us now turn to this higher level of biological complexity and organization, where we are concerned with organized behavior between cells rather than within them.

SOCIALIZING CELLS

Advanced organisms such as the one reading this book comprise many billions of cells that are organized into enormously elaborate structures during the process of development from egg to offspring. There are scarcely any mechanisms in this development that we understand well enough yet to be able to give a decent mathematical description. Nevertheless, nonlinear mathematics can provide a qualitative sketch of the self-organization of a community of cells, as we can illustrate with the help of a strange creature called a slime mold.

The slime mold falls halfway between a collection of single cells and an organism. Like the ant hive, *Dictyostelium discoideum* is a superorganism. At times it is multicellular (with around 100,000 cells), while at others, its cells wander independently. When the bacteria that make up its food are plentiful, individual cells feed voraciously, behaving like solitary wanderers and multiplying by direct cell division. Eventually, however, the colony runs short of food. Now the cells "notice" each other. For nonlinear reasons not yet fully understood, certain cells in the colony become active and act as pacemakers, "ringleaders" that send out rhythmic pulses of a chemical called cyclic adenosine monophosphate (cAMP). This is a ubiquitous molecule in biology that acts as a molecular message between neighboring cells. It is a glucose distress signal, announcing they have run out of food.

This clarion call to close ranks and organize travels at a few microns a second. Cells amplify and pass on the message, a form of feedback mechanism providing the nonlinearity that induces still more cells to hone in on the pacemaker centers. There are two additional ingredients: once a cell has released a burst of cAMP it cannot immediately respond to another signal, going into a "refractory state" before returning to an excitable condition. The cells also exude another enzyme—phosphodiesterase—that destroys cAMP, setting up a gradient of the chemical that provides a signpost. The starving cells slither toward the pacemaker cells, in the direction of increasing cAMP concentration. Aggregating populations can produce concentric and spiral waves that bear a compelling resemblance to the spiral waves occurring in the BZ reaction. This is no surprise: though the details are different, the positive and negative feedback processes are the same.

Once the cells have formed a slimy mass, they begin to differentiate and a tip forms that secretes cAMP continuously. The whole mass becomes organized into a glistening multicellular "slug," with a head and a tail, that wriggles in search of light and water. All in all, it takes several hours for these cells to form this simple organism. Between one and two millimeters long, it crawls along under the leadership of the pulsating source at its tip. It then rights itself to form a hard stalk above which perches a small head containing spores; eventually, the head breaks open and the wind casts its spores far and wide. If they settle in a suitable place, they can germinate and begin the cycle of this strange organism's life anew.

Remarkable biochemistry underlies this behavior, reminiscent of the sugar clock in glycolysis. The messenger molecule that organizes this wriggling mass, cAMP, is formed from ATP by the help of an enzyme called adenylate cyclase. Feedback occurs, just as in glycolysis: cAMP already present in the medium surrounding the cells switches on adenylate cyclase to produce more cAMP from ATP. In this way autocatalysis arises, an essential ingredient of self-organization. By employing largely the same nonlinear analysis as he used to model glycolytic oscillations in yeast cells, Albert Goldbeter was able to show in a detailed way on the basis of limit-cycles how oscillations of cAMP could be produced every few minutes.[74] This is an excellent example of self-organized behavior; moreover, chaotic cAMP oscillations are also now known. Indeed, in the mutant form of D. discoideum, we have observed temporal chaos in the form of cAMP oscillations and spatial disorder manifested in aberrant stalks and fruiting bodies, all of which can be returned to ordered behavior by adding phosphodiesterase.

THE CREATION OF BIOLOGICAL SHAPES

Nature has mechanisms for organizing cells into a dazzling range of forms and shapes extending far beyond the repertoire of the humble slime mold. A detailed description of the recipe for that diversity is a long way off. Nonetheless, the very fact that *Homo sapiens* is governed to a large extent by the same genetic programming as other creatures like fruit flies, dandelions, and chimpanzees fuels the hope that we can make sense of development in general terms. Some scientists simply want to throw a little light on the processes at work. Others dream of formulating biology's equivalent of Newton's laws. Earlier this century D'Arcy Thompson, professor of zoology at St. Andrew's University in Scotland, wrote in his study of growth of multicellular organisms that the development of living creatures must lie within a framework set by geometry. Many biologists would say that his approach led nowhere: the leaps and bounds in the field have not been made with computers but by observations in the laboratory.

Nonetheless, in parallel with the explosion of effort in biological research, theoreticians have made some progress in comprehending how these living patterns are woven in time and space. Our understanding of the evolution of form in living systems—i.e., morphogenesis— enjoyed its first significant advance more than forty years ago when Turing wrote his 1952 paper. Today, however, such studies can be taken much further thanks to the existence of fast computers for handling the omnipresent nonlinearities underpinning morphogenesis.

The hydra, a freshwater polyp measuring a couple of millimeters in length, is one of the most fascinating creatures to which we have applied this approach. Turing himself was introduced to the wonders of the hydra by the same book that roused his interest in machine intelligence, *Natural Wonders Every Child Should Know*, written by Edwin Tenney Brewster. If a small piece of tissue is taken from near the head of the hydra and put elsewhere on the body, a new head will grow within forty-eight hours. A part of hydra cut off from the rest will rearrange itself so as to form a complete new organism.

Turing idealized the tubular hydra, portraying it as a ring of cells. He found that by modeling two chemicals reacting and diffusing around this ring, he could set up chemical waves, defining patches where the tentacles would subsequently appear after the hydra had been chopped in two.[75] Developing this work, Hans Meinhardt of the Max Planck Institute for Virus Research in Tübingen concluded that the humble creature had a good chance of providing insights into how the development of a relatively simple organism is controlled.

Pattern formation in the hydra seems to depend on two components: short-range chemical activation (by autocatalysis) and long-range inhibition.[76] The resulting nonlinearities give rise to patterns with features common to many organisms. Typically, a small patch of tissue becomes slightly different from its surroundings and exudes a tiny amount of a molecular "activator" that rapidly builds up in concentration because it catalyzes its own production. The high concentrations in this region trigger the manufacture of an inhibitor signal—another biomolecule—that diffuses into surrounding tissue to prevent other regions making the activator. The concentration profiles of these so-called morphogens in effect tell cells where they are with respect to this landmark of special tissue, essential information for deciding if they are to evolve into a head or a body cell.

Activation and inhibition not only model the initial pattern development but are also believed to play an essential role in the spacing of repetitive structures. With the hydra, we are talking about tentacles. For other living things, it could be bristles, hairs, feathers, leaves, or body segments. In turn, other processes can follow. For instance, depending on which body segment of an insect is affected, legs or antennae can form. Other applications of Turing's theory explain many spatial features in numerous organisms, such as cartilage patterns in limbs, feather and scale distributions, the complex patterns on butterflies' wings, and animal coat markings.

In animals, such patterns are generally laid down on the embryo inside a shell or womb. The precise moment at which the patterns form and the size of the embryo at that instant are critical factors in determining the patterns that adorn the adult animal. James Murray of the University of Washington has produced detailed mathematical models of the kind that produced the diverse repertoire of the BZ reaction to support the observation that mice and elephants, at the two extremes of the mammal size spectrum, tend to have a uniform coloration. Similar models also show why the patterns of animals of intermediate size such as alligators, leopards, and zebras can be very exotic. By modeling an activator-inhibitor mechanism on tapering cylinders of varying widths, he also found that the tip of a leopard's tail is too thin to support spots, which coalesce into stripes, and why it is possible to have a spotted animal with a striped tail but never the other way around.[77]

HEARTBEAT

Morphogenesis explains how something as complex as a heart can develop, but we need other tools to understand the complexity of its beat. A human heart beats without stopping about 70 times each minute, 40 million times each year, and some 3,000 million times during a lifetime. It is robust to sudden exertion, yet its activity can be thrown off balance by only the slightest restriction of its internal blood flow, sometimes with fatal consequences. Because of the importance of a healthy heart for human survival, scientists have been interested in studying its properties for as long as it has been a potent symbol in art and culture.[78] Today, its global behavior is finally succumbing to understanding, thanks to the power of the modern supercomputer.

The Dutchmen van der Pol and van der Mark wrote down a simple nonlinear mathematical heart model in the 1920s and showed that several different breakdowns in regular rhythms—arrhythmias—could be generated from it as they adjusted various parameters; each arrhythmia was brought into play by a familiar mechanism in nonlinear dynamics—the bifurcations we described in Chapter 6. A major advance in the effort to produce more realistic mathematical models of nerve impulses came in 1952, with Alan Hodgkin and Andrew Huxley's Nobel Prize–winning work on the squid's giant axon, the long thread-like extension of a nerve cell that conducts impulses. The methods they employed led to a quantitative description now routinely used for studying the electrical activity of heart tissue.

For some time, there has been a single digital heart cell beating in a computer at Oxford University's physiology department, developed by Denis Noble and his collaborators. This group has stitched together a mathematical model incorporating the chemical processes taking place in the cell. The cell's beat is formed by the coordinated movement of filaments of protein. Just as a pendulum in a grandfather clock is kept moving with a slowly falling weight, special proteins called sodium pumps push positively charged sodium ions across the cell membranes to ensure that the heart is held away from equilibrium. During each beat, calcium ions surge into the heart cells. Interacting with the channels ferrying the calcium ions into and out of the cells, our old friend cAMP makes an appearance. As in the slime mold, it operates alongside the enzyme adenylate cyclase. In the heart the two act in concert in feedback processes that control how the calcium channels flip open and slam shut. The calcium ions that surge into a single heart cell are amplified within by a positive feedback mechanism to contract it by a kind of molecular "ratchet" mechanism. Passage of calcium into the heart is only half a heartbeat. Before the beat is over, the calcium must be extracted again by another transport mechanism so that the cell relaxes. Thus, each individual heart cell has its own internal clock, the calcium ion oscillator—a biochemical version of the BZ reaction.

The digital cell behaves in a very realistic way. If one type of cell, called the pacemaker cell, is given a shot of the heart drug Digitalis, the normal rhythm is disrupted, producing ectopic beats—additional heartbeats. The drug causes the overexcitation of the heart cell's calcium clock because of a surge of calcium ions. You can feel these palpitations when they happen, which is only rarely for most people. In a similar way, the effect of hormones such as adrenaline can be measured and various abnormal beats reproduced. Some people have suggested that there is little left to do, given the success of the single cell model. This is a "crass misunderstanding," according to Noble, whose computer simulations provide a convincing demonstration of how the properties of a slice of heart tissue are more than the sum of its cellular parts.

What is striking about Noble's approach is that he is scaling up a physiologically accurate model of a single heart cell, rather than using a stripped-down idealization. This approach is computationally very demanding. To simulate the dynamics of a single cell draws on a substantial amount of number crunching—it takes around one hundred seconds on Noble's workstation to work out one real-time second of heartbeat. To simulate even a small slice of tissue requires a supercomputer. Collaborating with Rai Winslow of Johns Hopkins University in Baltimore, Noble has been using the U.S. Army Research Center for High Performance Computing's Connection Machine located at the University of Minnesota. (See color plate 7.) Depending on the size of the simulation, each of the 64,000 processors in the computer may need to handle the dynamics of one or more separate heart cells; they have modeled up to four million cells, about one-hundredth the total number in a heart. Clearly, the larger the model, the more serial computations each processor must execute, slowing the calculations down.

One of the most interesting and medically important issues is how millions of individual cell beats are entrained to produce a globally coherent heart throb. There are known to be at least six different "conductors" of the heartbeat. One is the sinus atrial node, a small mass of muscle cells in the heart's right atrium. A particular tissue, known as the Purkinje fiber, contains cells that very rapidly convey the firing of the sinus node "pacemaker" to the great mass of ventricular heart muscle. The rhythms of these pacemaker cells have to be excited by a so-called "cardiac pacemaker current," analogous to the centers that initiate patterns in the BZ reaction. The reductionist approach would search for the cells that fire first, arguing that they must hold the secret of the beat. It turns out that the cells that fire first do not reveal the origin of the heartbeat, a counterintuitive discovery that was only made possible by modeling vast numbers of heart cells acting together.

This odd result was evident when the scientists managed to simulate one candidate pacemaker—the sinus node—in the Connection Machine. The graphical displays generated by the simulations are beautiful, since each of the 16,384 pixels on the monitor represents a cell that is colored according to its instantaneous voltage, ranging from blue (positive) to red (negative). Each cell is coupled to others electrically via tiny pores called connecting channels. Not surprisingly, the more intense the chatter between neighboring heart cells, the more the cells fire in unison. However, as Noble points out, this model of the sinus node gives "a totally wrong result."[79] A video display of the node shows a wave of excitation collapsing *inward* toward the center of the node, rather than rippling outward to trigger a beat. Winslow and Noble were bitterly disappointed. However, they found that when they wired the node to a vast network of surrounding cells that represent one of the heart's chambers, the wave of excitation that starts within the disc of sinus node cells washes *outward*, over the rest of the tissue, stressing how long-range interactions between cells are as important in the heart as in the brain's neural networks. "The overall rhythm of the heart is a function of the whole thing," observed Noble, "not just the sinus node alone and certainly not just the cells that fire first." It later emerged that experiments performed by a Dutch group on the isolated node *also* produced convergent waves—a pleasing confirmation of the veracity of their model. "We were absolutely delighted," Noble said.

The simulaticns have yielded other fascinating results. It is possible to determine the minimum number of cells within the sinus node pacemaker that can trigger the firing of atrial heart muscle. This turns out to be an island of ten thousand cells. In one simulation, Winslow and Noble found that by roughening the boundary of this island—as in the real node—the beat is conducted more efficiently to surrounding atrial tissue. In another, they found that a tiny chunk of damaged tissue, just 10 by 10 by 10 cells, is enough to trigger an extra, ectopic, beat.[80] "Just one thousand cells are enough to excite the lot," said Noble. Most ectopic beats are benign, but sometimes, if there is damaged tissue, a life-threatening disruption of heart rhythm called fibrillation occurs. Noble and his colleagues are now trying to simulate an abnormal beat invading the entire heart, perhaps from a larger critical mass of initially damaged tissue. For example, it could be that the ectopic beat then interferes with waves sent out by the pacemaker. The result would be spiral waves called rotors, similar to those formed in the BZ reaction, that have been observed during real heart attacks. This is a plausible scenario, since these rotors have already been made by clashing waves of excitation. "We may be closer to reconstructing fibrillation than we think," said Noble.

Winslow and Noble's work provides a rare but compelling example of the computer modeling of a complex macroscopic phenomenon, starting with a detailed microscopic description and building in complexity, step by step. In most of the examples we describe in this book, including cellular automata and neural networks, global behavior is computed by starting out from a so-called mesoscopic model, in which much of the microscopic (atomic and molecular detail) is ignored. Noble's work shows that, with contemporary supercomputers, the passage from a more microscopic description to macroscopic properties *can* be effected in some circumstances. Indeed, it is important not to overstress the shortcomings of reductionism. Noble admits that it can be difficult to understand what is going on in his simulations. Here certain key ingredients can sometimes be isolated by a reductionist approach, analogous to that which von Neumann adopted when he used computers to explore fluid dynamics.

THE RULES OF BEHAVIOR

Just as living things are in themselves patterns in space and time, so communities of creatures also display structures on a larger scale. There can be spatiotemporal organization, ranging from the waxing and waning of flour beetle populations[81] to the emergence of cooperative behavior between species in the struggle for survival. Structure arises in the way animal communities organize themselves: many examples of such organization spring to mind, none more graphic than those within the class of social insects known as *hymenoptera*, which includes bees, wasps, termites, as well as ants.

In colonies of these insects, one finds a plethora of intriguing phenomena: the presence of infertile workers and soldiers who heroically sacrifice themselves for the greater good of the colony in construction work and defense activities. This seems like "altruism," if one expects that life is ultimately a ruthless struggle of the individual, driven by its selfish genes, to survive. Indeed, an explanation for the existence of this kind of cooperation was one of the biggest problems for Darwin himself. However, we now recognize that these examples can be understood by viewing colonies of *hymenoptera* as "superorganisms" (rather like the slime mold), with each individual sharing the *same* pool of genes. We will not pursue these cases of kinship-based cooperation further, preferring instead to concentrate on cooperation between genetically unrelated individuals to show how reciprocal strategies of behavior can emerge spontaneously as a result of the same blind driving forces of survival.[82]

This subject has been investigated using a branch of mathematics that today is called game theory. It aims to determine the strategies that individuals or organizations should adopt in their search for rewards when the outcome is uncertain and depends crucially on what strategies others adopt. Von Neumann is the founding father of this subject, which weighs the risks and benefits of all the strategies in a game of war, economics, survival, or whatever. As with much of his other work, he was keen to use mathematics for analyzing what appeared at face value to be a nonmathematical subject. His first paper on the theory of games appeared in 1928. While at Princeton University, he collaborated in the late 1930s with the mathematical economist Oskar Morgenstern. In a style typical of von Neumann, this work not only led to important applications in economics, but also enriched pure mathematics through the advances simultaneously made in combinatorics (the theory of arrangements of sets of objects). Von Neumann and Morgenstern published their now classic tome on game theory, *Theory of Games and Economic Behaviour*, in 1944.

It is now increasingly apparent that the same principles can be applied to understand how cooperation emerges within human societies, "in a world of self-seeking egoists—whether superpowers, politicians, or private individuals—when there is no central authority to police their actions."[83] These principles therefore have relevance to cooperation between commercial companies, between individuals inside organizations, within government, in politics, economics, and international relations, as well as in biological science proper.[84]

Economists were fascinated by game theory because it offered to explain mathematically why Adam Smith's invisible hand can apparently fail to deliver the collective good. The theory helps us to understand how companies make business decisions in competitive markets. Political scientists, too, picked up on game theory because it shows how "rational" self-interest can then make everybody worse off. Game theory was introduced to biologists in the 1970s, mainly due to the work of John Maynard Smith.

Robert Axelrod, a professor of political science and public policy at the University of Michigan, is a leading worker in the field. He has modeled interactions between individuals on the basis of a simple game called the Prisoner's Dilemma.[85] The idea of this game is to simulate the conflicts that exist in real life between the selfish desire of each player to pursue the "winner-takes-all" philosophy, and the necessity for cooperation and compromise to advance that selfsame need. Like so many complex problems we have previously encountered, such as finding the lowest energy state of a spin glass, learning in neural networks, and the solution of the traveling salesman's problem, it is an example of an optimization problem that must be solved in the presence of conflicting constraints.

It works like this. Two individuals can choose to cooperate with one another or not. If both cooperate, each receives a reward of, say, three points. If one cooperates and the other does not, the defector gets a bigger reward, say five points. The reward for the "sucker" is nothing. Finally, if both defect, each gets a small reward, one point. Even though each player inevitably gains if both cooperate, there is always a temptation to defect, both to maximize profit and avoid being suckered. That is the dilemma.

It is easy to put flesh and bones on this game. Imagine that you and a friend have been caught with a stolen painting, spattered with blood. The police rightly suspect both of you of having committed another, more serious offense, of which they have no proof. You are being held in separate cells and not allowed to contact each other. A detective offers you a deal: if you inform on your friend and reveal his other crime, you will not be charged with stealing the painting. It is reasonable for you to assume that the police have offered your friend the same deal. What to do? If each of you refuses to give evidence, you both will be charged only with the lesser crime—a reasonable result. If each one informs on the other, the two of you will go to jail for the more serious offense, on the evidence of each other's testimony—a bad result. Here comes the dilemma: if you alone stay silent, you will be punished for both offenses while your accomplice walks free.

The Prisoner's Dilemma exercises mathematicians, social scientists, and biologists because it illustrates a widespread problem: how individual ambition can lead to collective misery. If the two players are never going to meet again, there is no reason for them to cooperate. But in real-world situations, which range from traffic jams to global wars, it is often more likely that they *will* encounter one another in the future. Consequently, different strategies emerge.

Robert Axelrod held a worldwide tournament for computer programs to play the Prisoner's Dilemma in an attempt to uncover the best strategy. He made the fourteen entries—some of which used very complex strategies—compete against one another. "To my considerable surprise," said Axelrod, "the winner was the simplest of all the programs devised, tit-for-tat."[86] Created by Anatol Rapoport, a psychologist and game theoretician at Toronto University, the tit-for-tat strategy is very simple: cooperate in the first round, and then do whatever your opponent does in successive rounds. It is a "nice" strategy, since it signals willingness to cooperate at first, and then retaliates whenever the opponent defects. Moreover, it has the property of "forgiveness" in that it does not bear a grudge beyond the immediate retaliation, thereby perpetually furnishing the opportunity of establishing "trust" between opponents: if the opponent is conciliatory, it forgives, and both reap the greater rewards of cooperation. Finally, it is not too clever. Highly complex strategies are incomprehensible: if you appear unresponsive, your adversary has no incentive to cooperate with you. Tit-for-tat's great success is its simplicity. Axelrod circulated the results and solicited entries for a second round that saw sixty-two entries from six countries, including some very elaborate programs. Tit-for-tat was again sent in by Anatol Rapoport. Again it won. "Something very interesting was happening here," remarked Axelrod.

After thinking about the evolution of cooperation in a social context, he realized that the findings also had implications for biological evolution and collaborated with the Oxford University biologist William Hamilton to investigate them.[87] In many scenarios, the same two individuals may meet up more than once. If an individual has a sufficiently powerful brain to recognize from memory another from a previous interaction, and remembers some of the previous outcomes, then the strategic situation becomes one known as the iterated Prisoner's Dilemma. The strategies now allow for the development of rules that take into account the history of the interactions of the two individuals in the game to date. Early in the 1970s the sociobiologist and former lawyer Robert Trivers of Harvard University had suggested that reciprocation of this form was the chief way in which animals not sharing the same gene pool achieve cooperation.[88] His discussion included the Prisoner's Dilemma; symbioses where one organism, such as the wrasse, cleans another, such as the grouper; the warning calls of birds and reciprocal altruism in human societies, which may be employed to avert possible revenge. One particularly colorful example was provided by the Bushmen of the Kalahari. They have a saying to the effect that "if you wish to sleep with someone else's wife, you get him to sleep with yours, then neither of you goes after the other with poisoned arrows."[89]

In the context of biology, one can interpret the rewards and punishments offered in the game in terms of the ability of individuals of a species (or of different species) to survive through reproduction: the reward can be taken to imply numbers of offspring produced during each breeding season. Many examples have been found. Tree swallows live in groups, but not all the birds involved are parents, so that elements of the Prisoner's Dilemma are present. Thus, by attaching themselves to a colony, nonbreeders might learn the characteristics of a good nest site. Parents might benefit by having extra birds around to challenge predators. However, the nonbreeders may kill young birds and take over a nest. Parents, however, generally do not chase off nonbreeders. Both birds benefit if they show restraint: the nonbreeders gain information and the breeders produce extra young. Evidence of how the tit-for-tat strategy had evolved in their relationship was found by Michael Lombardo of Rutgers University in New Jersey.[90]

On the coral reefs off Panama the tit-for-tat game is played by a small hermaphroditic fish called a hamlet.[91] During courtship each fish takes its turn in playing female, while the other plays male. Not surprisingly, it is easier to be a male than a female. Because of this incentive to defect, the "female" has evolved a strategy to lay only a few eggs each time, until the "male" has proved that he will not defect after fertilizing her eggs but will play female in his turn. As trust grows between them, more eggs are laid each time.

In terms of ecosystems, tit-for-tat is robust. It does well when playing a wide variety of other strategies. Though no strategy is evolutionarily stable, it turns out that tit-for-tat cannot be invaded and displaced by all-out defectors in a long-term relationship. The discovery that tit-for-tat is a common strategy sends an optimistic message to those who fear that human nature is founded on greed and self-interest alone, typified by the Hobbesian savage whose life was "solitary, poor, nasty brutish and short."[92] When working in human society at large, the tit-for-tat strategy means that a successful entrepreneur may be an opportunistic seeker of cooperation rather than a ruthless operator. Nice people do not have to finish last.

THE EVOLUTION OF STRATEGY

Because of the complexity of possible strategies in the Prisoner's Dilemma, Axelrod has employed a genetic algorithm to study their evolution. He represented each possible strategy as a string of genes on a chromosome, which undergoes the usual processes defined by Holland's genetic algorithm. The success of each strategy is determined in the context of the current environment, with mating, crossover, and mutation progressively favoring the better strategies. The specific class of strategies investigated by Axelrod in this way was the set of strategies taking into account the results of the past three moves and for which no mistakes are made. (In other words, the game is played deterministically.)

In fact, a huge number of strategies could be represented in this way.[93] An exhaustive search for the best strategies is clearly out of the question. As Axelrod points out, "If a computer had examined these strategies at the rate of a hundred per second since the beginning of the universe, less than one percent would have been checked by now."[94] Hence we are faced with a combinatorial explosion familiar from complex optimization theory, which is where the genetic algorithm scores its biggest successes.

In one series of runs, new strategies were sought to compete against an unchanging set of eight representative strategies Axelrod extracted from the second round of his Prisoner's Dilemma tournament. Remarkably enough, the genetic algorithm evolved (from a random start) a most dominant member whose rule base was just as successful as the tit-for-tat strategy that won the tournament. Indeed, most of its rules shared many features in common with tit-for-tat. But it also contained rules that did substantially better.

However, they may well not be very robust in different environments. As we have pointed out, the genetic algorithm as usually defined involves the use of a kind of sexual reproduction, in which the chromosomes of two parent solutions are recombined. Axelrod investigated what happened in the asexual case: populations still evolved toward rules that performed about as well as tit-for-tat, but they were only half as likely as in the sexual case to produce rules substantially better than tit-for-tat. This emphasizes the importance of sex. He then went on to consider the case where the strategic environment itself is *changing*: it was now taken to be the evolving population of chromosomes. He found that initially the population strategies moved away from one with significant cooperation. But after about twenty generations, these trends started to reverse. Participants evolved strategies of cooperating whenever possible, with a capability of being able to discriminate between others who reciprocate that cooperation and those who do not. Eventually, the reciprocators spread to dominate the entire population.[95]

Axelrod's game-playing provides fascinating insights into the evolution of strategy. In effect, his study depicts the game-theoretic problem as the search for high points in a fitness landscape corresponding to a multitude of gene. combinations.[96] The computer simulations show that sex helps a population to explore this multidimensional space and so find the gene combinations with the greatest fitness;[97] that there is a trade-off in evolution between the gains to be had from flexibility and specialization (flexibility is usually advantageous in the long run, but, in the short term, individuals have to survive); and that some aspects of evolution are entirely arbitrary. One of the most striking examples of this arbitrariness is the mass extinctions that have wiped out entire species at a stroke, a feature of evolution we will return to after considering more recent work in game theory.

UNCERTAINTY AND PAVLOV

Nice guys can finish last in the real world. The emergence of cooperation depends on several things: the "players" must repeatedly encounter each other; they must be able to recognize each other; and they must also remember the outcome of previous encounters. Other factors could influence the outcome, from the chance of an encounter to the probability that the genetic factors that shape behavior are passed from generation to generation. With such real-world uncertainties in mind, Robert May of Oxford University pointed out in 1987 that the groundbreaking work of Axelrod is highly idealized and unlikely to apply wholesale to the natural world.[98]

Attempts to find out how tit-for-tat copes with these complications were made by Martin Nowak and Karl Sigmund, working at the University of Vienna. They found that by adding a little real-world uncertainty, which might be caused by the all-too-human tendency to make mistakes, tit-for-tat is no longer the supreme strategy. To anthropomorphize, this uncertainty permits individuals to develop new strategies, while the addition of a little randomness to behavior allows "forgiveness," and a chance to test the behavior of another player. As a result, Nowak and Sigmund discovered that there were *no* evolutionarily stable strategies. Over thousands of generations, new strategies continually emerge, take over, and subsequently die off. Uncertainty does, however, allow cooperation. The optimistic message of Axelrod's work remains.

Two players deploying deterministic tit-for-tat can be locked into vendettas. If one started by cooperating and the other defecting, they would cycle through a never-ending alternation of behavior—cooperation, defection, cooperation, defection, *ad infinitum*. But by adding mistakes, another strategy eventually takes over from tit-for-tat after some 200 generations, called "generous tit-for-tat." As before, cooperation is met by cooperation. Sometimes, however, defection is "forgiven" with co-operation, rather than punished with Old Testament–style retribution. As a result, the cycle of mutual backbiting that plagues deterministic tit-for-tat can be broken.[99]

Nowak stresses that the success of the strategy depends very much on circumstances. In a population of defectors, for example, generous tit-for-tat is the best way for cooperation to evolve in the first place: Nowak calls the strategy a "catalyzer" because of its ability to kick-start reciprocal relationships. Moreover, the preferred strategy also depends on the way the game is set up. Generous tit-for-tat is rather restricted in that the strategy reacts only to the coplayer's previous move. But if the moves of both players are taken into account, four possibilities arise for each move. With this additional complexity, generous tit-for-tat is itself eclipsed by another strategy. Nowak encountered it while staying with Sigmund in a castle near Vienna. At first, he was irritated that generous tit-for-tat was being beaten as he ran his simulations on a laptop computer. "It took me two weeks to realize that the most interesting outcome for the Prisoner's Dilemma was a strategy that was different from tit-for-tat."[100]

Called Pavlov, the strategy is as simple as tit-for-tat and embodies a fundamental behavioral mechanism: "win stay, lose shift." This kind of behavior can be summed up by the more familiar maxim: "If it ain't broke, don't fix it." The strategy had been discussed and derided as the "Simpleton" strategy by Rapoport as long ago as 1965.[101] But its ability to correct for errors was only appreciated when Sigmund and Novak introduced a little randomness to represent mistakes.[102] Typically, Pavlov players can resume cooperation after two rounds of mutual defection. Sigmund likened it to a couple making up after a domestic quarrel. This sounds like generous tit-for-tat but Pavlov offers another advantage in that it has no qualms about exploiting suckers: a mistake that leads a Pavlovian player to defect then allows it to detect whether its partner cooperates in all circumstances.[103]

The new understanding has modified the conclusions of an important experiment conducted by Manfred Milinski of the Ruhr University in Bochum, who is now based in Berne. He thought he had found evidence that three-spined sticklebacks used the tit-for-tat strategy.[104] In the wild, sticklebacks often approach a stalking predator so as, it is thought, to identify it and gauge its readiness to attack. If the little fish do this *en masse*, they can get closer and, should it attack, they might be better protected by being in a group and confusing the predator. There are, however, choices to be made. Each time one fish swims closer, his companions can cooperate and go along, or defect and turn tail. A defector runs less risk of being eaten itself, and it may gain more information than the sucker as it watches that fish's fate.

Milinski put a stickleback in a tank from where it could see a large cichlid, *Tilapia mariae*, that resembled the perch, a common predator on sticklebacks. By the clever use of mirrors, Milinksi could either create a reflection of the stickleback that appeared to swim toward the predator with the stickleback or one that lagged behind and eventually vanished. "In both cases," he concluded, "the test fish behaved according to tit-for-tat, supporting the hypothesis that cooperation can evolve among egoists." But among sticklebacks that always defected, he also found that they attempted to cooperate on every second move. This is just what one would expect from a Pavlovian strategy.

SPACE, RANDOMNESS, AND THE PRISONER'S DILEMMA

Another elaboration of the Prisoner's Dilemma came in research by Martin Nowak and Robert May. They revealed the effect of spatial variations on the Prisoner's Dilemma using a deterministic cellular automaton model that permits large numbers of players to interact across a lattice. Without spatial structure, this model would allow a clear-cut win for the defectors. However, said May, "Without any remembering, or contingent strategies, or need to meet again, or discounting the future—just by dint of interacting with neighbors—you can get self-organized local spatial structures that enable cooperative behavior to persist." Thus, cooperation can be the child of repetition, of uncertainty, *and* of spatial structure.

Because the players in Nowak and May's model are located on a two-dimensional grid, there is no uncertainty about who an individual is dealing with at any instant of time: each square has eight neighbors. The players neglect all niceties of strategy or memories of past encounters—they are either cooperators or defectors. The only variable in this game is the payoff for defecting. After each round of play, each square is occupied by whoever won the most points, that is, either a neighbor or previous owner. Not surprisingly, cooperation is greater in this sedentary population. Defectors can thrive in an anonymous crowd, but mutual aid is frequent between neighbors.[105] Using color coding to display which cells have changed and which remained the same, these spatial games give an altogether new twist to the Prisoner's Dilemma, generating spatial patterns of extraordinary beauty and complexity.[106] The results are striking. (See color plate 8.) As the authors put it, "There is a new world to be explored."[107]

These evolutionary games generate irregular or regularly shifting mosaics, where strategies of cooperation and defection are both maintained. Even if the chessboard lattice is changed for a hexagonal array, or if the rules are altered slightly, or if the sites are updated at random, the mixtures of cooperators and defectors remain in proportions that fluctuate about predictable long-term averages. The outcome is an endlessly milling chaos, with clusters expanding, colliding, and fragmenting, with "both nice guys and nasties persisting." From a single defector, a fantastic kaleidoscope of gorgeous patterns, suggestive of Persian carpets, lace doilies, and rose windows can result. Though the rules are simpler than those of Conway's Game of Life, they can generate gliders, blinkers, rotators, and motifs that grow without limit, or exhibit fractal patterns.[108]

These studies are exciting for several reasons. First, they show that by introducing geography into the Prisoner's Dilemma, cooperators and defectors can exist side by side. In nature, this means that diverse populations of hosts and parasites or prey and predators can persist in communities, despite the instability of their interactions. Second, such studies can be extended to provide insights into the behavior of other spatially extended systems, such as two-dimensional versions of the spin glass models encountered in Chapter 4, artificial and biological neural networks described in Chapters 5 and 9, and the prebiotic soup of molecules (discussed earlier) from which life first emerged. As the scientists argue, "Catalyzing the replication of a molecule is a form of aid; a chain of catalysts, with each link feeding back on itself, would be the earliest instance of mutual aid. In this sense, cooperation could be older than life itself." This particular application may sound familiar: the spatial structures generated by this two-dimensional game are akin to the cellular automaton chemistry used by Marten Boerlijst and

Pauline Hogeweg of Utrecht University to argue that cooperative chains or "hypercycles" are not vulnerable to "cheating" by molecular mutants that are recipients of more catalytic aid than they give.

These computer experiments underline how cooperation can hold its own against the ever-present threat of exploitation, an apparently everyday feature of natural selection that thrives on the high premium placed on an individual's success. Throughout the evolutionary history of life, cooperation among a huddle of smaller units such as cells led to the emergence of more complex structures, whether cells with nuclei or multicellular organisms. In this sense, argue May and Nowak, cooperation and thus self-organization become as essential for evolution as natural selection: "Cooperation generates more complex structures, whereas natural selection chooses which of these can survive." However, as they emphasize, the twists and turns of the billion-year-old biological tug of war between cooperation and exploitation during Darwinian evolution have grown so complex that it would be unreasonable to expect any real-life examples to be reproduced faithfully by such a simple model.

CATASTROPHES

There are no "Darwin's equations" describing biological evolution in quantitative, mathematical terms. Nonetheless, as we have discussed in earlier chapters, biological evolution is a highly complex (i.e., nonlinear) dynamical system. In the next chapter, we will see how computer studies, based ultimately on Boolean logic, are helping to define the general features of evolution. While computer simulations are the only reliable way of investigating evolution's global properties, certain specific aspects can be approached using some concepts we have already encountered. One is the phenomenon of "punctuated equilibrium," proposed in 1972 by Niles Eldredge of the American Museum of Natural History and Stephen Gould of Harvard University, based on a study of fossil records suggesting that the evolution of individual species takes place in well-defined steps (the punctuation marks) separated by long periods of stability (equilibria or stasis).[109] During the 1980s, David Raup and John Sepkoski at the University of Chicago found from their studies of records of thousands of fossil *genera* that extinctions occurred in bursts.

Various suggestions account for these observations. Some regard the mass extinctions simply as a set of unrelated accidents. Others have sought a single mechanism to explain them all: the bombardment of the earth by comets, the destruction of shallow seas when continents collide and merge, a shortage of oxygen, or a dramatic climate change. The best known of such events was the impact of a huge meteorite, the widely favored explanation for the demise of the dinosaurs sixty-five million years ago. The applied mathematician Sir Christopher Zeeman has used "catastrophe theory"[110] to furnish a simple and general explanation of punctuated equilibria on the basis of variations in a species' environment leading to instabilities.[111]

Another evolutionary model has been proposed using the notion of self-organized criticality, which we discussed in the last chapter. Per Bak argues that life is a dynamical system that, far from ever existing in a stable steady state ("equilibrium"), organizes spontaneously into a characteristic and much more precarious critical state. Like Zeeman's suggestion, this model of evolution again predicts that life does not evolve gradually but intermittently, with long periods of inactivity or *stasis*, interrupted by spurts of change characterized by mass extinctions and the emergence of new species. What is significant about the idea is that it does not require extinctions to originate through some "external" cause. If life organizes into a critical state, catastrophes, no matter how large, can be an intrinsic, self-organizing feature of evolution requiring no external causes.

While the demise of the dinosaurs is likely to have been caused by a meteorite impact, there are plenty of other crises to choose from during the 600 million years following the Precambrian explosion, during which the earth became occupied by multicellular organisms. In the most severe of the mass extinctions, some 250 million years ago at the end of the Permian period, over ninety percent of the species in the fossil record were wiped out. Is it possible that such events are natural consequences of Darwinian dynamics?

In some highly idealized computer simulations of evolution, Bak and his colleagues found certain intriguing insights. During periods of frenzied evolutionary activity, the average fitness of species is low, as they relentlessly mutate in search of better fitness. The fitness of these various species is also low during mass extinctions, and high during periods of stasis associated with low evolutionary activity. While evolution takes ecology to a critical point by its self-organizing dynamics in these models, the fitness of species at that point is not particularly high. Thus, in their work, "survival of the fittest" did not mean evolution to a state where every species is well off: "On the contrary, individual species are barely able to hang on—like the grains of sand in the critical sandpile," they conclude.

According to their model, at the critical point all species influence each other. In this state they act collectively as a single "metaorganism," many sharing the same fate. This is highlighted by the very existence of large-scale extinctions. One corollary is that species with many biological connections and dependencies, resulting from such factors as food chains, predator-prey and parasite-host relationships— that is, those with a high degree of complexity—are more sensitive to fluctuations that perturb the dynamics and thus are more likely to be part of the next avalanche to extinction. As Bak puts it, "According to our model, cockroaches will outlast humans."[112]

THE GLOBAL ECOSYSTEM

The idea of communities of social insects acting together as a superorganism is familiar; as we mentioned previously, we can only make sense of kinship-based altruism in this way. But the application of self-organized criticality serves to emphasize once again that the *global* properties of biological evolution cannot be apprehended by breaking them into their individual parts and then analyzing these as if they were independent of one another. Clearly, such reductionism is only a first approximation to the truth, and while it may afford us many insights, it always behooves us to put the pieces together again.

In 1968 James Lovelock upset gene-centered proponents of Darwin's views by arguing that the earth was not a ball of rock with a green layer of life on the surface.[113] Biologists, following Darwin, see life adapting to its environment. The independently minded Lovelock viewed life and the environment as part of one superorganism in which creatures, rocks, air, and water interact in subtle ways to ensure that the environment remains stable. His concept was named *Gaia*, after the ancient earth goddess, by his then novelist-neighbor, William Golding.

The Gaia concept, as first described in 1973, has been attacked by many established mainstream biologists. Their main complaint is that the idea seems to be teleological—it implies action purposefully directed toward a specific goal—and is therefore unscientific. The problem is that Gaia looks like a return to the days before Darwin, when in the seventeenth century the Archbishop of Armagh in Northern Ireland, James Ussher, maintained on the basis of the scriptures that the date of creation was 4004 B.C. Lovelock and other proponents of Gaia now rarely use the word "organism" and prefer more bland terms like "system." They also reject the notion of preordained purpose but maintain that there are regulatory mechanisms that have kept the environment fit for life over the past three or more billion years. These feedback mechanisms are invoked to explain the relative constancy of the climate, the surprisingly moderate levels of salt in the oceans, the constant level of oxygen over the past few hundred million years, and why life forms are so diverse.

Like it or hate it, simply looking for Gaia can give new insights into the complex feedback systems that rule the planet. One recent example came from a study conducted in the Roaring Forties, the stormy seas off southern Australia, that supported the Gaian view that living organisms would help regulate the climate.[114] The work, led by Greg Ayers in Melbourne, followed the fate of a gas, dimethyl sulphide (DMS), released by the ocean's microscopic marine plants, phytoplankton. The interest in DMS dates from the 1970s when, during a round-trip voyage from Britain to Antarctica, Lovelock and some colleagues confirmed a prediction of "Gaia theory" by finding surprising quantities of the molecule in the surface waters of the middle of the ocean. Lovelock and colleagues at the University of Washington predicted that all the DMS would be changed chemically by the oxygen in the air to produce sulfur-containing particles (sulfates) on which clouds could condense.

Lovelock's Gaian hypothesis involved a self-regulating biological thermostat: clouds would shield the earth by reflecting excessive amounts of sunlight, limiting the growth of plankton that produce DMS and thus the sulfate particles that seed the clouds.[115] The Australian scientists led by Ayers measured DMS and the airborn particulates over Cape Grim, on Tasmania's northwest coast, from November 1988 to May 1990. As predicted, the DMS and particulate levels moved in step with the seasons, being high in summer and low in winter. Ayers was cautious: "Our work is a signpost in the direction of Gaia, but the results do not go far enough to vindicate the whole (thermostat) theory."[116] However, Lovelock points out that no one would have looked for DMS and its connection with clouds and climate without his concept of Gaia.

Lovelock claims that Gaia is a well-defined theory with a solid mathematical basis, able to cast light on such crucial issues as biodiversity.[117] Critics respond that the theory is at best vague, and at worst almost mystical. Indeed, Lovelock has had to rename it "geophysiology" because the peer-reviewed scientific journals will not allow the word "Gaia," except in denigration.[118] Many serious-minded evolutionary biologists recoil in horror at the thought that "in popular natural history, notably television documentaries, it is flourishing";[119] they hope that Gaia's influence will soon die out.

Yet, despite its flaws, complexity studies of the kind explored within this chapter are helping to make Gaia more plausible—at least in spirit—than reductionists would have us believe. By shifting the focus from individual molecules to global behaviors, it is possible to show in outline how natural selection in concert with self-organizing molecular processes led to the emergence of life: self-replication— effectively a form of autocatalysis—was at the basis of the nonlinear complexity that led to spontaneous self-organization at the molecular level; variation in self-replicators arose from mutations, caused by radiation or chemicals in the environment, or by errors that inevitably occurred during replication; molecular evolution then selected the self-replicating molecules that were best at harnessing chemical energy (food or "fuel") by self-organizing processes.

In short, there is more to the story of life than competition, inheritance, selfishness, and survival of the fittest replicators or genes. Concepts of self-organization and complexity that we encountered in the last three chapters also play an important role, though the relationship between evolution, complexity, and the spontaneous creation of order is still a matter of debate among biologists.[120] As we will see in the next chapter, the selfsame global features of complex adaptive systems may even enable artificial life to evolve.

Chapter 8

LIFE AS IT COULD BE

Perverse, all monstrous, all prodigious things,
Abominable, unutterable, and worse
Than fables yet have feigned, or fear conceived,
Gorgons and Hydras, and Chimaeras dire
　　　　　　　　　　　—MILTON

"Men will not be content to manufacture life: they will want to improve on it." With these words, the young Irish crystallographer John
Bernal anticipated in 1929 the possibility of machines with a life-like
ability to reproduce themselves.[1] He wrote of this "postbiological future" in *The World, the Flesh, and the Devil*: "To make life itself will be
only a preliminary stage. The mere making of life would only be important if we intended to allow it to evolve of itself anew."

Almost two decades later, von Neumann performed the first work
demonstrating the possibility of artificial life, in his efforts with self-
reproducing automata.[2] His so-called kinematic model, which we encountered in Chapter 3, aimed to isolate the logical content of biological
self-replication. Though von Neumann conceived his self-replicating automaton some years before the structure of the genetic blueprint (DNA)
had been unraveled, he laid stress on its ability to *evolve*. He had told the
audience at his Hixon lecture that each instruction that the machine carried out was "roughly effecting the functions of a gene"; he went on to de-

scribe how errors in the automaton "can exhibit certain typical traits which appear in connection with mutation, lethally as a rule, but with a possibility of continuing reproduction with a modification of traits."[3]

There were efforts to put flesh and bones on his ideas. In 1956, a proposal was outlined for "Artificial Living Plants," self-reproducing floating factories that could harvest important mineral and crop resources.[4] Realizing the dangers posed if one such machine ran amok on the planet, Freeman Dyson of the Institute for Advanced Study in Princeton proposed a more benign *Gedankenexperiment* in which self-reproducing machines seeded life in the solar system.[5] Today, however, von Neumann's realization of the logical nature of life has more significance than ever because of the ability of modern computers to evolve complexity.

We have already seen the trend to forsake rational design in biology and computer programming for techniques based on the blind forces of biological evolution. Imagine what happens when computer programs are evolved to solve hard optimization problems. Given that life is such a problem, is there any sense in which this might lead to new, computational forms of life? Indeed, is it possible to perform Darwinian evolution by "natural selection"—the single property defining life— inside a computer? While life on earth is restricted to carbon-based organisms, what we can create inside a computer is based on logical machine instructions; nevertheless, there is little to indicate that computer-based evolution does not have the potential for developing complexity on a par with that found in biology. Indeed, evolving machine codes should be able to generate *any* amount of complexity in that they may be capable of universal computation in Turing's sense.[6]

Just as biological life ultimately emerges from the complex interactions of a great number of inanimate microscopic units called molecules, so some believe that artificial life (ALife) may emerge from complex logical interactions within a computer. The analogy between the two arises because all the logical processes that take place within a computer are based on the atomistic building blocks of Boole's binary symbolic algebra (see Chapter 3). The new science of ALife is predicated on von Neumann's abstract vision and is beneficial in several ways. As we will see, it can help us to better understand biological life by a more abstract study of the emergent properties that honed and shaped it, through the processes of reproduction, competition, and evolution. In turn, by harnessing within a computer the problem-solving capabilities of Darwinian-like systems, it will become possible to effi-

Color Plate 1   Solutions of a simple nonlinear equation, called the logistic equation. The plot reveals the expected behavior of the living population, from generation to generation, as a function of certain parameters in the model. The "figures" in the foreground portray regular, periodic changes in the population of organisms, while the backgrounds correspond to irregular fluctuations (also known as chaos).

2(a)

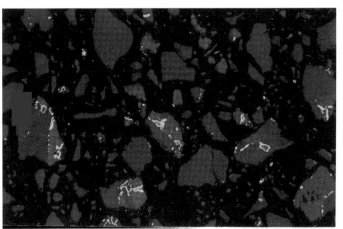

2(b)

Color Plate 2   Starting digital images for cellular automaton simulations of hydrating cement slurries: (a) two-dimensional image obtained from a combination of scanning electron microscope back-scattered electron and X-ray images; (b) computer-generated three-dimensional image of random grains. Different colors represent chemically distinct materials.

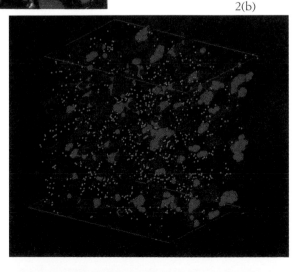

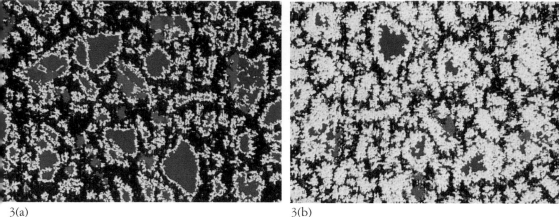

3(a)   3(b)

Color Plate 3   Digital images computed from cellular automaton simulations of hydrating cement slurries: (a) when 30 percent of the cement initially present has hydrated; (b) when 76 percent of the cement initially present has hydrated. The yellow regions indicate hydrated material.

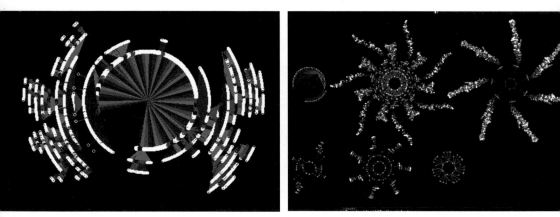

Color Plate 4   Examples of basins of attraction for one-dimensional cellular automata that can be used to summarize the behaviors we saw in Figure 4.4. On the left is a basin of attraction of a simple rule (such as that which arises in Wolfram Class I and II). On the right is the entire basin of attraction of a "complex" rule (Wolfram Class IV).

The attractors, each distinguished by a different color, are at the center of each spiral. Each step in time, shown as a horizontal string of pixels in Figure 4.4, corresponds to one of the colored points. As time ticks by, the points spiral clockwise into the attractor. The shape of the basins reflects the classes (I–IV) in the length and bushiness of their fleeting spiral arms. (The white points are so-called Garden of Eden States, that is, states with no predecessors.) [Images generated using Andy Wuensche's "Discrete Dynamics Lab" PC-DOS shareware, available by downloading the file ddlab.zip either by anonymous ftp via ftp://alife.santafe.edu/pub/SOFTWARE/ddlab or on the World Wide Web at http://alife.santafe.edu/alife/software/ddlab.html.

Color Plate 6 *(opposite)*   Joyce's unnatural selection of RNA molecules. Snapshots illustrate how much mutation (illustrated by vertical bars) has taken place at each point of the structure of the *Tetrahymena* enzyme. It shows that after twelve generations, the entire sample of molecules they analyzed had changed at four positions (green bars). Notice the rise and fall in the success of the red mutations. They are mutually exclusive to two of the green mutations.

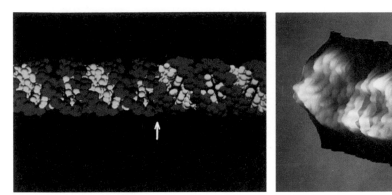

Color Plate 5   DNA. Computer-generated molecular model (left) and as observed by a Scanning Tunneling Microscope (right).

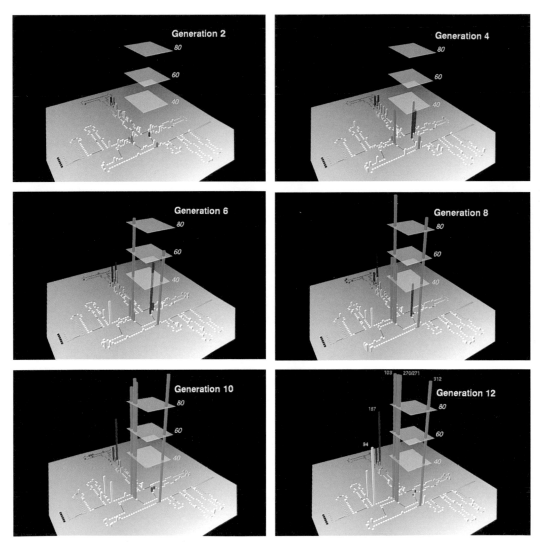

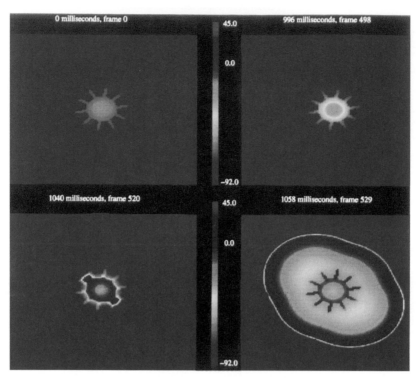

Color Plate 7   Rai Winslow and Denis Noble's computer simulations of a wave of excitation triggered by the firing of the sinus node pacemaker in the heart.

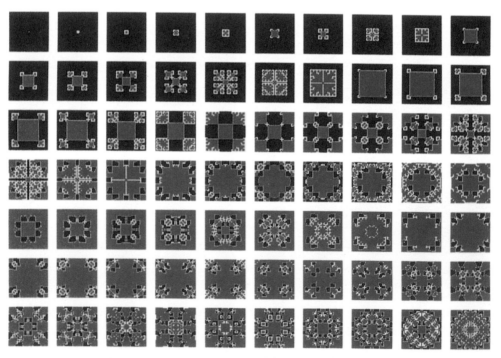

Color Plate 8(a)   The patterns generated within the spatial Prisoner's Dilemma by starting out with a single defector. See 8(b) for color code.

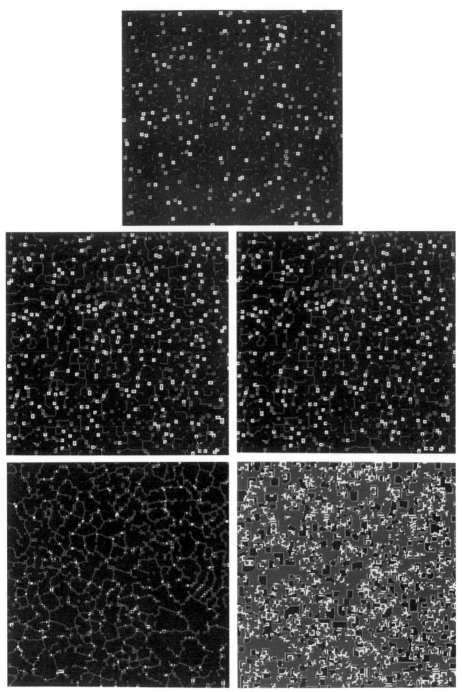

Color Plate 8(b) The spatial Prisoner's Dilemma generates a large number of different patterns, depending on the advantage for defectors, as defined by the parameter *b*. In the game, if two cooperators interact, both receive one point. If a defector "exploits" a cooperator, the defector receives the payoff *b* and the cooperator 0. The interaction between two defectors also leads to the zero payoff. These patterns are generated after 100 generations, with the value of *b* varying from 1.15 to approximately 2.0. Blue represents a cooperator that was already a cooperator in the previous generation; red is a defector following a defector; yellow a defector following a cooperator; and green a cooperator following a defector. The amount of green and yellow indicates how many cells are changing from one generation to the next.

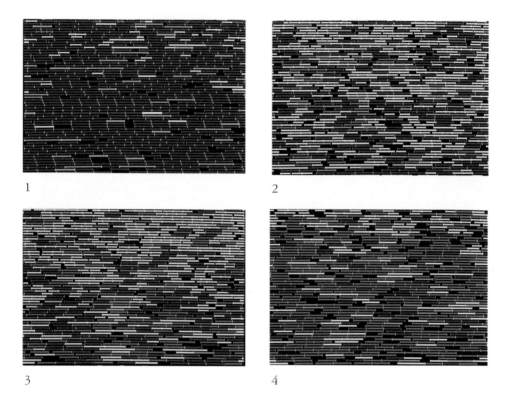

1

2

3

4

Color Plate 9 Evolutionary arms race between hosts and parasites in the *Tierra* Synthetic
Life program. Images were made using the Artificial Life Monitor (ALmond) program
developed by Marc Cygnus. Each creature is represented by a bar; the color corresponds to
genome size (e.g., red = 80, yellow = 45, blue = 79). (1) Hosts, red, are very common.
Parasites, yellow, have appeared but are still rare. (2) Parasites have become very common.
Hosts immune to parasites, blue, have appeared. (3) Immune hosts now dominate memory, while parasites and susceptible hosts decline. The parasites will soon be driven to
extinction. (4)

Color Plate 10 *(opposite)* Fish and silicon chips. Computer-generated shoals display a
host of realistic behaviors.

Color Plate 11 *(opposite)* William Latham's evolutionary art.

10

11

Color Plate 12 Karl Sims' evolved art using genetic programming techniques.

12(a)

12(b)

Color Plate 13 Karl Sims' evolved creatures. Rather than giving his creatures a goal such as walking, Sims made them compete. In one series of virtual experiments, two creatures were challenged to struggle for control of a green cube between them, as shown here. The creature in the foreground is attempting to drag the cube out of the grasp of its opponent. Sims found that weedy creatures with small, boxlike limbs were not necessarily inferior to larger creatures with greater reach. As they evolved strategy and counter strategy, some creatures displayed "clever" behavior, such as knocking the cube to one side, or attacking an opponent.

ciently solve many nonbiological complex optimization tasks, of the kinds discussed in Chapters 2 through 5.

ALife provides a new way of coming to grips with the tricky concepts of intelligence and artificial intelligence by placing emphasis on the environmental influences on learning, as Turing foresaw long ago. Most intriguing of all, some believe that ALife may lead to the emergence of radically new forms of life. The first step along this path occurred some time ago with the birth of various computer viruses, and now more truly "living" organisms are coming to life within computers. At its heart, ALife aims to discover the essential nature and universal features of "life": not only life as we currently know it, but life *as it could be*, whether on earth, within computers, or elsewhere, and in whatever shape or form that it may be found or made within our universe.

WEAK AND STRONG ARTIFICIAL LIFE

In a certain sense, artificial life research has been underway for decades, albeit by another name. Computers have been used to simulate a wide range of biological processes, which are often intractable by any other means, as we saw in the previous chapter. Typically, this use boils down to making a computer solve a specified set of equations that are believed to model the phenomenon in question, whether the pattern of limb growth or aggregation of a slime mold. Just as a computer model of the flash and mushroom cloud of a nuclear detonation is not itself a nuclear bomb exploding, so these simulations are in no way alive. Those who pursue such research can be called supporters of "weak" ALife, since they study computer models of biological processes in which the simulations would not be termed living. What distinguishes a certain strand of contemporary research in the field from the mainstream of study is an often unspoken belief in "strong ALife," according to which a suitably programmed computer may itself be deemed to be alive, or at least possesses properties of a living thing. This kind of distinction is quite familiar to researchers investigating artificial intelligence. There we find the "weak" artificial intelligence community, composed of people who use computers to model processes occurring within brains that could not be simulated without the power of the computer. The strong AI community, by comparison, has an even grander aim than the strong ALife group—to impute a conscious mind to a properly programmed computer. One way to distinguish these two AI positions is to translate the adjectives so that "weak" means modest and "strong" means daring.[7] Today, AI is a sub-discipline of ALife, since only some forms of life can be expected to manifest intelligent behavior. But there is one important difference between AI and ALife as subjects for contemporary research: as the philosopher Elliot Sober of the University of Wisconsin has pointed out, terrestrial life is in many ways far better understood than the human mind, so there is a firmer grounding for ALife than for AI.[8]

The term "life" is notoriously difficult to define, as we emphasized in Chapter 7. Since it is difficult to find two people who can agree on a definition of life, it might be thought that attempting to define artificial life would only increase the confusion. Petty arguments about definitions have not, however, discouraged the hubristic claims of the strong ALife community. One of the most provocative was made by Doyne Farmer of Los Alamos and Alletta Belin of Shute, Mihaly and Weinberger, in a manner reminiscent of the claims made by the proponents of strong AI. "Within fifty to a hundred years, a new class of organisms is likely to emerge. . . . The advent of artificial life will be the most significant historical event since the emergence of human beings. The impact on humanity and the biosphere could be enormous, larger than the industrial revolution, nuclear weapons, or environmental pollution. We must take steps now to shape the emergence of artificial organisms; they have potential to be either the ugliest terrestrial disaster, or the most beautiful creation of humanity."[9] It sounds outrageous. However, we need to examine the subject of ALife more closely before we glibly dismiss such views.[10]

VIRUSES

Some examples of "weak" ALife forms are already in the environment and widely known. They were born during Core Wars, originally a game played by computer addicts.[11] The idea was to create programs that compete against each other for processing time and space in computer memory—rather like animals competing for food and territory. Now there are unintended versions of Core Wars, better known as computer viruses, existing in PCs, workstations, and mainframes around the world.

The term "computer virus" is evocative; it was coined in 1983 for a short stretch of computer code that can copy itself into one or more larger "host" computer programs when it is activated. The threat posed by computer viruses is such that they are rarely out of the headlines and sport colorful names, such as Brain, Denzuk, Michelangelo, Elk Cloner, Festering Hate, and Cyberaids. They can also "mutate" when a malicious individual adds a minor adjustment to the viral code: for example, one of the most common, Jerusalem, has spawned Jerspain, Payday, Mendoza, Anarkia and Sunday, Fu Manchu, and Zerotime.[12] The viruses have also become smarter. The middle of 1990 saw the first virus to use two separate strategies for replicating itself, either of which it could employ depending on the circumstances. Overall, in the time between the first infection by the Brain virus in January 1986 and April 1, 1991, two hundred different viral strains infected the IBM personal computer alone.[13]

When these infected programs are carried out, the viral code is also executed. In the process, it may cause damage to the computer's operating system, overwriting important files, data, and instructions; moreover, the virus contains instructions for its own reproduction, thereby enabling it to spread via transfer to magnetic media such as disks and tapes. Other members of what is actually a menagerie of infectious codes also exist. For instance, "worms" are programs that can run independently and travel from computer to computer by the burgeoning global computer network, through the electronic equivalent of a chain letter. Other types of so-called vandalware include bacteria, Trojan horses, logic bombs, and trapdoors. Computer manufacturers have obligingly standardized the software in which they breed. Just as crop monocultures can be obliterated by a disease that would barely perturb a healthy mixed grassland, so cohorts of identical operating systems can be devastated by these seemingly insignificant computer infections.

Their proliferation begs an inevitable question: are computer viruses alive? This question is even more thorny than the same question addressed to bioviruses. The answer turns on one's definition of "life." Both natural and artificial viruses certainly possess some features of living things. They can reproduce themselves, they store information, and have a metabolism in the sense that they pirate the workings of a host, whether computer or cell. There are even examples of interactions between different computer virus species: the Denzuk virus will seek out and overwrite the Brain virus if both infect the same computer. However, one key difference can be highlighted with our understanding of hepatitis, a viral disease that has infected about half the world's population and is responsible for 90 percent of liver cancer.

There is not one "hepatitis virus" but many. Hepatitis B was the first discovered. Since then, A, C, D, and E have followed. Two new varieties have recently been suggested, and undoubtedly there are more. Surprisingly, these are not strains of one virus but arise from entirely different families, depending on their genetic makeup. But all are united in one respect. They are "solutions" evolved by quite different families of virus to the same problem: how to infect a human liver cell so as to reproduce. A and E enter the body in food and water contaminated with sewage; the other viruses exploit our enjoyment of sex and, like the AIDS virus, are blood-borne. This allows transmission from mother to child, via contaminated needles, and in such practices as ritual circumcision, tattooing, and ceremonial blood exchange.[14]

Although scientists dislike anthropomorphizing what is little more than a complex molecule at the borderline of life, one of the pioneers of hepatitis research, Baruch Blumberg, admits that the strategies the virus adopts are so clever that it is difficult not to endow it with cunning intent. A virus consisting of a handful of genes seemingly plans endless strategies to outwit the human body thanks to its ability to evolve: the reproduction and mutation of viruses are so rapid that among every population a few are always able to adapt to a change in circumstances or exploit a new opportunity.

Unlike natural viruses, no man-made computer virus evolves; indeed, to create one that does would be a major programming challenge, given the intolerance—also called "brittleness"—of most computer languages to errors or "mutations." This is just as well, considering the damage that viruses can already inflict on computer systems. As we described in Chapter 6, the most fruitful definition of a living system is one that is subjected to the rigors of Darwinian-style selection; at least by this definition such programs cannot be regarded as alive. Nevertheless, some strong ALifers, notably Farmer and Belin, still argue that although computer viruses need human beings to create them, many natural organisms cannot exist without the help of another. Thus, they argue, computer viruses have a symbiotic relationship with humans for their evolutionary development. Putting aside the sterility of terminology, it is clear that, although these artifacts comprise more or less disembodied information, they are disconcertingly close to being alive in the general sense.

ARTIFICIAL GROWTH

In some types of weak ALife research, computers have been used to model the processes of plant growth, a fiendishly complex enterprise. A few numbers can illustrate the problem posed by multicellular plants. Within each cell, thousands of genes are present. Not all are used within any given cell—the precise configuration of active and inactive genes depends on the type of cell, whether flower, stem, or seed, for instance. For n genes, the number of active combinations is 2^n. Even an unimaginably simple plant of ten genes would have 1,024 combinations of genomic states to spell out a cornucopia of plant designs.

Following in the footsteps of the British zoologist D'Arcy Thompson, in an attempt to capture the underlying universality of the genetic language used by all plants, the late Aristid Lindenmayer, a botanist at the University of Utrecht, developed a formal, mathematical description of plant growth. What grew from his interest over twenty-five years ago were intriguing simulations that are now dubbed L-systems in his honor.[15] He tried to model plant growth using the same sorts of grammatical techniques that linguists use to analyze sentences. Instead of "parsing" plants as one parses a sentence so as to work back from real words to abstract parts of speech, L-grammars are typically run in reverse by starting with abstract "parts of plants" and using the grammar to guide repeated substitutions until one gets to components such as bark, stem, leaves, and flowers. The beauty is that there is nothing in the grammatical rules about the overall plant shape. The structure simply emerges from the computation.

Lindenmayer's computer models represent the body of a plant by a string of symbols, one symbol per module (for leaf, stem, and bud). The body "grows" as one repeatedly applies the algorithms for manipulating the symbols. Even simple systems of rules can generate "plants" that look real. Another step was to introduce branching to this bottom-up approach so that "tree"-like forms could be made. His models worked in a manner not unlike that of a cellular automaton. Most spectacularly, L-systems have been used to generate extraordinarily realistic plant forms similar to ferns. Two of his graduate students at the University of Utrecht in the Netherlands, Ben Hesper and Pauline Hogeweg, took the digital results from an L-System rule set and displayed them graphically on a monitor. In this way, they have produced

Figure 8.1   Artificial plants. A lilac twig and a field of flowers, generated by Linden-mayer's methods.

realistic computer-generated images of ferns and of an aster, the plant that Lindenmayer was pictured holding at an Artificial Life workshop in Santa Fe in 1987. The image appeared in a dedication to his memory in a volume of papers from the second ALife gathering, held the year after Lindenmayer died.[16]

Some scientists have used Lindenmayer's approach to see whether real plant mutants can be accounted for in terms of a change in specified developmental rules.[17] Even film animation laboratories are now using L-systems to generate images of trees with computer graphics. His methods provide insights into morphogenetic processes, revealing how by the application of locally acting simple rules, whether genetic or algorithmic, branches, leaves, and flowers can evolve in parallel. However, L-systems are *not* designed to grow in the same way as biological cells. This is why they fall naturally within the broad field of "artificial" life studies.

UNNATURAL SELECTION

Core Wars programs, computer viruses, and computer worms are capable of self-replicating, but not of evolving in an open-ended fashion. L-systems do not even have self-replicating capabilities. The ability to evolve is the central aspect of life as we know it, and attempts are now being made to do just this artificially, within cyberspace. For there to be any chance of creating genuine artificial life within a computer, we must find ways to introduce novelty by mutations and then select the "fittest" objects.

In September 1987 the first conference on Artificial Life was held at Los Alamos in New Mexico. The ALife gathering attracted 160 delegates, ranging from zoologists and plant biologists to chemists, physicists, and computer scientists, taking in along the way researchers working on automated Lego sets for children and others, all of whom shared a common interest in the simulation and synthesis of living systems, whether using chemistry, software, or hardware. The organizer, Chris Langton, reported that "the most fundamental idea to emerge at the workshop was the following: artificial systems which exhibit life-like behaviors are worthy of investigation on their own rights, whether or not we think that the processes that they mimic have played a role in the development or mechanics of life as *we* know it to be. Such systems can help us expand our understanding of life as it *could* be."[18]

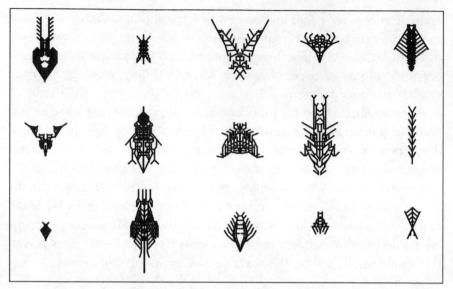

Figure 8.2   Breeding diversity in a computer. Dawkins' biomorphs.

Among those present was Richard Dawkins, the Oxford zoologist. He discussed how his *Blind Watchmaker* program could evolve "creatures" displayed on the screen of his Apple Macintosh personal computer. "Borrowing the word used by Desmond Morris for the animal-like shapes in his surrealistic paintings, I called them biomorphs," he explained.[19] "My main objective in designing *Blind Watchmaker* was to reduce to the barest minimum the extent to which I designed biomorphs. I wanted as much as possible of the biology of biomorphs to emerge."

The biomorphs were generated by strings of computer code rather like bodies are generated by genes. The computer carried out minor changes ("mutations") in the code describing a biomorph, and displayed the range of body shapes that resulted. Being unable to select biomorphs according to how well they performed in an environment, Dawkins picked some for aesthetic reasons, and then bred new generations from them. After a few generations, surprisingly life-like biomorphs resulted. "I was genuinely astonished and delighted at the richness of morphological types that emerged before my eyes as I bred," he remarked.

Dawkins used his program to show how a "blind watchmaker" could produce the diversity of living things, without recourse to God or a grand designer. Dawkins, following in Darwin's footsteps, argued that the intricate design of the human eye could result from evolution by natural selection, through the interplay of chance and competition.

This was elegantly demonstrated in later work by Dan Nilsson and Susanne Pelger at Lund University in Sweden, who showed that if selection always favors an increase in the amount of visual information processed, a light-sensitive patch of tissue will gradually turn into a focused lens eye through continuous small improvements of design over a few hundred thousand generations.[20] The study suggests that the evolution of something as complex as the eye could—in theory at least—have taken place in less than a million years, an eyeblink in terms of the vast span of geological time.[21]

EVOLUTIONARY ALGORITHMS

Dawkins' biomorphs provide a striking illustration of how a series of random mutations can turn a simple structure into a life-like object. However, they are still a product of *unnatural* selection. It was a "god"—Dawkins—who provided the selection pressure that led to complex shapes appearing, rather than open-ended evolution through competition with other objects in the environment. As we have repeatedly stated, living things have an innate ability to evolve by natural selection, that is, via "survival of the fittest." Those species best optimized to perpetuate themselves in the complex but finite environment furnished by all other species and energy resources will be the ones that survive. Less than optimal creatures will eventually die out. In Chapter 4 we saw that, couched in these terms, evolution is rather like the many other hard optimization problems we discussed there, with the added complication that the fitness landscape (valleys and mountains) is itself coevolving, due to the individual struggle of all other species to survive. This means that evolution follows highly nonlinear dynamics, involving massive feedback loops, leading to a system that arguably represents the apotheosis of complexity.

Nevertheless, based on the sheer power of modern computers, we can be optimistic that the secrets of biological evolution may be simulated by computational processes. Recall John Holland's genetic algorithms (GA) discussed in Chapter 5, which were inspired by Darwinian ideas. In the first twenty or so years following their creation, GAs have been largely used to solve complex problems within the inanimate world. However, even in the 1960s, Holland's group used GAs to investigate biological

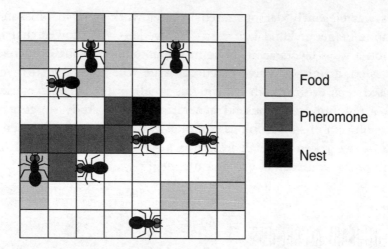

Figure 8.3a   Ant farm. At the start of a new generation, all the ants are in the nest, there are no chemical signaling molecules—pheromones—and food is distributed around the farm. The virtual ants evolve foraging strategies in their search for food, laying down pheromone trails as they go.

systems, starting with a simulation of single-celled organisms.[22] Although initially slow to take off, the work using GAs, particularly in recent years, has provided a splendid artificial laboratory within which to dissect evolution. Genetic algorithms use an interacting population of digital codes, with each representing an individual organism, to model a population of organisms. The evolving codes "exhibit counterparts to such phenomena as symbiosis, parasitism, biological 'arms races', mimicry, niche formation and speciation."[23] Other work with genetic algorithms has shed light on the conditions under which evolution will favor sexual or asexual reproduction, while Rick Riolo of the University of Michigan has observed genetic algorithms that display "latent learning," a phenomenon not dissimilar to learning in neural networks, in which an animal such as a rat explores a maze without reward and is subsequently able to find food placed in the maze much more quickly.

David Jefferson, working with an Artificial Life group at the University of California at Los Angeles, used a GA to develop a trailblazing "artificial ant," a digital insect existing within a computer, that consisted of a set of instructions inculcating it with a mission to "learn" how to follow a winding broken trail laid out on a grid, just as real ants follow the scent of a chemical messenger in nature.[24] The ant was represented as a "finite-state automaton"—that is, as a string of binary digits that could only access a finite number of ant behaviors, or as an artificial neural network.[25] The GA was used to evolve improved rules governing the next move of the ant once it had "sniffed" what was in the grid cell directly in front. The program started out with 65,536 (2 to the power 16) digital ants, an arbitrary number that suited the available computing power (one ant per processor). Within seventy generations on a massively parallel Connection Machine 2, it had evolved a significant population able to complete a twisting and tortuous trail of eighty-nine squares. This may sound like a lot of evolution but each generation "lived" for less than thirty seconds. In this way, the group demonstrated that it is feasible to produce, by evolutionary means, artificial organisms exhibiting complex, indeed life-like, behavior—behavior that would be very difficult to design *ab initio* by writing down a computer program. With colleague Robert Collins, Jefferson went on to develop AntFarm, a computer program that simulates the evolution of foraging strategies in artificial ant colonies (see Fig. 8.3a).[26]

Genetic algorithms show how such complex behavior evolves. In return, the ants again illustrate how the use of many agents obeying simple rules can produce complex foraging behavior. Steve Appleby at the British Telecom Laboratories in Martlesham Heath has developed a similar ant program to simulate the qualities of ants that give them robustness—their simplicity and ability to self-organize in the important job of gathering food. He hopes that these insects will inspire a solution to one of the most important problems facing any modern telecommunications company—how to route calls across a vastly complicated communications network.

Two extreme solutions are possible. In one, a supercomputer would sit like a spider at the center of this vast communications web. In theory the computer can always calculate the best way to route calls. However, in practice it may spend too long scratching its head and working out what to do next. At the other extreme, each telephone exchange could act independently, rerouting calls depending on how overloaded the neighboring exchanges are. This is fast, but never comes up with anything like the best solution because it lacks a global picture. British Telecom looked to the ants to provide a compromise between global and local control.

To persuade an ant colony to forage efficiently required just four rules. The first three exploited the way the ants communicate by leaving a trail of a chemical signaling molecule called a pheromone: (1) If the ant finds food, take it to the nest and mark the trail with pheromone; (2) If an ant crosses the trail and has no food, follow trail to food; (3) If the ant returns

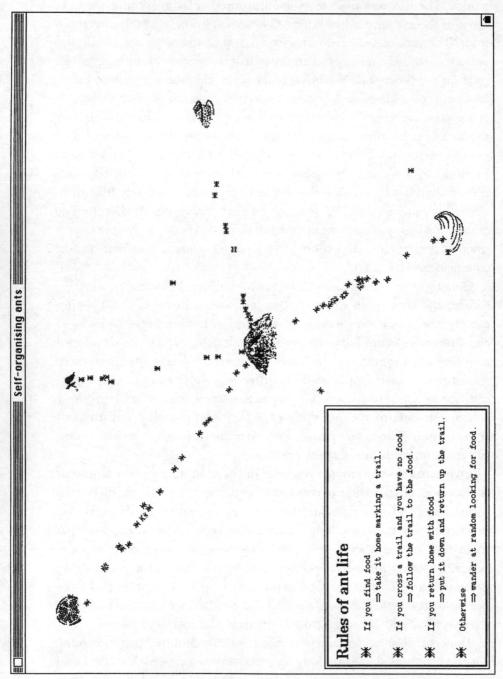

Rules of ant life

* If you find food
⇒take it home marking a trail.

* If you cross a trail and you have no food
⇒follow the trail to the food.

* If you return home with food
⇒put it down and return up the trail.

* Otherwise
⇒wander at random looking for food.

Figure 8.3b Self-organizing ants, the inspiration for a new way of managing telecommunications networks.

to the nest, deposit the food and wander back along trail. The final ingredient is a catch-all rule: if the other rules do not apply, wander at random
(see Fig. 8.3b).

"This pheromone trail is one of the keys to robustness," Appleby believes. "It is a message but is laid down for any old ant to stumble
across. This is quite unlike the way computers function, where a program working on one computer sends a message to another specific
computer." Robustness comes naturally as a result of not addressing
messages. One ant can take over the job of any other that perishes.
Working with Simon Steward, Appleby has now developed a swarm of
"ants"—mobile programs—that can help route calls through a telephone network, evening out the load to ensure that each exchange is
used efficiently and is not overloaded.[27] Like the ants, these pieces of
code do not communicate directly but leave messages for one another
in each exchange as they pass through. "If one program crashes, it does
not really matter because another one can come along later, pick up the
message, and carry on. That is what gives the system robustness," said
Appleby. BT is now exploring how to use such a system to manage dynamic networks, notably the networks that connect computers.[28]

OPEN-ENDED EVOLUTION

Conventional genetic algorithms endow the quest for artificial life
with some aspects of natural selection, but they suffer from numerous
limitations. One is the sheer size of a simulation needed to reproduce
the complexity of life as we know it. Natural populations can number
millions in the case of large animals, trillions in the case of insects, and
quintillions or more for bacteria.[29] However, as massively parallel computers become more powerful, it should be feasible to evolve digital
populations of increasingly realistic sizes.

Another limitation arises because the chromosomes that represent
encoded solutions to a given problem—the digital organisms—are all
the same size in a GA.[30] Many of the most important parts of the evolutionary recipe have already been designed into the simulation, rather
than evolving naturally. Moreover, the presence of "crossover" as a key
genetic operator implies "sexual" evolution, while in nature the earliest
and more primitive organisms reproduce asexually.[31]

Furthermore, genetic algorithms employ "task-oriented" evolution while the real world uses "environment-oriented" evolution. In other words, an environment seething with life poses its own problems to be solved, rather than having one imposed on it. Imagine that the object is to breed a creature that can run at 40 mph. There are two possible approaches. Realizing that horses are already rather fleet of foot, you could launch a selective breeding program by taking a herd of horses, racing them, discarding the losers, and breeding only from the winners. This is the approach most current GAs take. Alternatively, you could select an open grassy plain like the Serengheti, populate it with lions, release a menagerie of animals, and then wait a few million years to see what evolved. It could be a horse, or a particularly nimble rhinoceros.[32] However, all that matters is that the animal can consistently run at 40 mph. Very recently, it has become possible to simulate just this in GAs, so that evolutionary dynamics emerge spontaneously, just as they do in nature. Although less controlled, this approach is much more creative and some of its most enthusiastic advocates believe it may lead to genuine ALife.

Like biological life, genuine artificial life must be capable of evolving structures that are not designed or preprogrammed. To make the ALife evolutionary process open ended, it has to be random (stochastic); that is, it should embody an ever-present element of novelty that can cause change and evolution in quite unexpected ways. As the fitness of one organism alters, so the fitness landscape of its coevolving sister and brother organisms shifts accordingly. In this way, random simulations can produce solutions of a form never thought of or anticipated, imbuing a computer with originality and innovation. This aspect is essential for ensuring the plasticity of evolution—to prevent the whole process from running into a dead end.

The attempts at simulating evolution discussed so far have not been open ended; in other words, their "fitness functions" or landscapes are fixed once and for all. For instance, the effect of having genomes of fixed length in the GA is to tightly circumscribe the potential for innovation. In these models, a character string represents the organisms' genomes, which are mutated, recombined, selected, and replicated by unchanging design rules within the simulator. There is no mechanism of replication—the genomes are just copied if they survive the selection phase.[33] More must be done if full-fledged digital life is to emerge.

VIRTUAL LIFE: TIERRA

Thomas Ray from the University of Delaware has created the first example of artificial evolution based on Darwinian principles. His achievement is likely to rank as one of the most important developments in twentieth-century theoretical evolutionary biology. Inside his computer, digital organisms fight for memory space in the central processing unit (the analogs of food and energy resources). Until Ray's work, we had not been able to experiment with evolution, only observe it. Merely observing it is far from satisfactory, since a single human life is as nothing compared with the time scale—"deep time"—on which things happen in biological evolution. Ray's simulation of evolution, called *Tierra* after the Spanish for earth, provides a kind of metaphor for biological complexity, a tool to understand why it is seething with diversity.

Ray originally trained as a tropical ecologist and has spent much of his time studying vines in the rain forests of Costa Rica. He still lives there in his house-cum-laboratory set among forty acres of jungle. The inspiration for his computer simulation came not from the dizzying range of creatures that surround him, from humming birds to beetles, but the Chinese game of Go.

While a graduate student at Harvard, he once discussed the board game with a computer scientist, remarking on how the black and white stones used in the game form apparently self-replicating patterns, governed by the game's simple rules. The computer scientist made a fateful comment, saying it was equally possible to make a self-replicating computer program. That was new to Ray, who immediately imagined adding mutation and getting Darwinian evolution. He asked how this self-replication could be done and was told that it was "trivial." It took ten years before Ray was able to tackle this "trivial" problem.[34]

Mutation and the ensuing competition for finite resources are crucial to evolution—it is not enough for an object to simply reproduce. That would lead to a "virus" with the banal ability to clog up a computer with identical progeny. Thus, for evolution to occur in *Tierra*, Ray not only had to create a self-replicating computer program, but also mutate it. Once this occurs, "natural" selection can take place: varieties of "organisms" best suited to their circumstances breed more effectively. The ensuing artificial life would then be a combination of self-sustaining complexity and reproduction, the latter allowing the former to evolve. It would also be evolution subject to the laws of logic, rather than physical laws.[35]

There was, however, an important technical problem to overcome before this objective could be realized. Traditional von Neumann machine codes are not resistant to mutation: one period in the wrong place can cause a program to crash. This intolerance, the "brittleness" we have previously encountered, was originally seen as an almost insuperable problem by many ALife gurus. Farmer and Belin, for example, asserted that "Discovering how to make such self-replicating patterns more robust so that they evolve to increasingly complex states is probably the central problem in the study of artificial life."[36] "I was concerned but not convinced," said Ray. "Why should a machine language have that property of brittleness and the genetic language not?"[37]

Inspired by a high-level programming language ("C") compiler and an accompanying debugging program that clearly revealed the inner workings of his newly acquired laptop computer, in 1987 Ray decided to construct his version of evolution using assembly language. Recall that this is a low-level machine-specific, binary computer code whose commands directly invoke the instruction set inside the computer's CPU, as well as services provided by its operating system. Since assembly language is so closely connected to a computer's hardware specification, it was the natural language for Ray to use for keeping control of the artificial environment within which he hoped to breed his digital organisms.

Ray made his machine code more robust with respect to random mutations (bit flipping) by borrowing ideas from molecular biology. The biological genetic code is characterized by a very small instruction set: there are sixty-four instructions (codons) formed from the nucleic acid bases, which get translated into twenty different amino acids. The *Tierran* language has thirty-two instructions all told, and thus is of the same order of magnitude as the genetic code itself. This represents a dramatic distinction from the situation that exists in conventional computers: even the new generation of RISC (reduced instruction set) machines have associated assembly languages comprising many more instructions.[38]

Ray used three mechanisms for mutation, each reflecting parallel effects in biological evolution. Sometimes errors occur in the computations. Every now and then, a randomly selected memory location was changed, altering the binary code representing one digital creature;

and whenever such an organism reproduced, there was a chance of introducing a copying error into its offspring. To avoid clogging memory, which would ultimately freeze the artificial ecosystem, a routine called the "reaper" kills the old and the error-prone organisms to compensate for the lack of direct predators. Deaths of organisms within *Tierra* occur because greedier or more successful organisms have monopolized the resources.

Ray also introduced "addressing by pattern" as a means of permitting direct interactions between organisms. This allowed one organism to exploit the instruction sets—the genomes—representing other digital organisms located nearby in computer memory. While no organism in *Tierra* could overwrite another's genome, it could read and execute that code. Ray describes the importance of such interactions in nature by using the rain forest as metaphor: in some parts of the Amazon, the physical environment consists of clean white sand, air, falling water, and sunshine. Embedded in that physical environment is the most complex ecosystem on earth, with hundreds of thousands, possibly millions of organisms. These do not represent hundreds of thousands of adaptations to the physical environment but hundreds of thousands of adaptations to other organisms: the organisms themselves become the dominant part of the environment, with the physical environment almost fading into insignificance in comparison.[39]

The overall aim of Ray's work is clear: while life on earth is restricted to carbon-based organisms, the life we can create inside the computer is based on logical machine instructions; nevertheless, there is nothing to say that computer-based evolution does not have the potential for developing complexity on a par with the former. Indeed, the machine instructions within *Tierra* have been shown to be capable of universal computation, implying that evolving machine codes should be able to generate any amount of complexity, providing only that it is computable in Turing's sense.[40]

In describing the *Tierra* simulator, it is helpful to keep in mind the analogy between Ray's version of artificial life and life on earth. Evolution on earth follows the basic principles of self-organization we outlined in earlier chapters. It takes place far from thermodynamic equilibrium, and is maintained there by the continuing presence of energy and matter. The energy ultimately derives from the Sun; the available matter is the material resource (food) that all living things must consume to remain alive. Under these conditions, matter becomes self-organized; the

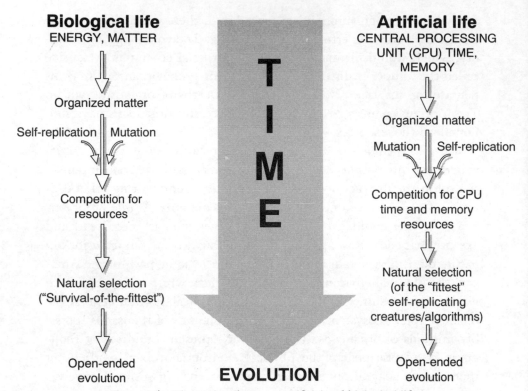

Biological life
ENERGY, MATTER

Organized matter

Self-replication | Mutation

Competition for
resources

Natural selection
("Survival-of-the-fittest")

Open-ended
evolution

Artificial life
CENTRAL PROCESSING
UNIT (CPU) TIME,
MEMORY

Organized matter

Mutation | Self-replication

Competition for CPU
time and memory
resources

Natural selection
(of the "fittest"
self-replicating
creatures/algorithms)

Open-ended
evolution

TIME

EVOLUTION

Figure 8.4   The analogy between artificial and biological life.

effects of self-replication and mutations, together with the fact that the
resources are finite, means that a competitive situation develops.

Ray designed *Tierra* as a parallel computer with the MIMD architecture we encountered in Chapter 2. He assigned a single processor—
that is, one with its own CPU—to each creature.[41] Parallelism is only
simulated on the serial machines Ray originally used, by assigning
each virtual CPU time to run in turn (however, now there are versions
that exploit *Tierra*'s inherent parallelism).[42]

Just as biological organisms use energy to organize matter, Ray's hope
was that digital organisms in *Tierra* would use their time in the central
processor to organize computer memory. A major difference between a
more traditional GA program and *Tierra* is that while in the former the
fitness function or landscape is controlled by the simulating code, in the
latter it is defined locally by the individual creatures in relation to the environment. The fitness landscape thus evolves with the organisms.

Ray hoped that the initial self-replicating digital organism would
evolve through mutations into distinct organisms that themselves might

be capable of replication. In this way, the organisms would compete for the finite computer resources: CPU time and memory space. Then the creatures that survive should be those that evolve most effectively to exploit the others present. (See Fig. 8.4.)

On January 3, 1990, Ray inoculated the virtual landscape in *Tierra* with life, starting with a self-replicating digital organism some eighty instructions long, *Tierra*'s equivalent of a single-celled sexless organism. This creature, the first Ray ever dreamed up, was simply a stretch of instructions written in assembly code. It identified the beginning and end of itself, calculated its size, copied itself into a free region of memory of the same size, and then divided. In terms of the analogy with biology, the machine instructions of the self-replicating digital organisms would be better regarded as amino acids than nucleic acids, since they are "chemically active"—they manipulate bytes, CPU registers, the instruction set, and operating system. But the digital organisms Ray envisaged are perhaps closest to RNA-based molecular-biological creatures— which died out long before the Cambrian explosion 600 million years ago that led to a riot of complex multicellular creatures—because they bear the genetic information *and* carry out "metabolic" activity; in short, they represent both genotype and phenotype.

To discover if he had been successful in producing a cybernetic community, his computer displayed how many creatures of any given genome size were multiplying. All he could do now was wait. Not long after Ray started the simulation, he saw a mutant appear. Slightly smaller than the original, its population grew until it exceeded that of the ancestor. Other mutants needed fewer instructions still to reproduce effectively, and increasingly grazed in the available cyberspace.

Ray called a halt to *Tierra*'s activities after several hours of processing time. It took a month-long follow-up study for Ray to tease out precisely what his digital organisms had been doing. *Tierra*'s artificial society had rapidly become more complex. A creature had appeared with around half the original number of instructions, too few to reproduce in the conventional way. *45aaa* (as it became known) was dependent on others to reproduce. It was the first of many parasites. Later, *Tierra* developed hyperparasites—creatures that force other parasites to help them multiply, although they can reproduce in their own right. They were able to drive the previous generation of parasites to extinction by sharing key operating instructions. With the absence of parasites in the soup, "social" behavior emerged, when each creature relied on at least one other to reproduce. Gradually the community "evolved into a corner," producing a minimal organism twenty-two instructions long, compared with the original eighty. *Tierra* even showed signs of Eldridge and Gould's notion of punctuated equilibria, with lengthy dull episodes of elapsed time when not much happened. (See color plate 9.)

It is interesting to note that, once the original soup had evolved to a certain level of complexity, parasites emerged that did not replicate perfectly because of their genetic design. Indeed, these parasites played a role similar to sexual reproduction in nature in that they mixed up pieces of the genomes between different creatures at random. From this point, it seems that random mutations are not necessary to furnish the genetic change that drives *Tierran* evolution.[43]

The result of all this was an effortless reconstruction of the kind of natural diversity that had seethed around Ray in Costa Rica. Just as the forest's rich diversity depends on "keystone predators," *Tierra* had keystone parasites without which artificial life fails. "I got all this ecological diversity on the very first shot," said Ray. "I was really excited—it was obvious I was getting my wildest fantasy. I thought it would take me five years of tinkering with parameters." The rich society of parasites adapted to each other rather than the environment in the computer. Once life starts, the evolutionary process feeds on itself. "It is an autocatalytic, explosive process that builds its own structures so that the simplicity of the physical environment becomes irrelevant."[44]

Although one might naively expect *Tierra* to favor the evolution of organisms that use less CPU time to replicate, which does indeed occur, much of *Tierran* evolution turns out to involve creatures discovering methods for exploiting one another. All of this emerges spontaneously—it is not preprogrammed—and is in full accord with the tenet of Darwinian evolutionary theory mentioned above: adaptation to the biotic environment (derived from the selection pressure provided by other creatures) rather than adaptation to the physical environment is the primary pressure driving the diversification of organisms.[45]

Tierra was originally unveiled at the second Artificial Life workshop at the Santa Fe Institute, after Ray frantically completed the analysis of the results from *Tierra*'s first successful run in his hotel room.[46] His presentation was, unfortunately, "a complete flop," as he himself put it; he had to wait until he won a prize in an IBM supercomputing competition before his work got significant attention. Today that work is widely recognized as a watershed in ALife research. Ray makes the staggering conclusion that "It would appear . . . that it is rather easy to create life. Evidently, virtual life is out there, waiting for us to provide environments in which it may evolve."[47]

There are, however, skeptics. Robert May of Oxford University's Zoology Department, who has contributed much to modern theoretical biology through his work on nonlinear dynamics and chaos, finds the work stimulating but has "slight reservations about the extent to which the conclusions are perhaps inadvertently built into the program" as well as doubts about the robustness of the findings.[48] Ray's response is to point out that "What we do in artificial life is create a universe, define the 'physics' of that universe and then set it in motion and observe what comes out. It is true that when you design the physics you predetermine what is possible in that universe." However, at the heart of these criticisms is the suspicion that Ray contrived the ancestral organism in *Tierra* to mutate in such a manner that communities would automatically follow. "That is definitely not what happened," he insists. "I am not clever enough to have 'built in' such a rich ecology and long chain of turns of the evolutionary race between hosts and parasites."[49]

PLATEAU AND C-ZOO

Despite the criticisms, Ray's work has been backed by other successful simulations of open-ended evolution. One was developed at Oxford University, inspired by Ray's success. Carlo Maley, then an American graduate student specializing in computer science and evolutionary theory, set out to test Ray's ideas for his master's degree at the Zoology Department. "I wanted to know whether it was hype or serious," said Maley. "It was important to me to write my own computer program so I could say something about whether Tom Ray had built things into his model or not, and whether it was a robust finding." He spent two years on one of the university's DEC workstations, attempting to reproduce Ray's work from scratch in a program called *Plateau*.

The name referred to the two-dimensional universe Maley created. While Ray used a one-dimensional string of programming instructions, Maley used a two-dimensional program so his creatures had shape: Loopy, which consisted of nested loops of instructions, was one of the two-dimensional creatures he set grazing on his computer memory. Unlike Ray, Maley found it difficult to get his model going because of bugs and because he lacked unpublished details of Ray's work. The two-dimensional programming language he used was also more brittle—less tolerant of mutations and creatures interacting with each other. But Maley did manage to iron out these wrinkles in his computer model. Then, within *Plateau*, a rich menagerie developed, albeit more slowly than in *Tierra*. "In any of these models, you will get some form of parasitism. That is the most robust thing about ecology in the real world and it was a surprising result that these models would show the same behavior," said Maley.

At British Telecom's laboratories in Martlesham Heath, another variant has been developed, based on a program called C-Zoo originally written by Jakob Skipper of the University of Copenhagen.[50] José-Luis Fernández explained that BT selected C-Zoo because it had the qualities of *Tierra* but was easier to use. One problem with *Tierra* is that so many instructions in each organism are devoted to reproduction that it is difficult to persuade them to evolve to achieve another goal, such as finding their way through a maze. Another is that the results, displayed as a series of multicolored bars on a screen, are a headache to analyze. And the *Tierran* programs are hard to interpret because they evolve error tolerance through redundancy. "The good thing about C-Zoo is that the biological ideas are the same as *Tierra* but it is much easier to understand," said Fernández.

In a typical C-Zoo demonstration, a series of ants scuttle about within a two-dimensional memory space, hunting for food, represented on screen by an apple. On the display an X marks a spot where there is more than one ant, and thus where a battle for memory space ensues. "What we did was to substitute these pieces of computer code with a bit map of an ant. When you see an ant, what you really see is a collection of cells, each cell containing computing instructions," said Fernández. The code for each ant consists of only thirty-two different instructions, grouped as four cells of eight instructions. From these, complicated behaviors emerge, just as in nature complex proteins can be built from a set of only twenty amino acids.

Four instructions are all that we need to tell the ants how to move: move forward, right, left, or stop. The aim is to find food, either apples or other ants. Those that fail perish. Successful ants breed and mutate to produce progeny, with each different "species" being distinguished by a different color. When an ant dies it deposits its food—its code—into a central pool of memory where the other ants go. As a final touch, the screen display was toroidal so that an ant that wandered off the top of the screen reappeared at the bottom, and those that wandered off one side reappeared on the other. Fernández demonstrates self-organizing ant behavior by depositing all the food in the middle of the screen.

First, he starts with ants that could only move forward. Those that survived constantly return to the middle of the screen, where the food is. Mutants that wandered off course would perish. After a while, green ants become the predominant species. Then he allows some ants to randomly mutate the instructions that control their movement. The ant army evolves to circle around the pile of food. "This feeding frenzy is an efficient solution and a very clear demonstration of evolutionary improvement," said Fernández. "The good thing about C-Zoo is that representation is much nicer compared with *Tierra*, which is difficult to analyze." As he discussed the program, a swarm of ants emerged that had their codes executed by nearby ants.

VIRTUAL BEES AND FISH

The computer simulations we have described so far are of virtual life, though no one would for one moment think that what can be seen on the screen of the computer bears more than a highly idealized resemblance to life on earth. Simulations are under development, however, that model real-world behaviors much more closely. At the Salk Institute in La Jolla, Terry Sejnowski, Read Montague, and Peter Dayan have used a neural network to model and thereby understand how honeybees learn which flowers give them the best return for their efforts.[51]

The focus of the biologists' interest is a brain cell called VUMmx1, which has connections extending throughout the bee's brain, according to studies by Martin Hammer of the Free University of Berlin. Some connections are stimulated by the bee's senses, as when the bee sips sugars, or detects aromas with its antennae. Others are connected to centers that control movement, or to the "mushroom body" responsible for learning. Through VUMmx1's connections, the bee learns to associate a stimulus with a reward, according to work by the bee psychologist Randolph Menzel, also of the Free University. What results is a memory linking flower scent to the nectar reward. Sejnowski reasoned that once these associations are set up, the bee can then use VUMmx1 to predict which flowers in a field are likely to yield the best rewards.

He then set about simulating a bee's brain. Though it has only about one million neurons, compared with our 100 billion, its tiny brain presents a formidable computing challenge: it can tackle about 10,000,000,000,000 floating point operations per second (10 teraflops), when today's most powerful computers usually work at less than one-tenth that speed. Sejnowski's neural network model contains enough detail to be biologically realistic, without oversimplification. "The model does not go down to the molecular level but it is not so completely general that it loses track of the individual neuron," he said.[52]

The artificial neural network was trained on the basis of responses from "senses" to detect color and the taste of nectar. Depending on the timing of these responses, and a memory of the possible nectar reward, the artificial VUMmx1 neuron predicted whether a particular flower was worth investigating, or whether to flit to another bloom. This network could be tested against the real thing because studies of foraging bumblebees had been conducted using artificial flowers by Leslie Real of Indiana University: the insects preferred blue flowers, which had been arranged to give a more consistent return of nectar, even when yellow flowers provided the same average return, albeit in an erratic way. Sejnowski's artificial bee had exactly the same preferences.

Another splendid virtual creature—a shoal of them to be precise— can be found in the Department of Computer Science at the University of Toronto. There, one can glimpse a striking "tank" of "virtual fish." The fish offer all the advantages of the marine world without any worries about feeding, cleaning out the tank, or disposing of the occasional victim of disease and aggression.

To enjoy the fish, all you need is a high-powered Silicon Graphics workstation worth a few tens of thousands of dollars and a copy of the "Artificial Fish World" program. The simulations were designed to emulate real fish as much as possible, capturing their form, motions, and behavior, by Demetri Terzopoulos, working with doctoral students Xiaoyuan Tu and Radek Grzeszczuk.[53] The fish swim gracefully through water, scatter when pursued by a leopard shark, and compete for morsels of food. They even produce elaborate courtship displays. And yet they do not exist physically. Each is described by an individ-

ual computer program nested within a larger program, which generates a simple underwater ecosystem. "We have demonstrated realistic-
looking artificial fish that are capable of some astonishingly lifelike
behaviors," said Terzopoulos.[54] (See color plate 10.)

To write the program, the Canadians first used images of the real thing
to give the fish coloration and texture. Next they gave the fish "brains,"
rules patterned after the real thing that control its twelve muscles, and
"eyes" that enable them to perceive and react to their surroundings. The
program took account of the mass and elastic properties of each fish, modeling it so that it is able to deform as it swims through simulated water.
To coordinate the complex action of all the muscles, the fish learn to swim
"pretty much the same way a baby learns to walk," said Terzopoulos.

The fish tries random combinations of muscle actions, using an algorithm to refine their use. With the help of simulated annealing, the best
combination is chosen depending on speed and, most important, swimming efficiency. After ninety annealing steps a virtual leopard shark
hardly moves at all, because its muscles twitch randomly. After several
thousands such steps, it can swim gracefully. "What comes out is very
very natural," Terzopoulos said, "what ichthyologists call caudal locomotion because it depends mostly on the rear, caudal, fin." And, just as the
undulating swimming motion is not programmed but emerges naturally,
so the behavior emerges from simple rules: the researchers program each
fish's affinity for darkness, coolness, and schooling, plus motivations such
as its level of hunger, fear, or desire to mate. "Then we can start to develop predators and prey," explained Terzopoulos, "where the prey form
schools, take evasive action, and scatter—as real fish do."

To model the elaborate courtship rituals found in the real world, the
Toronto team studied the literature. "For some fish, the female displays an ascending behavior and the male goes underneath and nuzzles
her belly," he said. "There are also courtship behaviors where the female and male circle around, chasing each other's tail." To create these
displays, the team chained together such primitive behaviors as looping,
ascending, nuzzling, in a sequence that depends on various events—for
instance, the female has to witness the male's mating dance for a certain time before she responds.

Although the behavioral repertoire of the fish is programmed, what
happens is highly complex and unpredictable because it depends on
other fish in the neighborhood and what they are doing. "If the male
gets interrupted by a predator, then the mating may not get consum-

mated." Terzopoulos hopes to take the work forward by allowing the fish to mate by mixing the genetic components of male and female fish in forming offspring: "We may be within reach of computational models that can imitate the spawning behaviors of the female and the male, hence the evolution of new varieties of artificial fish through simulated sexual reproduction."

ANIMATS

Another flourishing area of artificial-life research is that on *animats*, or physically real, yet artificial creatures. While the fish are virtual animats, efforts are underway to construct animat robots. Early attempts to build intelligent robots, in the late 1960s and early 1970s, centered around good old-fashioned AI (GOFAI). As we saw in Chapter 5, GOFAI is a top-down approach that compartmentalizes intelligence into discrete "modules" dealing with specific types of knowledge. One classic example was the "Shakey" project at the Stanford Research Institute, another "copy-demo" at MIT. The first robot navigated around obstacles in a room; the second piled blocks according to a model it was shown in highly structured environments.

In the new bottom-up approach to artificial intelligence, also described in Chapter 5, complex behavior emerges from the interaction of simple reflexes, making use of adaptive, learning algorithms that have been developed in the study of connectionism, that is, of artificial neural networks. The basic premise is that interesting robots are too complex to design. The pioneer of this approach in robotics is Rodney Brooks of the Massachusetts Institute of Technology, who inspired the field in 1989 with a prototype animat called Genghis that was based on a cockroach. Instead of being directed by a GOFAI computer program, Genghis had six independent legs that communicate with each other and operate by a few simple rules combined with an ability to learn on the job.[55]

In general terms, Brooks' approach has two strands: a distributed, parallel-processing architecture, and the abandonment of the "compucentric" approach—explicit programming of a robot's brain with a preformed mental model of the world—in favor of real time control. Brooks wanted his machines to generate their own models of the world *adaptively* through their sensory experiences. "It is the coupling of the machine with the environment that is important," he maintains. The artificial nervous systems in these mobile robots is layered. As with the legs of Genghis, each layer contributes to behavior in its own right, although it may implicitly rely on others. For instance, an "explore" layer is not concerned with obstacles, because the existing "avoid" layer will take care of it. Building more and more sophisticated robots through progressive layering is analogous to the long-term results of evolution.

An animat clambering over an irregular terrain has no need for a pre-specified computer program. "When the robot goes at different speeds, different gaits will automatically emerge," said Brooks. "It does not worry about which gait is the best or make an explicit decision." By adding senses, reflexes naturally emerge through adaptive control techniques. As one example, we can add a heat sensor so that if the animat senses more warmth on its right side, compared with its left, the range of the legs on one side can be altered so that it turns away. "The actual path the robot takes is not precomputed. It emerges."

Such is the speed and simplicity of the design that IS Robotics, a company set up by Brooks, has now commercialized Genghis with a Canadian company called Applied AI Systems. A swarm of twenty of these small robots was used to simulate the behavior of a termite colony. Large versions of the robots are being considered for trimming the undergrowth around trees in Canadian forests.[56] A budget version of Genghis, called Marv, has been developed to negotiate rugged terrain of the kind encountered in places like Death Valley, California, in a project led by Chris Melhuish at the University of West of England, Bristol. The U.S. space agency NASA is also interested, in the wake of a 1989 paper written by Brooks for the *Journal of the British Interplanetary Society* entitled "Fast, Cheap and Out of Control: A Robot Invasion of the Solar System." It argued that rather than sending a single heavy-weight rover to explore a planet, a swarm of smaller robots would offer a cheaper, lighter, more robust alternative. NASA liked the idea of robots being fast and cheap, but it could not quite handle their being out of control, according to Brooks.[57] NASA is currently considering a six-wheeled explorer for Mars.

There is a great deal of related work on animats. Neural networks that control robots are being evolved at the University of Sussex by Dave Cliff, Inman Harvey, Phil Husbands, Nick Jakobi, and Adrian

Figure 8.5   Some robots from Applied AI Systems.

Thompson.[58] Instead of designing fixed robot control programs, they rely on genetic algorithms to carry out simulated evolution on a random population of between sixty and one hundred dynamic recurrent neural networks. The networks control the behavior of simulated robots; their sensors and actuators closely model those of a real robot. During artificial evolution, selection pressure is controlled by evaluating the performance of each robot: the better the robot performs its task, the more offspring its cognitive architecture has. In one experiment, the Sussex team could evolve the networks to use visual sensors, "eyes," so that a palm-sized robot can avoid obstacles and seek light.[59] Another wastepaper basket–sized robot used sonars, whiskers, and bumper bar-sensors to find its way around the Cognitive Science Department at Sussex University. "We apply selection pressure so that those that are better at doing the job we want them to do are more likely to breed," said Dave Cliff, whose colleagues have now turned to evolving control programs within a real robot.[60] "We have some very promising results which suggest that we could evolve visually guided robots, which would find their way around a room," he said. "As far as we know, we are the first people in the world to do this."

Traditional AI researchers have scoffed that this kind of work has little to do with intelligence, because such animats do only what insects do. Brooks has responded with an ambitious plan to reproduce human evolution using a humanoid robot called Cog. Under construction since 1993, Cog has human form, with a head, arms, and even a voice, though it lacks legs (it was modeled on an MIT graduate student). "And we want it to have behavior like that of humans," he said.[61] Sensing in the hands and arms is carried out with conducting rubber that allows touch. Strain gauges, heat sensors, and current sensors will allow Cog to feel its arms being used and how they are performing. Cog will also have two eyes, each of which consists of two tiny cameras, one giving a broad field of view, the other a central field. The eyes saccade—that is, dart back and forth like those of a human. Cog will have three microphone receivers for locating the origin of sound, a trick we manage with two ears and fancy signal processing in the brain. Such details are crucial, because the fact that we have bodies matters. Our bodies define, constrain, and enable us to interpret the world. The way the brain evolved and human cognition developed was predicated on our interaction as individuals with the world. "Any intelligence that we will really be able to communicate with better have a human-like body, otherwise it will be an alien," he said. "I did not want to build aliens but things that we can know and love."

Cog sits in a public area in the AI laboratory so that it can interact with people, rather as a baby might, perhaps by playing with toys, stacking objects, passing things back and forth, and so on. Although Cog will have an ability to learn and its view of the world will be built through experience, it will still have some preprogrammed responses, just as a real infant is equipped with a sucking response and the tools for language acquisition. Cog has a sense of balance, and coordinates the movement of its head to keep the face of any nearby human in its field of view. "It wants people to pay attention to it," said Brooks. "That is its inner drive and we are hoping we can use that motivation for lots of developed behaviors." The information-processing power is located in an external computer that is connected via an umbilical cord. "My eight-year-old son was very disappointed when he found the brain was not in

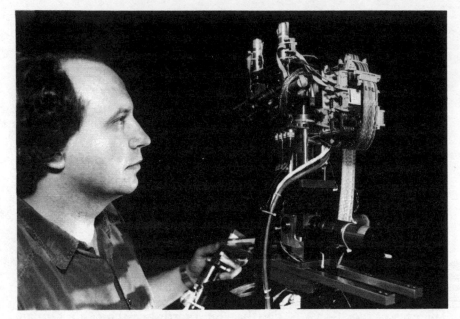

Figure 8.6   Rodney Brooks and his humanoid robot, Cog.

the head," he said. Understandably, Brooks is coy about when he expects Cog to evolve human-like behavior. "The six-month- to two-
year-old level is the capability we are aiming at." He does not expect it
to evolve a voice but only guttural responses. "We could have put a
speech synthesizer on it but that is cheating," he said.[62]

EVOLVING BETTER PROGRAMS

Clearly, the evolutionary approach to programming computers (and
robots) is already dawning. Evolving communities of digital codes
could be useful for the task of programming highly complex systems.
British Telecom has already used the experience with *Tierra* and C-Zoo
to calculate when we may expect to see people breeding software rather
than writing it. Chris Winter, Paul McIlroy, and José-Luis Fernández
at British Telecom started out by assuming that a good programmer
can write up to thirty lines of debugged code every day. By their reck-
oning, a machine of an assumed processing speed of ten MIPS (million
instructions per second), such as a present-day Mac Quadra, would

take 100 days to achieve the same result. However, given the leaps and bounds in computing speed, they estimate that by the year 2000, desktop computers, able to crunch through about 3,000 MIPS, will generate computer code at approximately the same speed as a good human programmer. Indeed, it might even be possible to leave this software to breed in a computer so that it can evolve to cope with new challenges, such as infection with a computer virus.

Of course, tricky technical problems do exist. For instance, problems can arise because of shortcuts used in testing each generation of evolved software. It is computationally expensive to test every organism in every generation for a given task, say, routing calls through a telephone network. Instead a simpler test is used, measuring the organism's performance when one particular type of network traffic jam occurs. The danger is that the successful organisms are only good at solving that particular network problem and no other. One cunning way to overcome this problem is to evolve the population of problems as well as solutions, so that the tasks we find most difficult are constantly selected. Developed by Danny Hillis at Thinking Machines, the method of evolving both problems and their solutions is called the "co-evolving parasite model"—a reference to the similar evolutionary race that occurs in nature between parasites and their hosts.[63]

An illustration of how software may one day be evolved rather than expressly written down has been proposed by Tom Ray, as part of the ongoing development of his *Tierra* ALife simulator at the Evolutionary Systems Department of the Advanced Telecommunications Research Institute International in Kyoto. Ray wants to set up a "biodiversity preserve" by repeating the *Tierran* evolutionary computer experiments with volunteers on the International Network of Computer Systems, the Internet, which connects an estimated twenty-three million users worldwide. In collaboration with Kurt Thearling, he has developed software that would enable "digital wildlife" to multiply on the network.[64] With the help of a few thousand volunteered computers linked across the world, the programs would be faced with a more complex environment that should encourage them to evolve yet more sophisticated strategies for survival and reproduction. Among the creations in this digital menagerie may lurk commercially useful software suited to spreading tasks among many different computer processors or doing jobs that Ray and Thearling have not even thought about. For example, it has been found that digital organisms can spontaneously evolve programming tricks, and the set of processes involved in *Tierran* evolution could be regarded as a general optimization technique for parallel programming. In Ray's words, "In the end, artificial evolution may prove to be the best method of programming massively parallel machines."

Because they will be chased off any given machine when its rightful user wants to get on-line, the organisms on the Internet will be constantly forced to try out new strategies, shrinking to the minimum size required for copying, learning quickly how to find little-used resources on the network. "They should start migrating around the globe, staying on the dark side of the planet," said Ray.

The creatures may start to cooperate, creating "multicellular" organisms. This adaptation of *Tierra* to a more collaborative multicellular form (as opposed to the "unicellular" one used to date) should lead to constructive cooperation between organisms in the solution of a single problem, rather than the fragmented effort on a variety of distinct ones typical in the unicellular case.

Various communication mechanisms are needed to coordinate composite organisms in *Tierra*: one operates like a hormone, sending a message from the mother cell to all its daughters; a second acts like a nerve, running between two specific cells. However, certain control tasks must only emanate from a single cell. For instance, when a mother cell divides to make a daughter cell, only the mother program should be executed if the pair need to take offensive action, or search for a resource or a mate. If mother and daughter cooperate successfully in this way, the first artificial multicellular *Tierran* creature will emerge. That could prove a milestone in the quest for artificial life and intelligence. As Ray puts it, "We are living examples of this kind of parallelism on an astronomical scale, with trillions of cells and hundreds of cell types that are beautifully coordinated. I am arguing that evolution has a proven ability to achieve that, just as I would argue that evolution is the only proven technique for generating intelligence."

CELLULAR AUTOMATA AND THE EDGE OF CHAOS

Various ALife groups have latched on to Ray's work, and today numerous people are working along similar lines. One variant on *Tierra* is *Avida*, developed by Chris Adami and C. Titus Brown working at Caltech.[65] While *Tierra*'s organisms exist in a spatially nondimensional cyberspace, *Avida* is designed to introduce spatial dimensions into artificial evolution. It retains most features of *Tierra* but is based on a two-dimensional array of cells. The digital organisms occupy sites on the lattice, and in a manner characteristic of cellular automata they may only interact with their nearest neighbors. However, the update rules that determine how any site in the lattice is to alter from one time step to the next are not fixed once and for all prior to execution. Rather, the rules are determined by the genomes at the closest neighboring sites. These genomes change randomly via point-mutations of their strings of instructions, thus providing the impetus for evolution. As with *Tierra*, *Avida* can be used to perform tasks specified by a programmer. For example, if we wanted to evolve a method of performing multiplication of two integers, then we could reward every digital organism that accomplished this task by allowing it more slices of CPU time in which to execute. Over time, the artificial organisms will evolve to deal with this task. In principle, as with *Tierra*, such a system should be able to evolve to solve any task, no matter what its complexity, provided only that it be computable in Turing's sense.

Some scientists have argued that living things must be able to perform computations of essentially arbitrary complexity to survive the evolutionary arms race. Moreover, certain types of cellular automata (CA) are known to be capable of performing universal computation—in other words they can simulate a universal Turing machine. (Such a capability was first established for the Game of Life encountered in Chapter 4.) This claimed requirement for universal computation has inspired some people to revisit Wolfram's classification of CA dynamics, to see whether it furnishes any clues as to whether certain types of CA are better suited to artificial life.

As we have seen, Wolfram conjectured that his so-called Class IV CA would support universal computation.[66] Lying between periodic and chaotic regimes, Class IV *appears* to support the most complex dynamics of CA. It might therefore seem reasonable to suppose that in ALife systems based on CA—in reality only a small subset of all such possible systems—the dynamics would have to evolve into Class IV when "life" arises.

YIN AND YANG

These ideas on the relationship between dynamics and computation are part of a larger effort, notably in the field of statistical physics, to find complex behavior in "critical regimes" between order and deterministic chaos. The search for complexity in such regimes is also intellectually appealing since living things appear to capture an elusive mixture of yin and yang. Biological life seems to occupy a zone between regularity and turbulent chaos, where randomness coexists with creative adaptation. Organisms combine the ability to change and innovate with the stability of feedback systems that ensure a well-defined structure and metabolism.[67]

Suggestive evidence for this kind of balance between chaos and order can be found within ant colonies. By using video recordings, Blaine Cole, working at Houston in Texas, was able to reveal that individual ants behave chaotically. For a while, an individual scuttles about. Then it takes a rest. And so on and so forth, in behavior that rattles around on a strange (chaotic) attractor of the kind described in Chapter 6. But at the colony level, the ants show quite rhythmic behavior. Nigel Franks, at the University of Bath, observed that the ant colony is active for a while, takes a rest, then becomes active again, with a cycle time of around twenty-five minutes.

Experimental studies by Cole showed that the pattern of behavior depends on the density of ants. If there are only a few in a territory, their behavior is chaotic. But if the number increases beyond a threshold value in a given area, the whole group becomes rhythmic. It is a bottom-up organizing process driven by ant-to-ant contact. The ants excite one another, so that an active ant gets an inactive one moving when they encounter one another. This behavior was simulated by Octavio Miramontes and Richard Solé, working with Brian Goodwin at the Open University. They developed a cellular automaton model of ant colonies in which the activity of individual ants is propelled by a Hopfield neural network.[68]

"At a particular density of the computer ants there is a sharp transition in which what was a collection of individuals each doing its own chaotic thing suddenly transforms into a single whole—the colony becomes a superorganism with a well-defined rhythm and, at the same time, spatial order appears," according to Goodwin.[69]

Observations from Nigel Franks' laboratory suggest that real ant colonies adjust their densities so that they live near this transition point, at the "edge of chaos." Ants regulate the size of the territory within which they make their nests, with the queen at the center and the developing ant embryos and larvae arranged around her. Given grains of sand, the ants define the boundary of the brood chamber. But if a malicious scientist pushes the sand grains in to cut the size of the chamber, the ants push them out again. Likewise, if the territory is increased, the ants reduce it again. "The colony has a sense of density and spatial order," said Goodwin. "So it appears that ants may indeed adjust their colony density such that they are near the edge of chaos."

These examples are what makes the "edge of chaos" idea so seductive. For Goodwin, it is "almost a theorem about life, the universe and everything that is complex and nonlinear (which is *nearly* everything). Speaking more anthropomorphically, the edge of chaos is a good place to be in a constantly changing world because from there you can always explore the patterns of order that are available and try them out for their appropriateness to the current situation. What you don't want to do is get stuck in *one* state of order, which is bound to become obsolete sooner or later (remember the dinosaurs, or the British Empire, or IBM before the shake-up). So complex systems that can evolve will always be near the edge of chaos, poised for that creative step into emergent novelty that is the essence of the evolutionary process. At least, that is the conjecture."

EVOLUTION AND UNIVERSAL COMPUTATION

Chris Langton at the Santa Fe Institute, New Mexico, has been an ardent advocate of these suggestive ideas. He has devoted much time to an attempt to apprehend how life balances the yin of apparent mayhem with the yang of self-organization: "In living systems, a dynamics of information has gained control over the dynamics of energy, which de-

termines the behaviour of most nonliving systems. How has this do-
mestication of the brawn of energy to the will of information come to
pass?" he asked.[70] He claimed that cellular automata capable of performing "nontrivial computation"—including universal computation—
are most likely to be found in the vicinity of transitions between order
and chaos; that is the Class IV dynamics that lie "between" the ordered
Class II behaviors and chaotic Class III behaviors.[71]

Langton argued that if living systems perform complex computations in order to survive, evolution under natural selection would tend
to favor systems near the border between ordered and chaotic behavior—at the "edge of chaos," where the ability to process information
would be maximal. From a study of a large number of CA simulations
using different dynamical rules, Langton maintained that the conditions for life are optimal when apparently random behavior coexists
with more regular dynamics. Here "life" should be interpreted as either
real or artificial.

Langton believed that he had found a critical transition region in
the parameter space of all two- and one-dimensional cellular automata
where "one observes a phase transition between highly ordered and
highly disordered dynamics, analogous to the phase transition between
the solid and fluid states of matter."[72] This was the Class IV region between Class II and Class III. Phase transitions in physical systems,
from ice to water or water to steam, are able to establish correlations
between molecules over arbitrarily large distances in space and time.
Since universal computation in Turing's sense can only work in systems with memory and communication over essentially arbitrarily
large distances in space and in time, it appears that the complex (Class
IV) category of CA dynamical states would be the most likely to support nontrivial and possibly even universal computation. Put another
way, Langton was arguing that information processing within a parallel-processing network could be maximized at "the edge of chaos."[73]
One might then expect systems such as living things to evolve toward
this region if they are to perform the many complex tasks their environment demands of them.

To bolster the "edge of chaos" concept, like-minded colleagues have
tried to tie in other ideas. For instance, Stuart Kauffman has argued
that the concept provides "a powerful new framework to understand
evolutionary biology."[74] In the natural ecosystem of a rain forest, a co-
evolutionary system depicted by a rubbery fitness landscape, the suc-

cess of one species (such as a frog) may spell doom for another (a fly) that it prefers to dine on. Kauffman has claimed that the entire ecosystem may coevolve to a state poised at the edge of chaos, linking the idea with Bak's concept of self-organized criticality discussed in Chapter 7, who in turn proposed that a number of systems, including some cellular automata,[75] exhibit evolution to a critical state.[76]

One crucial issue remained: what is the *computational* capability that was supposed to blossom at the edge of chaos? This was addressed during the same period, but quite independently, by Jim Crutchfield and his colleagues at the University of California at Berkeley. Crutchfield and Young showed that a system's computational capability increases dramatically at the onset of chaos, which is a kind of phase transition.[77] Their work was published in 1989 at the culmination of a decade-long line of research involving Ditza Auerbach, Remo Badii, Peter Grassberger, Bernard Huberman, Gyorgii Sepfauluzy, Robert Shaw, and others.[78]

Crutchfield's ideas concerned a neat method for describing the yin and yang of complexity in a statistical way.[79] His definition of complexity led to a counterintuitive result. A string of ones represents pure order. A random string of ones and zeros would seem to be infinitely complex. However, Crutchfield and Young argued that, given a source of such random binary digits (e.g., from a chaotic system), it is as easy to generate random bits strings as a string of zeros, so that utter randomness is as simple as pure order. Discounting the computational cost of randomness in this way, we can reveal higher levels of complexity between these two extremes: statistical complexity is maximized somewhere between order and randomness. In fact, Crutchfield and Young showed that there was a jump from finite to infinite memory at the onset of deterministic chaos. So in terms of their measure of complexity, computational capability is maximized in the regime intermediate between order and randomness.

For reasons that will become apparent, it is unfortunate that much more media attention has been focused on the efforts of Langton at Santa Fe and also the work of Norman Packard at the University of Illinois, who put forward the suggestion of the biological relevance of the edge of chaos and evolved cellular automata to perform a range of computational tasks. Using a genetic algorithm, Packard had investigated which cellular automata rules are "naturally" selected on the grounds of their computational efficiency[80] and concluded that cellular automata rules associated with Class IV are the ones most likely to be capable of performing complex computations; he believed that if cellular automata rules are allowed to evolve so that they perform complex computations, those within this region would tend to be selected. Packard's work was promptly hailed—by a few scientists and the science media—as a landmark concept in the science of complexity.

DISSENT AT THE EDGE OF CHAOS

Criticisms of the work of Packard and Langton date back to at least 1988.[81] "Serious flaws in its general reasoning and in the technical details were pointed out then and continue to be," commented Jim Crutchfield,[82] whose important contributions have been overlooked in most popular accounts. Langton's work on the edge of chaos was qualitatively different from Crutchfield's in one respect. While Crutchfield and Young dealt with continuous time dynamical systems, the cellular automata studied by Langton exist only in discrete time. More important, Langton used a crude measure of complexity for his studies. The overall consequence, argues Crutchfield, "is that Langton's work cannot be construed as giving results on how evolutionary processes produce structure."[83]

In a room opposite Langton's at the former site of the Santa Fe Institute, Melanie Mitchell also decided to investigate whether the media accounts of the "edge of chaos" had been inflated. She, too, was irritated by the vagueness of some of its claims. With student intern Peter Hraber along with Crutchfield, she tried to reproduce Norman Packard's important claim that his evolutionary CA simulations "adapted to the edge of chaos." They concluded that he was wrong.[84] His "landmark" finding was almost certainly an artifact that said more about how his computer was programmed than anything else. In Mitchell's words, "To the extent that one can make sense of what Packard and Langton meant by the 'edge of chaos,' their interpretations of their simulation results are neither adequately supported nor are they correct on mathematical grounds."[85]

In redoing Packard's computer experiments, Mitchell and her colleagues found no evidence for the idea that cellular automata with values close to the border between Class II and Class III dynamics had any particularly enhanced computational capabilities. Mitchell and her collaborators came to an important conclusion: "it is mathematically important to know that some CAs are in principle capable of universal computation. But we argue that this is by no means the most scientifically interesting property of CAs. More to the point, this property does not help scientists much in understanding the emergence of complexity in nature or in harnessing the computational capabilities of CAs to solve real problems."[86]

High levels of computation at a *cellular automaton* phase transition between order and chaos have never been shown to exist.[87] This topic is surrounded by confusion because of the suggestive yet ambivalent nature of the word "chaos." In everyday language chaos is synonymous with randomness, making people contrast it with ordered behavior, and thus think of some kind of precarious balance between opposites. But its scientific usage is quite different; there, as we have pointed out, the term masks the fact that chaotic dynamics is actually exquisitely organized.

EXPLORING COMPLEXITY

In the final analysis, there is no simple nor lofty mathematical theory of life, whether real or artificial. However, we may draw strength from the ancient paradox that although the world is complex, the rules of nature are simple. The universe is populated by a rich variety of physical forms, from bacteria and rain forests to spiral galaxies, yet they are all generated and sculpted by the same underlying laws. Thanks to fast and powerful computers, biologists, physicists, and computer scientists pondering complexity can now explore complexity in its full glory, throwing light on questions that once lay exclusively in the province of philosophy and mysticism.

The most important contemporary ALife system, *Tierra*, has an evolutionary dynamic of such great complexity that the only way to investigate its behavior is to perform experiments on it within a computer, the equivalent of experimental biologists' field studies. One might anticipate that, given enough time and a sufficiently powerful supercomputer, it might be possible to evolve within a *Tierra*-like system digital organisms endowed with intelligent capabilities and even consciousness. The work of Brooks and others underlines that, for this to be possible, digital life must be exposed to and interact with the complexity of a rich and varied environment.

For us to pursue the quest for genuine artificial intelligence, we must now turn to the supreme manifestation of complexity in nature: the human brain. This is an object that fascinated von Neumann and Turing and provided the inspiration for much of their most significant work. Now more than ever before, it is setting the research agenda for scientists worldwide. The reason is easy to understand: it poses the ultimate challenge to the science of complexity, and to lay bare its secrets we will have to draw on many of the concepts and systems we have explored in this and previous chapters, from mathematical logic and spin glasses to Belousov-Zhabotinski reactions and evolution.

Chapter 9

THE ENCHANTED LOOM

The spirit within nourishes, and mind instilled
Throughout the living parts activates the whole mass
and mingles with the vast frame[1]

—VIRGIL

It is wrinkled, weighs around three pounds, and has the consistency of a ripe avocado. It is sophisticated enough to harness the motion of atoms and molecules, give split-second instructions to the fingers of a concert violinist, or conjure up three-dimensional images from rays of light dancing on a pair of two-dimensional retinas. It can also dream, spout poetry, and devise jokes. It is unmatched in its ability to think, to communicate, and to reason. Most striking of all, it has a unique awareness of its identity and of its place in space and time. Welcome to the human brain, the cathedral of complexity.

The brain presents us with a baroque hierarchy of dynamic and static structures. At one level, we need to understand how a few cells in the early embryo can divide and self-organize to generate a living adult brain. At another level, we need to explore the structure of the brain and how interactions between vast numbers of neurons can give rise to many emergent properties, including the ability to distinguish a fine claret from vinegar, or the smile on the face of a baby from the fixed grin of an advertising executive. The omega of all these emergent properties, indeed of all complexity, is human consciousness. If we ever come to fully comprehend the brain's complexity, the ultimate test of that understanding will be an attempt to simulate it within a computational machine.

The nub of the problem of consciousness is admirably expressed by John Searle, a philosopher at the University of California, Berkeley: "The secret of understanding consciousness is to see that it is a biological phenomenon on all fours with all other biological phenomena such as digestion or growth. Brains cause consciousness in the same sense that stomachs cause digestion and in neither case are we talking about something spiritual or ethereal or mystical or something that stands outside ordinary physical processes in the world. The two biggest mistakes, and at bottom they are both the same mistake, is to think that consciousness, because it is private, subjective, touchy, feely, ethereal, etc., cannot be part of the ordinary sordid physical world of drinking beer and eating sausage. The second big mistake is to think that it is all a matter of computer programs."[2]

Astonishing progress has already been made in our knowledge of the brain, for many reasons. First, we are armed with unprecedented understanding of its chemistry. Second, we now have a range of tools to watch the living brain at work. Third, neural network simulations of the brain have managed to capture some of its emergent properties *in silico*: these range from the way a damaged brain behaves to how it processes the signals from the retina for vision. It is becoming clear that obstacles to creating artificial consciousness may not be as formidable as we had thought.

A HISTORY OF THE MIND

Our understanding of the brain has come a long way since the father of modern philosophy, René Descartes, put forward his ideas about it. Born in Touraine, France, in 1596, Descartes is said to have had divine revelation of his mission in life—the unfolding of the general principles of science (*Scientia mirabilis*)—while shut away one day in a "stove-heated room."[3] Thirteen years later, in 1632, Descartes launched the scientific study of the brain with publication of *The World*.[4] He rejected the prevalent view that biology could only be explained by invoking special "vital" principles of life, claiming that there was nothing about the human body that could not be explained by the same laws that governed the behavior of stars and rainbows.[5]

Descartes believed that all operations of the senses and the nerves both began and ended with the tiny pineal gland at the base of the brain. The pineal influenced the flow of "animal spirits," the term he used for a refined form of blood that supposedly surged through nerve and brain. Information from the senses, meanwhile, was transmitted to the brain by cords, which ran inside the same nerves. A cut to the skin of a finger would tug a glorified bell-pull, which would open a valve inside the head, sending a flock of animal spirits to the muscles to pull the hand out of harm's way. But Descartes did not pursue this reductionist picture to its ultimate conclusion. He claimed that the soul ultimately ruled the pineal. Modern brain scientists beg to differ. The brain is in charge, not the soul. And the brain consists not of spirits, plumbing, and bell-pulls, but cells.

The structure and function of many of the cells within the brain are beginning to be understood, as are many of the molecules that shuttle signals between them. On a computer screen, we can display a colorful image of the receptor sites on the surface of brain cells, revealing the precise docking site where these messenger molecules act. We can even mimic the ripple of electrical activity generated when a single brain cell fires. Yet when all this detail has been accounted for, the most important feature will still be lacking. This is the big picture of how emergent properties such as memory result from the brain's structure. While it is essential to understand the molecular detail of brain chemistry, only the science of complexity enables us to make sense of the higher level of organization at which networks of billions and billions of neurons act together apparently miraculously to handle not only memory, but also vision, learning, emotion, and consciousness.

Complexity arises in the brain through self-organization at several levels. First, during development, self-organization fashions the brain from a series of feedback and selection processes between neurons. Second, within the complex chemistry of living brain cells, vast interacting networks of molecules self-organize to create spatial and temporal order, which can be seen as dissipative structures similar to those we encountered in the Belousov-Zhabotinski reaction (Chapter 6). Third, self-organization constantly rewires the huge numbers of neurons in the brain to store memories, tailor its performance to the environment, and create a host of other emergent properties. Beyond all this, we should not forget that many aspects of the brain have been optimized through millions of years of evolution.

The physiologist and Nobel laureate Sir Charles Sherrington, writing *Man on His Nature*, described brain self-organization in his characteristically evocative prose: "It is as if the Milky Way entered upon some cosmic dance. Swiftly the head-mass becomes an enchanted loom, where millions of flashing shuttles weave a dissolving pattern, always a meaningful pattern, though never an abiding one: a shifting harmony of sub-patterns."[6] Attributes like intelligence and consciousness, over which philosophers puzzled long before and ever since the death of Descartes in 1650, are now themselves coming within the scope of scientific analysis. No wonder that scientists today are filled with similar excitement to that felt when Ernest Rutherford revealed the structure of the atom, and when Francis Crick and James Watson established the molecular structure of DNA.

THE EVOLUTIONARY ART OF BRAIN DESIGN

The brain is a child of the rich environment it experiences through its senses. Its origins lie in the self-organization of the central nervous system, over billions of years of evolution. Faced with the necessity for survival, for making order in a buzzing booming chaos, as the American philosopher and psychologist William James called it, the nervous system evolved to extract ever more useful snippets of information about its surroundings. As a result, a profound harmony exists between the organization and structure of the brain and the world in which we live.

This harmony had its origins in the earliest primitive sensing cells. First there were simple cells to sense a chemical gradient or the direction of the Sun. Then multicellular organisms evolved that relied on several of these cells. The hydra, the freshwater polyp we encountered when describing how self-organization occurs during development (Chapter 6), needs a diffuse network of nerve cells in order to dine on the small organisms that brush past its tentacles. The flatworm boasts an even more complex neural wiring plan. Its ladder-like nervous system is not evenly distributed within its flat, black body: its sensory apparatus is located at one end and nearby are clusters of nerve cells called ganglia.[7] These clusters are the primitive forerunners of what we would consider a brain.

To improve the coordination of sensory information in more advanced forms of life, various strategies evolved to alter the connections between the nerve cells within and between ganglia.[8] Initially, connections became more sensitive if stimulated, a temporary effect that one can observe by tickling the gill at the front end of a sea slug. At first the simple creature withdraws the gill, a response controlled by a cluster of just six neurons. But if you tickle it again and again, the slug gradually learns to live with the nuisance through a simple form of memory, called habituation. Our brains are much more than billions upon billions of sea-slugs's worth of neurons. More sophisticated possibilities evolved, notably associative memory, when connections between a fixed number of neurons were snapped and woven in response to coincidence and correlation in the firing of nerve cells stimulated by the senses.[9] In this dialogue with the environment, simple learning processes could wring more value from the senses: for an animal, the memory of the bitter taste of a yellow berry could usefully remind it of a stomachache. This enabled the brain to adopt strategies for enhancing survival probabilities. Driven on by the engine of evolution, the brain has turned into an organ of richly interconnected processors that are able to predict as well as react to challenges from the senses. As the Oxford neurophysiologist Colin Blakemore put it, "The interesting parts of the brain are driven by the senses, right through to language, which surely evolved from sensory categorization."[10]

During the evolution of species, both the size and the sophistication of animal nervous systems have increased, culminating with our brain, the largest any primate has possessed.[11] The secret of the human brain's remarkable power lies in its extraordinary complexity. There are something like one million million cells (10^{12}) in the brain. The majority are small glia whose function was traditionally thought to be to support, nurture, and protect the all-important signal-processing cells, nerve cells called neurons. Although the glia are increasingly suspected of playing a role in information processing, the main focus of interest remains on the brain's neurons.[12] The total length of "wiring" between the neurons is roughly one hundred thousand kilometers. The number of connections between the neurons is about one thousand million million (10^{15}). The total number of neurons is one hundred thousand million (10^{11}). To put this last number in context, it rivals the number of stars in our galaxy.[13]

One may think that to build an organ of this complexity would require at least a million million instructions, one for each brain cell. Perhaps a need exists for many more, if instructions are also required to lay down every connection. Perhaps far fewer, if we assume that brain cells are not all different and come in distinct families. It turns out, however, that the design and function of the entire body are constrained by no more than 100,000 genes, one-third of which provide the wherewithal for building the most complex known object in the universe. To bridge the gap between the genetic ingredients and the resulting organ, all stages of brain development depend on an element of contingency as well as self-organization—that is, on creativity.[14] However, the Nobel laureate Francis Crick has warned against empty sloganeering: "Who or what else is organizing it if it is not doing it itself? There are any number of ways that the brain can self-organize. What we want to know is which way it does it."[15]

NATURE'S BLUEPRINT

The brain grows according to a recipe that depends partly on its genetic blueprint, partly on the environment. Let us start by concentrating on the blueprint. A cascade of genetically programmed proteins acting in a range of feedback processes turns a spherical, fertilized cell into the billions of cells in an adult, triggering not only cell division but also cell migration, adhesion, and death. Differentiation, the process that makes a brain cell distinct from a liver cell, is similarly controlled.

Recent advances in molecular biology have made it possible for us to investigate the detailed genetic and cellular mechanisms taking place during brain development when as many as 250,000 cells are made every minute.[16] The key stages of brain development are similar whether they occur in chick, fly, or human. The first phase is to make nerve cells in the right place. With a chick, this begins to occur during the formation of a blastoderm, a plate of around 100,000 cells. Three separate layers—the ectoderm, mesoderm, and endoderm—then form in the process called gastrulation, one of the phenomena originally modeled by Turing, as described in Chapters 6 and 7 (see Fig. 9.1). The central axis of the plate pinches off a structure called the neural crest atop a linear fold called the neural tube. Cells flow out of the neural crest

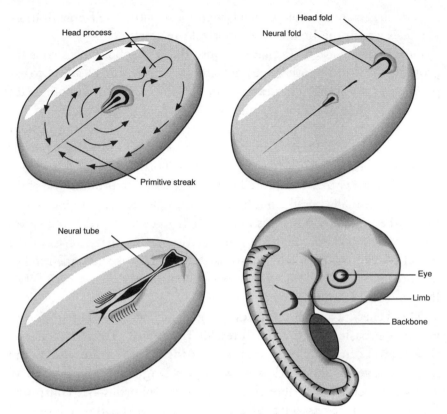

Figure 9.1   Key stages in brain development.

into a series of pouches known as branchial arches, the evolutionary remnants of gill arches—relics of our fishy ancestors—that develop into the various structures of the face and neck. Although some cells from the crest and other parts of the embryo become nerve cells, the most interesting structures—the central nervous system, brain, and eyes—arise from the neural tube itself.[17] By four weeks, the major regions of the human brain can be recognized, including the forebrain, midbrain, hindbrain, and the optical vesicle from which the eye develops. The brain gets its characteristically wrinkly appearance after about six months.[18]

During early development, different lineages of brain cells evolve. Previous work on fruit flies, the geneticists' favorite experimental organism, had shown that the role of *all* cells, including neurons, in a developing fly is controlled by a special class of programmed instructions called *Hox* genes. *Hox* is an abbreviation of "homeobox," a stretch of DNA enabling one gene to "talk" to another. This latter interaction is the key to gene regulation, which is vital for structural self-organization, just as it was in the Belousov-Zhabotinski reaction.

Genetic feedback takes place because the homeobox genes code for proteins that bind to other specific gene sequences, thus leading to expression or suppression of nearby genes on the same chromosome. *Hox* genes can set in motion interlinked cascades of genetic activity that control developmental pathways. The result is the genetic equivalent of painting by numbers: a process called segmentation that controls the form of an embryo, whether human or fly. Broadly speaking, segmentation refers to any repeated structures that occur along the main axis of an animal's body, from head to tail, such as the ganglia. In humans, the most obvious signs of segmentation are the ribs, the vertebrae of the backbone, and the roots of the nerves that run from the spinal cord to the peripheral nervous system. This is the same kind of self-organization Jim Tabony observed in the experiments with tubulin we described in Chapter 7.

This genetic control organizes the general "wiring diagram," providing the basic structural features for the brains of each species, so that spiders can effortlessly spin webs, owls possess acute vision, and humans have an innate capability to use language. The growth of axons and dendrites—fibers sent out by the neurons—is controlled by their interactions with variations in concentration of chemical markers.[19] Some of these markers are attractors, others repellors. The conflicting nature of these interactions is qualitatively similar to the ingredients that organized the hydra.

Half or more of these neurons die around the time their fibers reach their target, even though they all appear healthy. This process is thought to run along the principles of natural selection: they are all competing for a limited amount of a survival chemical.[20] If they do not get enough, they commit suicide by a process called *apoptosis*. The result of all this cellular sacrifice for the greater good of the brain is a rough "map," where neighboring nerves in a sense organ such as the retina plug into neighboring areas of the brain. The map is then fine-tuned even before the retina starts to respond to light. Evidence shows that waves of spontaneous organizing activity sweep across nerve cells of the retina whose fibers make up the optic nerve and help the termini of adjacent retinal nerve cells to plug into adjacent regions of the brain.[21]

The resulting adult brain is a contrary mixture: neurons span large regions and connect with thousands of other neurons. They are organized into functional maps so that different regions of the brain do different things. Yet the result is a single consciousness. This conflict between integration and segregation is a paradox central to the effort to provide a theoretical understanding of the brain. As Giulio Tononi of the Neurosciences Institute in La Jolla commented, "There has been an incredible tension between the localizationists on one side and the holists on the other."[22] To help understand the apparent conflict between local and global functions, he has tried to develop a numerical measure of neural complexity, called C_N, which works in a way akin to the statistical complexity measure developed by Crutchfield and Young mentioned in the last chapter. Tononi uses an example from physics to show how this works: C_N would be low both for a gas consisting of molecules moving randomly *and* for a crystal containing endless neat ranks of atoms. But for the cortex, where local specialization jostles with global integration, C_N is very high. Indeed, Tononi has predicted that this complexity measure will be higher for the billions of neurons working in a conscious brain than for those in a sleeping one, or during the electrical storms that occur within the brain during epilepsy.[23]

BIOLOGICAL NEURONS

Before we can deal with the emergent properties of billions of neurons acting together and thus understand how brains can learn from and adapt to their environments, we need to find out more about the design and function of individual nerve cells. Across the animal kingdom, neurons have a generic appearance, further confirmation—as if it were needed—of the Darwinian thesis that we are descended from common ancestors. Brains, like computers, pass electrical signals along "wires": each mass at the center of a neuron—the soma—receives nerve signal inputs via highly branched extensions called dendrites and sends signals racing along a thick extension called an axon, which can stretch from one end of the body to another (see Fig. 9.2). In general, each cell receives inputs from about ten thousand other cells, typically via three or four contacts but sometimes hundreds with each individual neuron. Conversely, the axon of a single neuron can be branched or unbranched, sprouting as many synaptic "buttons." Each provides the point at which a signal is sent to another cell. The brain is thus a dense jungle of synapses.

Although they have a similar generic structure, neurons come in breathtaking variations that reflect how self-organization through development has given them defined roles.[24] Their appearance can be revealed like "trees in a winter mist" using silver salts, a staining method discovered by chance at the end of the nineteenth century by the Milanese anatomist Camillio Golgi.[25] Because of his skilled use of Golgi's method, the Madrid researcher Santiago Ramón y Cajal (1852–1934) became the father of modern neuroanatomy.[26] The silver stains revealed that the brain was made up of discrete units rather than being a continuous web. Cajal thought these units, the nerve cells, to be the "mysterious butterflies of the soul." As varied as butterflies, they range from the Purkinje cell that looks like a frond of coral to one type of neuron that resembles a rope frayed at each end. There are at least five types of neuron that process light in the retina alone, with the variety steeply increasing within the brain so that there may be hundreds of types overall.[27]

Just as a spiral pattern is an emergent property of the BZ chemistry, so the firing of a nerve cell is an emergent, nonlinear property of its design and interactions with other neurons, which depends on the structure of its surrounding membrane. Recall our discussion of the cross-catalator in Chapter 6, where we saw how a simple nonlinear chemical

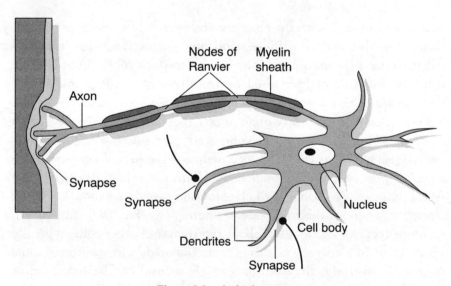

Figure 9.2   A single neuron.

scheme could generate oscillations of chemical activity not dissimilar
from those electrochemical patterns observed during the firing of nerve
cells. The membrane that cloaks each and every cell in the body is electrically charged because of a difference in numbers of electrically charged
atoms—ions—inside and outside; this ensures the cell remains far from
equilibrium, like the crosscatalator.[28] Neurons, unlike other cells, are
able to alter their electrical properties. When a signal is received from
another neuron, the properties of the cell membrane change, allowing
ions to travel across. Sodium ions flood in through large protein-based
ion channels and the membrane potential changes rapidly from -70
thousandths of a volt to +40 thousandths of a volt. Although the local
membrane properties change again to restore the status quo, this triggers similar changes in adjoining membrane so that a spike of electrical activity is sent out. This impulse, called an action potential, ripples
along the length of the nerve. To travel any distance in the body the
action potential must charge up the capacitance of the axon, and overcome the resistance and leakage to the extracellular fluid it encounters
along the way. To this end, the axon is heavily myelinated to reduce its
capacitance and the attenuating action potential is boosted along the
length of the axon by repeater stations called nodes of Ranvier. The
maximum speed the action potential can travel is 100 meters per second—one-millionth of the speed at which an electrical signal moves in
a copper wire inside a computer.[29]

The signal processing that neurons carry out is an emergent property
of these ion channels and the stimulation of other cells. Some signals
boost the probability that the neuron fires; others inhibit it, depending
on the details of the wiring. Axons connect not only to dendrites, but also
to other axons—usually to suppress the signal the other axons send. Dendrites do not just receive signals; they may also send them to neighboring
dendrites or axons. And pieces of a dendrite may affect each other, to veto
competing signals. The output of a neuron depends crucially on the timing and type of signals it receives. The processes occurring at a single
neuron are so complex when studied in detail that many scientists spend
their entire working lives studying only these wonders.

One example is provided by Rodney Douglas, a neurophysiologist
working at the Medical Research Council's Anatomical Neuropharma-
cology Unit in Oxford.[30] In his laboratory, painstaking work has taken
place to map out the myriad connections and responses of just one
among the many neurons from the vision-processing part of the brain;

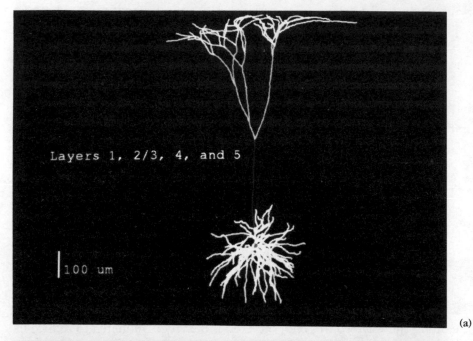

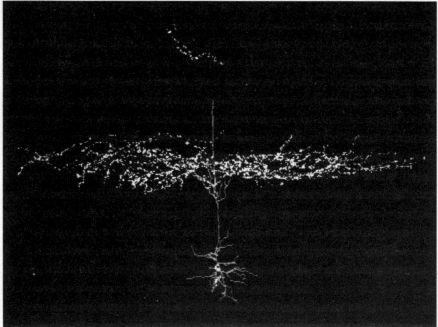

Figure 9.3 Neural architectures. (a) The J4 neuron found in the vision processing centers of the brain. (b) A reconstructed neuron from layer 6 of the visual cortex of a monkey.

that neuron is called J4, after John Anderson, who pioneered its study. J4 has been simulated by countless groups around the world because its overall shape, its behavior, and the way it is wired up to other neurons are highly consistent.[31]

On the screen of a workstation in his laboratory, a three-dimensional image of the J4 neuron looks a bit like an uprooted tree (see Fig. 9.3). Its apical dendrite branches are bent over, corresponding to the surface of the brain, while a thicket of basal dendrites reaches out to the "white matter," a layer a few millimeters below the surface where vast bundles of axons sweep information from one region of the brain to another. In the simulation of J4, the neuron is divided up into around 150 discrete compartments, and differential equations are used to calculate how an electrical signal propagates through its branches, trunk, and roots. A mind-boggling number of inputs are possible: any one or more of 5,000 neurons can feed signals into J4. The way the neuron responds is the result of a war between signals that can range in amplitude from one unit to 10,000 units. A typical simulation reveals the action potential as a ripple of red and yellow passing over the neuron. Another simulation shows how the signal can fizzle out before it reaches the axons. These responses are two of the huge repertoire that J4 uses to amplify signals that originate from the retina.

NEURONS IN SILICO

Armed with such detail, efforts have begun to construct "neuromorphic" devices—hardware implementations that process electrical signals in precisely the way real neurons do. The ambitious synthetic idea underpinning this work is based on the view that to understand the properties of a brain, one must first build a solitary silicon neuron and then construct a silicon brain by connecting 100 billion such neurons. Along the way, it is hoped, it may be possible to build analog devices that mimic biological hardware such as the retina, which is capable of image-processing feats way beyond the ability of the most powerful supercomputers available today.

Carver Mead at Caltech has been a pioneer in this field. His group began by emulating the senses, building an artificial version of the ultimate organ of hearing, the cochlea. His cochlea-on-a-chip can be used to locate an object from the sounds it makes, simulating the way in which barn owls rely on a rustle in the undergrowth to locate a field mouse.[32] An owl infers this spatial information from the time lag between a sound registering in the cochlea of each ear. Mead's group has also built a silicon chip inspired by the neural architecture and function of the retina of the mudpuppy, an amphibian.[33] Unlike a camera's single snapshot of an image, a retina devotes itself to reporting *changes* in the image perceived (see Fig. 9.4).[34] The output of the silicon retina was remarkably similar to the real thing and even generated optical illusions: for instance, a grey square appeared darker when placed against a white background than when against a black background.

The next step was to build a silicon neuron to process information from silicon senses.[35] In 1991, success was reported by Misha Mahowald, a biologist working with Mead, and Rodney Douglas.[36] Douglas, collaborating with Kevan Martin and others in Oxford, had provided the detailed information on the behavior of the J4 neuron we encountered above. Mahowald then developed an analog silicon chip to mimic in every detail the way a single real neuron generates input and output voltages. This in turn meant reproducing the action of tiny protein channels embedded in the membranes of the neuron that control the flow of electrical current into and out of the cell, and hence the voltage across its membrane. The ebb and flow of ions through these channels creates and shapes the output impulses enabling the neuron to communicate with others. Indeed, according to Mahowald, the "neuron itself is an emergent property of the ion channels."[37]

The mimicry is far from trivial to implement: the thin membrane around each neuron contains a vast number of ion channels. Some open like little trapdoors to permit the passage of sodium ions, others to allow potassium through. Some open and close quickly; others act relatively slowly. Different types come into play during the various stages of a nerve impulse. Some channels open and close in response to a change in the voltage across the membrane; others respond to ion and/or messenger molecules. In the silicon neuron the analog response of a channel is represented by a transistor. These channels have to stay shut while their input voltages are low, yet snap open to produce a surge of current as soon as the input voltage reaches a threshold value. The cell's membrane is represented by a capacitor: the fatty molecules in the membrane store charge—acting like a memory—in a similar way to a capacitor in a chip.

Figure 9.4   Artificial sight. How Misha Mahowald appears on the artificial retina.

When it was tested, their artificial neural creation was found to behave in a biologically realistic way. For example, it produced output
voltage spikes at regular intervals until the input voltage was stopped.
This is how a real neuron computes: it converts a constant input signal—the current—into a series of voltage spikes. The silicon neuron
could alter the rate at which it fired voltage spikes; and it was able to
discriminate between different levels of input, adapting its output
accordingly. It is also capable, in principle, of operating a million
times faster than the cells you are using to read this sentence. When
they announced the artificial neuron in the journal *Nature*, Douglas
and Mahowald concluded: "The silicon neuron represents a step towards constructing artificial nervous systems that use more realistic
principles of neural computation than do existing electronic neural
networks."[38]

However, the all-important step of connecting such neurons has
proved difficult. The group has succeeded in wiring thirty-six silicon
neurons together on a single 4mm × 6mm chip. Yet this is only the
very beginning: determining the rules that allow a neuron to talk to
another is one big challenge. Another problem they face is developing

a way to dynamically alter the connections between the neurons during learning. This is a fundamental hurdle. The connections underpin the emergent properties of networks of neurons, which in turn underpin the emergent properties of networks of networks—that is, the global properties of the brain itself.[39] And an entirely separate effort to build a brain by culturing networks of living neurons also has an extremely long way to go to mimic the complexity of the real brain.[40]

NEURAL MESSENGERS

One could be forgiven for thinking that we will never be able to capture the details of the brain's interconnections *in silico*. For the communication medium of the silicon neuron, indeed of any computer, pales in comparison with the complexity of that used within the brain. When an action potential sends a signal across a synapse, the firing of the cell does not consist of electrons racing along a wire but instead of molecules called neurotransmitters that diffuse across the junction between nerve cells. Since the first neurotransmitter was identified in 1921, some fifty more have been labeled.

Dopamine, one of this list, is as good as any for highlighting their importance. Found primarily in a region of the brain called the *substantia nigra*, this chemical messenger is used during cognitive processes, emotional states, walking, and running. As has so often been the case in brain research, its role has been emphasized by individual misfortune. Studies of patients suffering from Parkinson's disease conducted in the 1950s showed that they had unusually low levels of dopamine. A factor in the environment, perhaps combined with a quirk of brain chemistry, can trigger the loss of the nerve cells that manufacture dopamine in the *substantia nigra*. Losses of over 90 percent of these cells start to deprive the patient of mobility and muscle coordination.[41]

The brain's complexity is further increased by the variety of sites where chemical messengers like dopamine act: at protein receptors within the neural cell wall. One recent study provides intriguing evidence that schizophrenia, the most common psychotic illness, is linked to a disorder in the way dopamine acts on receptors.[42] Schizophrenics have normal levels of dopamine in the brain but an excess of sites where it acts.[43] One site, called the D4 receptor, was discovered to be six times more abundant in the brain tissue of schizophrenics than in normal people. For every bombardment of dopamine, schizophrenics received six times as much message. The discovery that dopamine over-stimulates sufferers seems to go along with the symptoms of hallucinations and delusions, though it is unlikely that this finding alone will explain such a complex condition.[44]

WEAVING THE LOOM

Reductionists, among whom may be included many molecular biologists and biochemists, believe in the dictum "God is to be found in the details." Fortunately, we do not have to understand the monstrous detail of neurotransmitters and receptors for an overall grasp of the way the brain works. The most important emergent property of such detailed brain activity is that experience of the real world strengthens and weakens synaptic connections between brain cells, along the lines described by Donald Hebb, whose work we briefly encountered in Chapter 5. Building on the ideas of Eugenio Tanzi[45] and Ramón y Cajal, Hebb proposed that, during learning, synaptic connection strengths would be increased if a neuron fires at the same time as one or more neurons connected to it. Hebb wrote: "The most obvious and I believe much the most probable suggestion concerning the way in which one cell could become more capable of firing another is that synaptic knobs develop and increase the area of contact between [nerve cells]."[46]

In other words, action potentials not only carry signals between neurons, their metabolic wake also alters the circuits over which they are transmitted. Within the brain are thousands of millions of neural connections of varying strength that change in this way with use. If, as a result of stimulation of the senses, interesting things happen simultaneously, frequently, and to neighboring neurons in the brain, these neurons tend to be connected by the network. Because of this *plasticity*, or adaptability, of neural connections, instruction by the world enables the brain to self-organize networks for recognizing objects (e.g., a pen), whether viewed end-on, from the side, or at any other angle. The result of these underlying molecular mechanisms is that the structure of the brain adapts to reflect the connections between events in the real world. The process of creating memories provides a good example of such networks.

By comparing the brain with a pocket calculator, one can easily mis-represent the brain's stunning memory abilities. All it takes is a simple exercise in which individuals are shown a number for a brief time and then asked to recite it a few minutes later. It is easy with a four-digit number but trickier when the number of digits is extended to eight, which is the limit for most. On this basis the typical human memory possesses a mere 41.86 bits of information, while the memory of a sim-ple pocket calculator is about 1,000. It is a surprise that we can re-member anything at all. Yet a child learns new words at an average rate of more than ten each day and may eventually build a vocabulary of up to 100,000 words. The best storytellers among the ancient Celts knew some 350 epics by heart, and the neurologist Oliver Sacks once treated a patient who had memorized all nine volumes of the 1954 edi-tion of *Grove's Dictionary of Music and Musicians*, not to mention the music of 2,000 operas.[47] There is less mystery about these abilities than one would expect.

We remember things on the basis of meaning—that is, through as-sociation, a process described in Chapter 5. To act as a "memory," a network of neurons must be able to "recall" a specific pattern of electri-cal activity on demand. Suppose that each neuron in the cluster is con-nected to all the others by synapses that can be strengthened according to Hebb's rule: strengthen connections with those neurons you are connected to if they are active at the same time as you. (In fact, things are a little more complicated than this, as rival networks compete to store the memory.) Once all the synapses in the most successful network have put the rule into effect, the memory has been stored. Neurons that were once active together are now linked by stronger synapses; during recall, they will tend to fire simultaneously and help to re-create the original pattern. A neural network of this type could cradle many memories simultaneously, with each synapse taking part in several memories with each memory being "encoded" by numerous synapses. Now we can understand how such a network's powers of recall can be so spectacular. If prompted with only a small fragment of a memory, its synapses would ensure that it regenerates that memory in its en-tirety—a phenomenon known as completion and closely related to the inorganic concepts of information storage and retrieval.

Hebb's suggestion was a theoretical explanation of how associative memory works at the level of neurons. Supporting experimental evi-dence came in 1973 from work by Tim Bliss at the National Institute for Medical Research, London, and Terje Lømo, at the University of Oslo. They used microelectrodes to study electrical activity in the brains of anesthetized rabbits, focusing on one region of a seahorse-shaped structure called the hippocampus. Impulses enter the region, called the dentate gyrus, through an array of input fibers and then spread through synapses to a network of neurons nearby. Bliss and Lømo found that sending a brief volley of artificial impulses along these input fibers increases the strengths of the synapses.[48] Once these changes have been triggered, they can last for weeks or even months— enduring enough to make the process a likely candidate for the physical changes that underpin the formation of recent memories. This process of strengthening connections is commonly called long-term potentiation.

Although the Bliss and Lømo experiments were somewhat artificial, they behaved in a pleasingly Hebbian manner, only taking place at synapses connecting neurons that are both simultaneously active. And they demonstrated a cellular mimic of classical conditioning, the variety observed in Pavlov's pioneering experiments on dogs, which learned to salivate in response to a bell regularly rung just before they were fed. (In the psychological jargon, they had become *conditioned* to *associate* a response with a specific stimulus.) Bliss found that weakly activated synapses do not become potentiated. But—and this is where the analogy to Pavlov's experiments becomes so striking—later workers showed that if the weakly stimulated synapses are active at the same time as another strongly activated input, then the weak input does become potentiated.[49]

Subsequent work has supported the significance of such cellular changes for the learning process; if long-term potentiation is blocked pharmacologically, then rats have difficulty in learning to navigate mazes, a task known to require the hippocampus.[50] Rats are not the only creature helping with this research. Some scientists hunt for the genetic basis of memory in the humble fruit fly (with dim-witted strains such as *Dunce* and *Rutabaga*) and others specialize in studies of the sea slug, which has conveniently oversized brain cells. At the Open University, a team led by Steven Rose studies the chick.[51]

Chicks are not very smart but offer various advantages nonetheless: they start learning as soon as they stumble out of the egg, they learn reliably, and they have a large and accessible brain. Most important, they can be shown to memorize by a simple test. One peck of a chrome bead coated with a bitter substance and the bird rejects similar beads: it has memorized that the chrome beads taste awful. The bitter taste triggers changes in the brain to lay down the memory. These changes involve the shapes and sizes of nerve cells and, most important, the connections between cells. As a result of such an experiment, dendrites in certain key neurons sprout 60 percent more connections in a manner predicted by Hebb. This is learning. Where the chick once had an urge to peck the bright bead—its normal response—the new web of connections formed in the brain ensures that this is shut off. "It spells out: No, don't peck it."

While memory is important to us, so, too, is forgetting. Luria's shocking story about a man whose brain never forgets anything and who ends up overburdened with memories of all he has ever done, is a compelling reminder of this fact.[52] So it seems that there must be some mechanism analogous to that discovered by Bliss and Lømo, which is responsible for *reducing* synaptic connection strengths. This is known as long-term depression. One might expect to induce such forgetfulness by arranging an anti-Hebbian correlation between pre- and post-synaptic activation states.[53] By this is meant a diminution of synaptic connection strengths when two connected neurons fire concurrently. Such phenomena have been observed in certain parts of the brain, including most notably a structure called the *cerebellum*.

AMMON'S HORN

Most neuroscientists—experimentalists as well as theorists—use Hebb's rule as an idealization of synaptic reinforcement during learning. As we have seen in Chapter 5, Hopfield demonstrated the power of this rule in his artificial neural network models for associative memory. The problem facing neuroscientists is determining whether Hebb's rule and Hopfield's net are indeed unifying principles or not. Further complicating matters is the fact that memory is a diverse collection of talents, each depending on different learning mechanisms and involving different brain areas.

To appreciate the distinctions that can be made within memory, imagine that you are talking with a colleague about the day you first learned to drive a car. Your conversation involves *semantic* memory, your knowledge about language and the world, including the concept of a car and what you know about cars and how to drive them. Second, it relies on *episodic* memory, your memory of the day itself, including the near miss with a pedestrian, or the feeling of elation when you succeeded in a three-point turn. Researchers sometimes group episodic and semantic together as *declarative* memory—that is, memory we can bring to mind and reflect upon. Your conversation is about another type of memory: the skill of driving a car. Skills are classified as part of *nondeclarative* memory along with other phenomena, such as classical conditioning. And your conversation also depends on another brand of memory: *working* memory, which enables us to hold material in our heads for long enough to build and understand complex sentences. This type of short-term memory would also come into play if you were to read off the serial number of the car chassis, or repeat your colleague's phone number in order to write it down.

The hippocampus plays a central role in various aspects of memory. Its role was first suggested by studies of patients whose hippocampus had been damaged. From such work, it seems probable that the hippocampus acts as a buffer store, relaying short-term memories to parts of the cortex in order to lay down a long-term memory during sleep.[54] Indeed, it was proposed as long ago as 1971 by David Marr, in a remarkable paper utilizing many ideas drawn from computer science, that the hippocampus is an associative memory store, consolidating input data from various locations prior to passing on fully formed memories to the cortex.[55]

Armed with modern theories of neural network function, Edmund Rolls of the University of Oxford hunted for likely candidates for memory networks in the hippocampus. Intense study of the neural architecture of one part has yielded particularly promising results. This region is called CA3, where CA refers to *Cornu Ammonis* (Ammon's horn), the archaic name for the hippocampus. In rats each CA3 neuron receives inputs from 12,000 other CA3 neurons. For example, the components of one recollection, forced onto different CA3 cells, might be what the rat ate for lunch yesterday, where, and with which fellow rats. The CA3 cells would then link these components to form an episodic memory, the whole of which could be recalled by any of the parts.[56]

A close correspondence exists between the architecture of the Hopfield net and the structure of the hippocampus.[57] Indeed, it seems that the axons of neurons within the CA3 region have such an extended reach in all directions that they can potentially make connections to almost all other CA3 neurons. (Recall that in the Hopfield net, the neurons are completely connected to each other, regardless of distance.) Building on anatomical work, Rolls collaborated with physicist Alessandro Treves to develop a mathematical model of the CA3 region along the lines of a Hopfield net. The results are fascinating. They show that the connectivity within the CA3 region does not have to be complete for associations to be set up: a 5 percent chance that any one neuron is connected to another is good enough to provide memory. The analysis revealed that the most important parameter affecting memory is the number of connections per neuron from other neurons in the network. This underlines the role of the hippocampus in episodic memory, when it must be capable of linking various items within a memory—for instance, the bouquet of a white burgundy with a particular lunch. That means widespread connectivity: it is no good having CA3 divided into separate blocks because that would mean you could never associate the bouquet in region A with the lunch in region B.[58]

Treves and Rolls' analysis showed that the more sparse the representation—that is, the smaller the proportion of neurons firing at any one time—the more memories can be stored and recalled. Intuitively, this can be understood by accepting that the brain must make efficient use of its synapses and not invest too many per memory. Developing this kind of analysis led Rolls to estimate that the CA3 region of the rat hippocampus could store 36,000 "memories."[59] Though appealing, such approximate mathematical analysis of the complex web of connections in CA3 does not *ipso facto* provide proof that a cluster of neurons functions as a memory network.

To liberate themselves from the necessity of making mathematical approximations, Rolls and Treves have started to simulate the CA3 on a workstation. Using their understanding of how the structure operates, Treves and Rolls employed Hopfield-type networks to investigate how the hippocampus acts as a memory buffer store.[60] First, however, they had to generalize the Hopfield model to make it more complex and thus more biologically plausible. One problem with Hopfield nets is that they have symmetrical interactions between neurons; an afferent (firing) neuron is affected in the same way as an efferent (receiving) one by the firing process. This is biologically unrealistic.

There are other deviations from the real thing in a Hopfield net. In

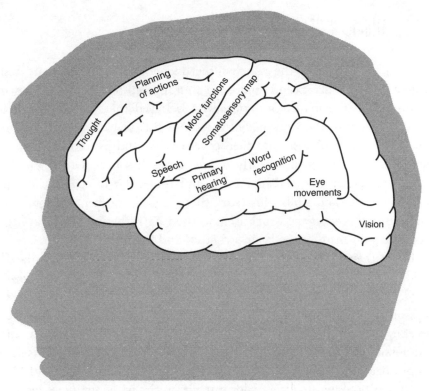

Figure 9.5   How tasks are divided among different regions of the human brain.

the net, neurons are usually represented as binary nodes (either on or off), while in the hippocampus neurons have continuously variable firing rates and only about 1 percent of them are doing anything meaningful at any instant. To make the Hopfield model more realistic, the Oxford team introduced variable firing rates, asymmetric connections, and also reduced the connectivity of the Hopfield network, so that each neuron is no longer connected to every other. Despite these alterations, the simulations show that the network preserves the ability to store and recall information. It can create attractors, the valleys where memories reside. The work suggests that from just 5 percent of an input—say, a fleeting glimpse of a face—we can reconstruct an entire memory. So that their ideas about recall can be put to the test, Treves and Rolls have even made some predictions of how the hippocampus should behave in some hitherto unobserved circumstances, for instance, when drugs are used selectively to block synapses.[61]

COMPUTATIONAL NEUROSCIENCE

For over two centuries, the principal methods for studying the brain and animal behavior had been founded on an approach originally laid down by philosophers such as Descartes, who believed that the only way to gain insight into the workings of the brain was via introspection. Following this, neurologists and then psychiatrists and psychologists took up the challenge; most recently neuroscience has been largely in the hands of biochemists, cell biologists, and molecular biologists. As the work of Treves and Rolls illustrates, there is now another way forward, *computational neuroscience*, which is concerned with the use of neural networks to gain insight into brain structure and function.

The surge of interest in computational neuroscience was stimulated in large part by the publication of Rumelhart and McClelland's *Parallel Distributed Processing* in 1986, a two-volume edited collection of essays on neural computing. Centered around the work carried out over several years by the Parallel Distributed Processing (PDP) group at the University of California at San Diego, it became an academic bestseller on publication. Although, as we saw in Chapter 5, Rumelhart and McClelland's book has made a major impact on the wider artificial intelligence community, their overriding aim has been to account for neural information processing through the so-called connectionist paradigm— in other words, using the principles of neural network computing. Some biologists are wary of the lack of biological realism of these endeavors.[62] But they have missed a key point: throughout this book, we have shown how highly complex behavior can result from simple rules acting on large collections of simple units. To understand something as complex as the emergent properties of the brain, we must start off with simplified systems in order to distill its key features and only then build in the fearsome details of neuroanatomy.

Artificial neural networks mimic the brain's complexity through their nonlinearity and a high degree of interconnectivity among their nodes. Like the brain, they are inherently parallel devices that can do many things concurrently. Qualitatively, their processors behave like neurons and the connections between them act like synapses. Any "programming" of such a net consists of rules to alter the all-important strengths of connections between processors. And the "programs" that solve various problems emerge within such networks spontaneously.

There are, however, important differences between the many types of artificial neural networks in existence today. Some are more biologically realistic than others, depending on their architecture and the methods they employ to alter connections between nodes.

Rumelhart and McClelland's multilayer perceptrons, discussed in Chapter 5, are composed of feedforward connections between distinct layers of neurons, and a supervised learning algorithm that relies on a process called the backpropagation of errors, in effect a "teacher." The multilayer perceptron network is trained on data—for instance, to distinguish round geometrical shapes from square ones—and its performance is continually corrected using the backpropagation algorithm until it reaches a satisfactory performance. It is then in a state capable of "generalizing" so that, for example, it will class a football as a roundish shape and a house as a squarish one. The concept of supervised learning from a teacher is one that does exist in nature—it is similar to how we are educated at school. However, backpropagation *per se* does not occur in real brains, nor are most real neural networks layered in this way.

On the other hand, as we have already described, John Hopfield's recurrent network, which was born out of the theory of spin glasses, and in which every neuron is connected to every other, *does* have a ring of limited biological authenticity, since real neurons are also highly interconnected over extended regions in some parts of the brain. These networks are usually involved in supervised learning activities to ensure that the error landscape (Chapter 5) is appropriate to the task at hand.

There is, however, another kind of neural net we have not yet discussed, but that throws further light on the processes within the brain. This is a neural network in which the emphasis shifts from supervised learning to finding out through self-teaching. Teuvo Kohonen, a computer scientist at the University of Helsinki in Finland, used these kinds of networks to investigate *unsupervised* learning. No correct answer to the problem at hand is provided to the network; the computer algorithm causes the network to self-organize in solving a problem as information is taken in from the external world. This is thought to be a possible way in which various networks form within the brain to handle different types of tasks, such as vision, motor control, olfaction (smell), and so on. Such self-organization is often referred to by biologists as *localization* of brain function. Self-organizing computer networks were the main focus of Kohonen's research long before the enormous growth of interest in artificial neural nets that occurred in the mid-1980s and have raised the profile of his work. Kohonen's approach throws light on the processes involved in the self-organization of the brain itself. It is worth discussing some aspects of what is known of brain localization before describing Kohonen's computational contribution.

CEREBRAL CARTOGRAPHY

Studying how the brain dedicates different regions to different jobs is common to much of brain research: anatomists examine the structures of the brain and psychologists study its response to various types of illusion; pathologists may pore over a microscopic slice of brain tissue, while physiologists examine its many and varied mechanisms; molecular biologists watch how genes are turned on in different structures— all seek to pin down the brain's functions to particular locations.

What we now understand as localization was elaborated in Paris in 1861, when the French physician Pierre Paul Broca described to the Société d'Anthropologie a study of a patient called "Tan" because that was the only syllable he ever uttered to hospital workers. An autopsy on Tan revealed a cavity large enough "to accommodate an egg" in his left frontal lobe. Broca concluded: "Everything allows us then to believe that, in the present case, the lesion in the frontal lobe had been the cause of the loss of speech."[63]

Cerebral cartography thrived on such misfortune through studies of people unlucky enough to lose or damage parts of their brains as a result of surgery, disease, or a blow to the head. Researchers investigating the localization of memory were fascinated by H.M., an American assembly-line worker, who had surgery in 1953 to relieve him of epilepsy. During the operation surgeons removed the hippocampal region, together with nearby tissue, from deep inside his brain. As a consequence, H.M. suffered a severe loss of long-term memory, though his so-called "short-term" memory, defined as his recollection of events occurring within about one minute, was normal.[64]

Over the years, it has become clear that the grand syntheses of mental life take place in the corrugated surface of the two symmetrical massive cerebral hemispheres that dominate the brain's appearance (see Fig. 9.5). This cell-rich surface is only two millimeters thick and is given the name *cortex*, after the Latin for bark. There is a surprising

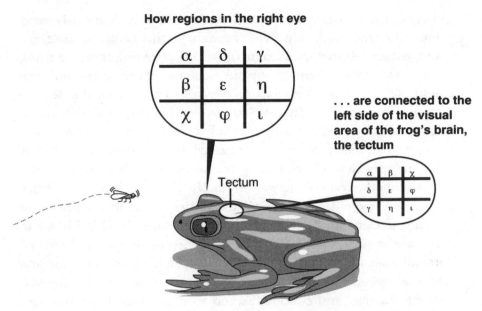

How regions in the right eye

α	δ	γ
β	ε	η
χ	φ	ι

. . . are connected to the left side of the visual area of the frog's brain, the tectum

α	β	χ
δ	ε	φ
γ	η	ι

Tectum

Figure 9.6   Mapping the eye and its visual fields to the tectum, part of the frog's brain.

amount of room for cerebral activity. If its convoluted surface were flattened, each hemisphere of the cortex would have the surface area of a large napkin. Vision is processed in up to one-third of the thin, wrinkly rind at the back of the cortex. An area in the left hemisphere deals with language. The frontal lobes play a role in the control of attention, planning, and social skills. Tucked under the back of the cerebral hemispheres is the *cerebellum*, a structure concerned, among other things, with the coordination of fine movements. Close by to the hippocampus lies the *amygdala* responsible for emotional memory (e.g., the ability to recognize fear in a facial expression).[65] The *thalamus* is located at the top of the brain stem (there is one for each brain hemisphere); known as the gateway to the neocortex, it is responsible for coordinating all sensory input to and output from the cerebral cortex, except smell. The *hypothalamus* is situated beneath the thalamus and regulates hunger, thirst, pleasure, and pain.

For many years, controversy raged among neurophysiologists about whether individual neurons in the brain are intrinsically coded— perhaps genetically—to perform specific, localized jobs, or whether the function of individual cells arises from spontaneous self-organization. The former approach is that adopted by proponents of the "Grandmother cell" hypothesis—the notion that there is a specific cell whose job it is to recognize grandmother, along with an unbelievably large number of other such cells. Yet throughout this book, we have encountered endless examples of emergent collective properties resulting from interactions between many simple units under nonequilibrium conditions. The kind of interactions described by Turing that lead to structure in a Belousov-Zhabotinski reaction can also produce patterns of connections in the brain. In view of our knowledge of how neural networks function, it seems highly probable that localization emerges spontaneously in a self-organizing neural network, without the need for grandmother cells. This was originally glimpsed in 1973 in work by Christoph von der Malsburg, then at the Max Planck Institute in Göttingen, Germany.[66] He later collaborated with David Willshaw in Edinburgh to develop computer-based models of neurons in the visual cortex.[67] They focused on connections between the retina of a frog and its visual center, called the tectum. What they were investigating was how the neurons in the retina map on to neurons in the tectum (see Fig. 9.6). What they discovered was an astonishing degree of flexibility.

Fascinating insights into how self-organization controls the mapping of the senses to the brain have come from the frog. If, for example, we remove half a retina at a particular time in its development, the other half will spread its neural connections across the tectum to compensate. But if we remove half the *tectum*, the entire retina will map to the remaining half of the tectum. This spontaneous ability to develop a "topographic map" is found for all the senses: for instance, the local regions of the brain that receive signals from neighboring fingers of the hand will rapidly reorganize, even in the adult animal, if a nerve linking a digit is cut.[68] If a monkey is trained to discriminate a rough surface with a specific fingertip, the brain's map of that digit will spread accordingly. At the cellular level, neurons deprived of an input will manufacture a wide range of receptors, as they hunt for new neurotransmitter inputs.

Willshaw and von der Malsburg developed "top-down" artificial neural network-type algorithms that displayed similar flexible behavior to the self-organizing cortex. This phenomenon of self-organization has been further investigated in the work of Kohonen, who effectively used a simplified version of Willshaw and von der Malsburg's algorithm.[69]

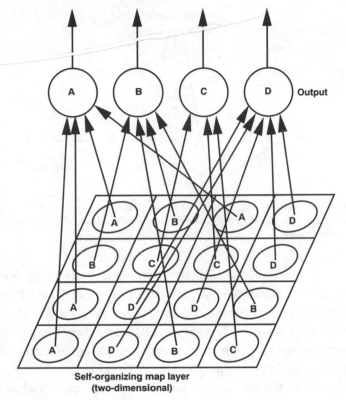

Figure 9.7   Architecture typical of a self-organizing Kohonen neural network. Circles represent neurons; arrows, the connections.

SELF-ORGANIZING BRAIN MAPS

Kohonen's self-organizing neural network differs from the multilayer perceptron and Hopfield nets because it is a single layered, two-dimensional collection of neurons; each input neuron is completely connected to the neurons in this layer (see Fig. 9.7). When taught a task, such as converting speech into text, it automatically generates a feature map within the network.

The network's key ingredient for self-organization is the same as that for spatial organization in the Raleigh-Bénard cell and the Belousov-Zhabotinski chemical reaction: nonlinear feedback.[70] However, while molecules in the cell or a BZ reaction interact with their neighbors,

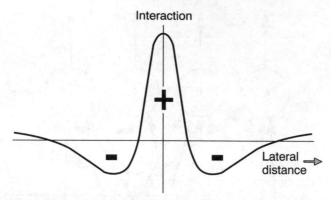

Figure 9.8   The Mexican hat function. The excitatory (+) and inhibitory (-) activation regions are shown as a function of distance from a given neuron. This pattern is typical in mammalian brains.

neurons can interact with each other over huge distances. Kohonen's network does not have the extreme connectivity of Hopfield nets: the feedback is restricted to interactions between processors within a small distance of each other in the two-dimensional layer. The nodes in this layer generate output information, and are also involved in establishing the self-organized mapping of the input data.

Biology provided the inspiration for the rules used by Kohonen to ensure that his networks could learn without supervision. In fact, a rather close analogy exists between his inspiration and the work discussed in Chapter 7, concerned with morphogenesis in hydra. Just as in that work, the key to the development of self-organized neural structures lies in a combination of excitatory and inhibitory influences, in this case the connections between neurons. Those close to an active (firing) cell are themselves encouraged to fire, while ones further away are actually inhibited.

Kohonen justified his choice of architecture and learning rules because most neural networks in the brain, especially in the cortex, are essentially two-dimensional layers of processing units in which the units are densely interconnected through lateral feedback. To account for the connectivity of his model, he pointed out that, as we noted earlier, in the neocortex there are some 10,000 interconnections emerging from and converging on every principal cell. And there is experimental evidence—for instance, from studies of primates—of those all-important competing interactions: there are short-range lateral excitations reaching up to a radius of 50 to 100 microns. This excitatory area is then surrounded by a penumbra of inhibitory action reaching up to a radius of 200 to 500 microns. This variation of nerve cell activation with distance follows the form of a Mexican hat (see Fig. 9.8).[71]

The kind of localization of function seen in real brains emerges naturally in this model. Kohonen's networks have since been used in many ways; he himself was primarily interested in building a "phonetic typewriter," an electronic word-processor that could produce text from dictation in real time. This represents a particularly difficult problem in speech recognition. As we mentioned in Chapter 5, simulating natural language processing is tough. In addition to the basic interpretation of sound waveforms, as human beings we are effortlessly aware that speech recognition involves a whole raft of other aspects, including context, syntax, parsing, and inference. Our hearing apparatus contains clever ways to filter out speech in noisy environments such as cocktail parties.[72] Speech differs from one person to another, and a great deal of effort has been expended in trying to address all this complexity using conventional methods that rely on specific models.

Kohonen's solution combines the best available solutions for the processing of speech. Only the unsupervised recognition of units of sound—*phonemes*—is performed by his neural network. After training on a range of speech, the performance of the "typewriter" is between 92 to 97 percent correct dictation (in Finnish, Kohonen's predominantly phonetic native tongue).[73] It was further adapted to cope with speakers who did not help to train the network. Kohonen developed a variant of the self-organizing feature map that fine-tuned a trained feature map.[74] Such tweaking of the network to allow for a new speaker requires around 100 words and is complete within ten minutes.

This work has been highly influential in raising interest in neural networks for the analysis of complex problems, both within and outside biology. The Adaptive Resonance Theory networks we described in Chapter 5, which overcome the stability-plasticity dilemma, are an extension of Kohonen's concept of a self-organizing network.

BRAINY ALGORITHMS

Of course, these "unsupervised" neural networks also have their drawbacks. Many lack multiple layers, as seen in the real thing. Worse still, they can be slow and imprecise.[75] According to Geoffrey Hinton, a computer scientist at the University of Toronto, everything we know about the cycle time of neurons that fire when we discern depth, a whisper in a noisy room, or a hidden agenda in a political statement, suggests that current artificial networks are too sluggish.[76] It is likely that, in general, biological networks use learning algorithms that are a combination of unsupervised and supervised methods. The unsupervised element enables the broadest features of a given set of data to be discerned; supervised learning hones the network's structure through determining which features are the important ones that must be handled.[77]

In the long term, however, we will probably need new algorithms. Hinton is heartened by one aspect of neurobiology: the most interesting part of the brain, the cortex, has a *homogeneous* structure, unlike more primitive centers. This suggests that the rules governing the operation of the cortex are similar, whether it is processing speech, language, or social skills. "My bet is that there are one or two algorithms in there," said Hinton. "One for building models of the world and another for associating responses with those models." He speculates that perhaps puzzling cerebral activities, such as dreaming, could be a natural consequence of their importance for building up brain structure and function. Hinton is confident that a confluence of algorithm development and empirical data will crack the problem in the next few decades. This effort will thrive on the dialogue between those who think that the brain will only succumb to deep, fundamental, and generic principles and those who want to document only the *minutiae* of its operation.

NETWORKS FOR VISION

Fundamental understanding of information processing has already helped Hinton to come up with one candidate algorithm for vision. He is considering a neural network that takes raw data about the world—say, an image on a retina—and boils it down to a minimal representation for the brain that can still be used for re-creating the original image. It is akin to data compression, or the kind of signal processing used by TV broadcasters to squeeze more stations into the same bandwidth: information is only transmitted on the *changing* parts of an image so that it can be described in a more compact way. This is similar to the work of

Mead's group on artificial retina. Hinton has developed an algorithm for a self-organizing network based on this approach that uses information about the expected error in the compressed representation as a kind of internal benchmark to judge how successful that representation is. He calls the resulting network the Helmholtz engine, after the German physiologist and pioneer of thermodynamics, Hermann von Helmholtz (1821–94); it is the successor to his Boltzmann machine we encountered in Chapter 5 and builds on work by Ralph Linsker at IBM that has already shown some success in modeling vision. By coupling two Helmholtz networks together, Hinton has carried out a fascinating study that suggests how backward connections in the brain generate dreams and fantasies which help train the brain to recognize images.[78]

The visual plasticity of the brain is greatest just after birth, as experiments on cats have shown. Born blind, they open their eyes after seven days; over the following three weeks of life their vision is perfected by a dialogue between the visual world and the brain.[79] As David Hubel and Torsten Wiesel first discovered in their Nobel Prize–winning work in the 1960s, most neurons in the primary visual area of the cortex in adult cats (and monkeys) respond when an edge of particular orientation appears in the visual field. Different cells respond to different orientations, as if the whole population is decomposing the image to represent the shapes of objects on the basis of their component orientations. However, when a kitten's eyes first open, only about one-fifth of these cortical nerve cells are already selective for edges, and most of them respond best to horizontal or vertical orientations. Presumably, the connections that give these prespecified cells this property are laid down before the eyes respond to light, either on the basis of strict genetic instructions, or through self-organization stimulated by spontaneous activity reaching the cortex from the retina.

After the eyes open, the prespecified cells, which tend to occur in clusters in the middle layers of the cortex, may act as "teachers," according to Colin Blakemore of Oxford University.[80] Within a couple of weeks, nearly all the cells respond selectively to orientation and form a system of "columns": all the cells above and below each prespecified cluster coming to prefer the same orientation. In this case, the supervision would not depend on backpropagation but on the provision by the prespecified cells of a "conditioning signal," enabling the initially unspecified cell to acquire the same selectivity for orientation by detecting coincidences between the activity arriving from the retina and the conditioning signal reaching them simultaneously from the pre-specified cells.

Early use of the eyes is crucial to the cortex's development. The influence of visual experience is clearly demonstrated by the fact that the instruction process does not occur in an animal deprived of vision and can be biased by changing the visual "diet." Colin Blakemore and his colleagues showed twenty-five years ago that kittens exposed to a visual field comprising only vertical black and white stripes early in life subsequently had difficulty in seeing horizontal lines, and this was explained by the fact that the majority of cells in the primary cortex had acquired a preference for vertical orientations.[81] If vision is restricted to one eye alone, even for just a day or two, during a crucial month-long period of early development, the other eye becomes dysfunctional because most cells in the cortex stop responding entirely to stimulation of that eye. These permanent changes in the visual cortex are a form of memory: indeed, they may utilize the same molecular and cellular mechanisms of long-term potentiation seen in the hippocampus.[82]

There is, however, a difference between the memory-based approach to processing images and memory itself: while the latter uses a neural network to recall an input, the neural networks of the visual cortex rely on this form of memory—early plasticity—to learn how to turn the retinal image into a representation. One set of brain cells could, for example, "remember" the component of upward motion of a contour. With a real object, say, an aircraft, the upward motion of the wing's flat surface would often be recognized in association with the upward motion of the vertical tailplane. Sensitivities to corners, round wheels, and cylindrical engines would gradually emerge to build up a representation of the object in the cortex.[83]

This kind of "memory" effect also occurs for the other senses, and in a wide range of animals. Monkeys are born with more prespecified training cells in the primary cortex than cats, so that they arrive in the world with more matured vision, but are also dependent on early visual stimulation. This suggests that "active" rewiring of the visual cortex, when the animal is conscious, aroused, and attempting to make sense of the world, is as important as the passive rewiring that takes place within the womb. Such activity-dependent self-organization, as it is called, seems to confirm that there are general principles governing synaptic strengths, such as those put forward by Hebb.[84]

To get a handle on how the cortex self-organizes by such memory effects so that it can process vision, Ralph Linsker of IBM's laboratory at Yorktown Heights, New Jersey, developed a feedforward network consisting of several layers, each composed of hundreds to thousands of artificial neurons. By comparison, the visual system of the brain consists of many millions of neurons.

Each neuron communicated with hundreds of others, mostly its neighbors. Starting with surprisingly few but biologically plausible rules that governed how neurons in a network were connected, such as Hebb-type modification, Linsker watched these connections evolve as the network optimized its overall properties, its "sight." He wanted to find out whether the network developed properties that were biologically important for the analysis of features in the patterns of light cast on a retina. The answer was "Yes": the network self-organized from one with connections of random strength into one whose connections gave it properties that bear a resemblance to the visual cortex, as the neurons took on specific jobs.[85] This occurred even when there was no visual input.

Recent artificial neural net studies have suggested possible reasons for aesthetic appeal, whether it be the allure of a gemstone, the legs of Marilyn Monroe, or a painting by Claude Monet. Work by Anthony Arak and Magnus Enquist suggests that the reason we find something appealing is a side effect of the way our brains handle the everyday problem of pattern recognition.[86] In the case of a neural network designed to represent the visual recognition system of a female bird, Arak and Enquist found that this artificial "brain" responded more strongly to certain novel images of male birds than to the male birds it had been trained to recognize. In particular, the female brain responded strongly to males with longer tails than those seen during training. Moreover, when the male tail length and the female visual system were allowed to evolve by mutation, progressively longer tails were preferred by the female, with a decline in responsiveness to the original tail length. The female preference for an exaggerated trait (in this case, long tails) could drive the evolutionary growth of the latter until a serious deterioration in male survival fitness becomes important. These findings support Darwin's original view, quoted by Arak and Enquist, that "When we behold a male bird elaborately displaying his graceful plumes or splendid colours . . . it is impossible to doubt that [the female] admires the beauty of her partner."[87] Extrapolating somewhat, this research suggests that innate preferences, whether in personal relationships or in the arts, may be an indirect result of the way the brain has evolved to interpret sensory information.

Indeed, our disdain for the irregular, distorted, and lopsided seems to be inherent in the way our brains recognize patterns. We may at last understand why it is that we love the symmetry of a snowflake, a beautiful face, or William Blake's Tyger, with its "fearful symmetry." Subsequent work by Arak and Enquist, complemented by independent work by Rufus Johnstone,[88] has shown that neural networks have an inherent preference for symmetry when trained to recognize visual patterns because symmetrical patterns are easier to ascertain from a variety of viewing angles—think of a sphere compared with a cube. These findings are corroborated by the discovery that our own love of symmetry is shared by other creatures—for instance, crows and monkeys.[89]

INTEGRATION

There are other examples of how our senses can be understood on the basis of artificial neural network simulations. However, we are still faced with the important but difficult question of how these networks interact with one another; specifically, the roles of integration *versus* specialization need resolving, along with the associated problem of net-net synchronization. We touched on this issue earlier in this chapter in our discussion of a particular, though restricted, measure of neural complexity developed at the Neurosciences Institute in La Jolla that attempts to express and quantify the subtle link between local and global functions.

There are centers in the brain that can recognize a face, while others detect movement, colors, and expressions. How do we reconcile the existence of a unified mental scene and, ultimately, the unity of consciousness with the astonishing specialization of the brain, often called the *binding problem*? This may sound esoteric but it is important in conditions such as schizophrenia, which occur when the process breaks down.[90] The same group in La Jolla—Olaf Sporns, Leif Finkel, Giulio Tononi, and American Nobel laureate Gerald Edelman—focused on binding in the visual cortex. For example, when we gaze at a red picket fence, how do the cells within the cortex that register the vertical orientation of the fence posts know that it is the selfsame stimulus (the fence) that makes other cells register the color red?

The team drew on the work of Charles Gray and Wolf Singer of the Max Planck Institut für Hirnforschung in Frankfurt who had found high degrees of synchrony of neural activity in studying the primary visual cortex of the cat.[91] Gray and Singer's experiments suggested that different processes in the brain were bound together by the fine temporal structure of neural activity. This inspired a bottom-up model by Sporns and his colleagues on a supercomputer that exploited the temporal properties of discharges between networks of about 200,000 artificial neurons. The neurons were arranged in three separate streams for form, color, and motion, analogous to those of the mammalian visual system. To process images from a video camera, the team connected the units in a biologically plausible way via several million connections, most of them arranged to link individual visual maps in a reciprocal fashion.[92] From this and subsequent models that integrated up to nine cortical areas there emerged a dynamic alternative to the traditional idea of a static grandmother binding cell. The group used this approach to segregate a moving figure from various backgrounds and applied it to give an account of visual illusions and *Gestalt* phenomena: an image of shapes and symbols that appears meaningless when viewed in close up can reveal a face, cube, or pattern when we stand back and see the image in its entirety.[93]

Using a more idealized network, John Taylor of Kings College, London, has been modeling a part of the brain, called the *nucleus reticularis thalami* or NRT, which acts as the playground for competition between many distinct activities in separate cortical areas.[94] Taylor envisages it as a gateway linking primitive centers that govern emotion, as well as inputs through the eyes and ears, with the cortex, the outer layer of the brain responsible for memory, language, thought, and intellect. In his neural caricature, Taylor has mimicked this process by allowing competition between different activities in an artificial network of inhibitory neurons. What emerges is a single wave of electrical activity across the net that, he claims, provides global correlation of cortical activity. Similar waves of activity have been observed in vivo using magnetoencephalography, a brain-scanning technique we discuss in the Appendix. "These fit exactly with what I would expect from my model," says Taylor.[95]

The jury is still out on the significance of this work, though it does complement in some ways Edelman's Darwinian model of thought processes, in which ideas compete for "workspace" within the brain. Perceptions of the thinker's current environment and memories of past environments may bias that competition and shape an emerging thought.[96] We should not forget, however, that no one has yet succeeded in providing a plausible description of such higher-level cognitive functions as awareness—the basis of consciousness—let alone the multitude of emotional states such as happiness, pleasure, pain, and sadness.

NEURAL NETS AND DAMAGED BRAINS

Artificial neural networks have further demonstrated their realism by providing deeper understanding of the effects of brain damage. Tim Shallice of University College, London, working with Geoffrey Hinton and David Plaut, used a neural network to model how the damage resulting from a stroke can lead to visual errors and difficulty with certain abstract words that superficially appear to present a random collection of behaviors.[97] Efforts to retrain the neural network after damage show that some strategies are better than others. Doctors can now begin to try rehabilitation procedures based on this conceptual framework.[98]

Neural networks have also been used to model a form of amnesia called *prosopagnosia*, caused by lesions with the boundaries between the occipital and temporal regions of the cortex. A sufferer loses the ability to recognize the faces of friends and family, even a photograph of his or her own face. Such patients often have difficulties in distinguishing between individual members of a given class of objects. For example, they can usually classify objects such as cars, dogs, cats, and kettles correctly while being unable to identify individuals within these classes. Since prosopagnosia affects only the visual recognition of faces, and memories of other classes of objects, it may still be possible for the afflicted individual to identify the person or object from other cues, such as posture or gait, or by using another sense, such as a telltale sound. A sufferer may not be able to recognize the face of his pet cat, Pushkin, but if he hears her plaintive meow he may immediately know the beast.

As ever, these unfortunate patients help reveal how information on individuals and objects is stored within the brain. Are there many representations of a single object, each derived from a different sense, whether smell, sound, or sight? Perhaps one abstract representation can be accessed in numerous ways via different senses? This idea can be readily grasped from the understanding we gained in Chapter 5 of the way in which recurrent neural networks act. Inputs from the different senses may all converge on some global attracting network state that represents Pushkin. The global nature of this state means that it is very likely to be spatially highly distributed—that is, the neurons whose collective activation serves to represent Pushkin are spread throughout the brain. Thus, if this interpretation of prosopagnosia is correct, it implies that only the visual stimulus route to the concept of Pushkin is cut by the lesions, since it can still be reached by other sensory routes. Indeed, it is possible that a patient, given such alternative routes to the neural representation of Pushkin, may even be able to describe her visual attributes accurately.[99]

The hypothesis is that there is a hierarchy of recurrent networks nested within recurrent networks nested within recurrent networks. It is to be expected on the basis of such a hierarchy that, when synaptic connections—connection weights in the language of Chapter 5—are destroyed by lesions, fine-grained learned patterns (more specific, individual recognition patterns) will degrade first, while the broader classes will tend to survive.

The retrieval of stored classes of objects (e.g., faces) from neural networks, as opposed to individuals within each class (one person's face), was investigated by the Argentinian physicist, Miguel Virasoro, at the Universita degli Studi di Roma "La Sapienza," and now director of the International Center for Theoretical Physics, Trieste, Italy. He found, using the Hopfield type of neural network model, that the stability of the class was much greater than for individuals. In other words, brain damage would indeed tend to wipe out sites that distinguish the faces of individuals, precisely the effect observed in prosopagnosia.[100]

CONSCIOUSNESS

The study of consciousness was shunned in learned scientific circles until recently. It was widely believed—and still is by many—that the phenomenon lies beyond the reach of scientific explanation. One key element of consciousness is its subjectivity—each of us can only know of his or her own conscious state. And the realm of subjective experiences is normally regarded as a strictly private affair. But brain-scanning techniques (see the Appendix) can now glimpse this private world and artificial neural nets offer the means to model it. The realization of artificial consciousness is a tall order and has not yet been attained; however, we should take heart from the many examples described in this book of how we can re-create extraordinary real-world complexity using computers. Given this kind of progress, it is no wonder that today a lively debate involving scientists and philosophers is attempting to find out what it is about that three-pound lump of grey and pink cells in our heads that is responsible for consciousness.[101]

On one matter most are agreed: the effort is among the most challenging and exciting ever undertaken. The challenge is rooted in the complexity of the brain's endless tangles of neurons and synapses. The excitement of the quest rests on the claim by some that our future survival and that of the planet could depend on a more complete understanding of the human brain.[102] This urgency is heightened by the increasing numbers of scientists who believe an explanation of consciousness, whether neurobiological or neurocomputational, is now feasible. As Francis Crick has stated, "I believe the problem of consciousness is now open to scientific attack. . . . The flavors (qualia) of what we see (such as the redness of red) may be private but it should be possible to discover the general type of activity in the brain that corresponds to consciousness."[103] To find out what sort of activity that may be, Crick and his collaborator Christof Koch are studying the visual illusion posed by the Necker cube, a line drawing that either appears to be going into the page or popping out, depending on how long you stare at it (see Fig. 9.9). "Aside from eye movements, what is coming into your eye is constant but your percept is changing," said Crick. "What we want to know is which neurons in the brain are changing when your percept is changing."[104]

Controversy abounds because there is no concrete definition of consciousness. It is a property of an undamaged human brain, just as a particular type of yellow glow is a property of a sodium atom. Unlike the wavelength of the yellow light emitted by sodium atoms, however, consciousness is not a conventionally observable property.[105] Then again, it is by no means invisible to the form of observation we call introspection. This elusive quality, on which most of us can agree, is captured in a typical dictionary definition of consciousness as "the waking state of the mind: the knowledge which the mind has of anything: awareness: thought."[106] Unfortunately, the "mind" itself is an elusive immaterial concept, so we are in danger of circularity unless we

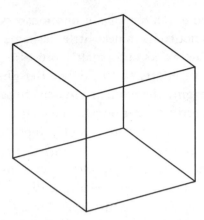

Figure 9.9   Conscious illusion. A Necker Cube.

accept Cartesian dualism—the rigid distinction between mind and matter. We can retain some scientific sense if we replace "mind" with "brain" in this definition. But we still face the problem of whether to assign consciousness to some kind of *homunculus* ("little man") in our brain before whom all experience is played out in some kind of "Cartesian theater" or, indeed, whether it is possible to depict consciousness as a physiologically consistent and unified concept.

Gerald Edelman is convinced that the mystery of consciousness will never be solved at any single level of description, whether molecular, neuronal, or psychological. Instead, he places great emphasis on the effects of evolutionary selection at a range of levels, starting with the myriad neural connections within the brain. Faced with the necessity for survival, for making order out of a chaotic world, the brain is highly plastic and adapts itself, mapping sensations, categorizing and recategorizing them constantly. "Nerves that fire together, wire together," he says.[107] Every neuronal map, every part of the brain, is dynamically, or, to use Edelman's term, "reentrantly" connected with every other, evolving and integrating itself in continuous cross-talk.[108] Thus, the brain actively represents and maps the world, and compares these mappings with one another. Crucially, however, this evolution of self is made possible by selection, strengthening of existing neuronal groups, and the constant emergence of new neural networks on the basis of "value systems" derived from evolution, such as reflexes, taste, and appetite.[109] All these processes naturally develop a diverse and degenerate repertoire of connections, no two of which are alike, even in identical twins. Understanding how the brain categorizes the world is a key problem in the search to explain consciousness. As Edelman puts it, "The world does not come in neat little packages with labels."[110]

Artificial neural networks and parallel processing consistently feature in many current efforts to pin down the phenomenon of consciousness. By envisaging the brain as a massively parallel device, there is no need for a Cartesian theater on which events are enacted before an all-seeing homunculus. Since there is no homunculus, we do not need to maintain the notion of Cartesian dualism, that mind is separate from matter. As an example, the philosopher Daniel Dennett maintains that "It is beyond serious argument that the brain is a computer. It is not a serial computer of the familiar sort, but a *parallel* computer, an architecture alluded to in the name for my theory of consciousness, the 'multiple drafts model.' I envisage the mind to work rather like the Reagan presidency—lots of sub-agencies and coalitions and competitive functionaries working simultaneously to create the *illusion* that one Boss agent is actually in control."[111]

This connectionist picture of consciousness creates problems for the traditional reductionist mission to dissect the workings of the brain. "Science has always had its great triumphs when and where it succeeded in subdividing complex phenomena into very simple paradigms. Doing the same to the brain we are in danger of being left with bits and pieces in our hand. Using a simile, in order to understand the value of money, we shouldn't stare at dollar bills," wrote von der Malsburg. "We should rather try to understand the system of beliefs and habits that make money do what it does. What I am saying here is that none of the isolated components of the brain can be expected to hold the essence of consciousness. That resides in the modes of interaction of all parts of the brain, and maybe even in the way the brain is integrated into the social world and the world at large."[112]

ARTIFICIAL CONSCIOUSNESS

The ultimate test of our understanding of the brain will come with the design and simulation of an artificial one, which displays such attributes as intelligence and consciousness. We have seen how neural networks are today providing insights into memory, pattern recognition, and the way the brain is organized. With these more realistic models of brain function, as well as our knowledge of artificial life, we can begin to see why intelligent behavior—and consciousness—may not necessarily be restricted to biological beings alone.

Perhaps the most useful early contribution to the debate over artificial intelligence was made by its founding father Alan Turing, who took a pragmatic "operational" view. An operationalist would say that a computer has a human attribute so long as the computer's attempts to imitate that attribute are indistinguishable from the real thing. Turing gave a description of this, the Turing test, in an article entitled "Computing Machinery and Intelligence" that appeared in the philosophical journal *Mind* in 1950. Turing reasoned that a computer must be said to be capable of thinking if a human being, conducting a dialogue by electronic typewritten messages, cannot tell whether he or she is communicating with a machine or with another person. Such issues have a substantial philosophical content, and have stimulated the growth of a huge body of literature.[113]

The mathematician Sir Roger Penrose, in his thought-provoking and widely read book *The Emperor's New Mind*, delivered an interesting critique of the entire enterprise of artificial intelligence (AI). He has developed his views in a recent sequel, *Shadows of the Mind*.[114] His argument turns on the significance of Gödel's undecidability theorems in mathematical logic, which we encountered in Chapter 2. Penrose maintains that human brains have an ability to "see" the truth or falsity of Gödelian statements whose truth-values cannot be decided within the formal axiomatic framework of the logical system concerned. This does not mean he believes that human brains are essentially different from those of many other animals. The important point is that, for Penrose, this ability is "a clear-cut *instance* of noncomputability—a noncomputability which must be present in conscious processes generally and [is] not at all unique to human brains."[115] According to Roger Penrose, because we can step outside these formal axiomatic frameworks and gain insight into the truth value of such undecidable statements by reasoning that is nevertheless mathematical, it follows that our brains cannot operate algorithmically. Since computers merely execute programmed instructions, they are algorithmic; hence computers cannot be as smart as we are.

If Penrose's argument were correct, the ambitious goal of strong artificial intelligence—namely, to build a *conscious* computational device—would crumble into dust. This argument has considerable force, as well as a certain mystical appeal. For, like many mathematicians, Penrose is committed to the notion of a Platonic reality, existing independently of us, yet one which we can make contact with through mathematical insight.[116] Thanks to Gödel, at least a part of this abstract reality is forever veiled to "dumb" algorithmic computation. Our conscious brains—or at least those of mathematicians—are what we need to reach into the Platonic cosmos and divine answers to these uncomputable problems.

Roger Penrose takes his argument still further. As we saw in Chapter 2, it is possible that the laws of physics—or at least their mathematical representations—may have consequences that are not computable. Indeed, we described this possibility when discussing the mathematical work of Pour-El and Richards, whose relevance to science remains unclear. For reasons wholly unconnected with AI, Penrose claims that the many problems that bedevil quantum mechanics and gravity will be resolved by the explicit incorporation of noncomputable elements in some more successful but hitherto unknown theory of quantum gravity.[117] From this, he speculates that it is precisely the supposedly nonalgorithmic new physics that lies behind the intelligent properties of conscious brains.

To provide a bridge from this currently unknown quantum gravity to the neurons within a conscious brain, Penrose draws on the ideas of Stuart Hameroff on microtubules.[118] Most neuroscientists agree that microtubules provide a "skeleton" for the neuron with two functions: to control the neuron's shape, and to transport molecules back and forth between cell body and synapses. Penrose goes beyond this consensus, suggesting that the network of microtubules might exhibit behavior that would correspond to a quantum measurement and that this could yield the noncomputability he believes he has shown is necessary for consciousness. However, microtubules would provide only one part of the overall setup envisaged by Penrose, which would still require many cells acting together in concert.[119] "The neuron level of description that provides the currently fashionable picture of the brain and mind is a mere *shadow* of the deeper level of cytoskeletal action—and it is at this deeper level where we must seek the physical basis of *mind*!"[120]

The basic Gödelian argument that Penrose and others have used to attack AI has been widely criticized.[121] Crick believes that, by now extending his ideas to microtubules, Penrose has moved far out of his depth.[122] The scientist who elucidated the three-dimensional structure of microtubules, Sir Aaron Klug, is unimpressed by their proposed new role in consciousness[123] while Gerald Edelman points out that "There is an old fashioned drug to treat gout and arthritis that dissolves your microtubules—what happens to your soul then?"[124] Yet despite numerous criticisms, Sir Roger Penrose's argument about the elusiveness of computational consciousness is an important one, and we cannot easily dismiss it.

However, it can be argued that Penrose's position is based on a somewhat restricted view of what constitutes a computer. For Gödel's theorem is a theorem of logic, concerning mathematical systems of axioms; it does not apply to machines. Michael Arbib, a computer scientist at the University of Southern California, concluded many years ago, as others have also, that, though fascinating, Gödel's theorem is an irrelevant technical statement. "Those of us who model human intelligence know that people do not argue from axioms all the time. We argue by analogy. We keep learning new things. We make mistakes. We are not consistent, unlike the axioms," says Arbib.[125] Gödel's theorem would indeed limit artificial intelligence if it were as restricted as a GOFAI system.[126] Penrose regards the potential of modern machines to learn mathematical axioms and rules of thumb as irrelevant to the way humans understand mathematics.[127] But for most people, it is precisely the ability to learn that enables modern AI to escape Gödel's clutches, as Turing argued long ago.[128] Any artificial consciousness would have the ability to incorporate new "axioms" into its structure as a result of experience with sensory or other data. The neural computing machines inspired by the brain are ultimately not intended to be logical inference engines but machines that interact with and explore the world, and can learn from their mistakes. In the jargon, these machines are "situated" so that they can constantly match their behavior to that of the world. In Arbib's opinion, "Gödel's theorem has absolutely nothing to say about that"[129]: it is a red herring.[130]

Most of those currently working on the simulation of intelligence and consciousness feel that the real challenge lies with the biochemical machine we call the brain. The demise of GOFAI hegemony has led to a proliferation of computational strategies, as we have discussed throughout this book. What is happening now is that a menagerie of competing computational approaches is evolving that can only enrich artificial intelligence research. At this early stage, these efforts will continue to thrive, regardless of Roger Penrose's arguments against strong AI. For instance, though many details of a human brain and that of a slug are similar, they are wildly different in ability, reflecting their relative complexity. It seems highly probable that different degrees of complexity lead to different degrees of consciousness. Therefore, the quest for AI is a quest for complexity. Earlier in the chapter, we described efforts to construct the complex "wetware" that brains are made of; such efforts are still at a primitive stage. Other approaches try to re-create that complexity in an artificial neural network by computer hardware, software, or a combination of both. Although the representation of artificial neurons is far simpler than their realization in wetware, they do at least capture several important features crucial to the way the brain functions. Most important, the complexity of such networks results from the collective action of a large number of simple units, just as the brain's complexity rests on its myriad neurons.

What is novel about the neural net approach is that it does not involve explicit programming. To be sure, some kind of algorithm is always present in all software neural network simulations in order to specify the dynamics of learning, in the same way as the DNA code "programs" the brain's architecture and learning processes. But such nets, once established, learn by experiencing the world with which they interact. Therefore, it seems entirely conceivable that sufficiently complex types of machines could also learn to "see" solutions to certain types of Gödelian undecidable statements. Indeed, this process is the same as that by which some humans develop an ability to resolve Gödelian problems—through a sufficiently deep education (i.e., a lengthy and highly specialized learning process) that enables them to stand outside any given formal logical system. And even if, as Roger Penrose contends, there do turn out to be significant limitations to the intelligent capabilities of digital neural networks, analog recurrent neural networks possess the ability to perform "super-Turing" computations, rendering computable that which for a finite state (Turing) machine would be noncomputable.[131]

Whatever laws of physics the brain obeys, one thing is certain: they are the laws governing the behavior of any physical object, whether a neuron or silicon chip. Most scientists feel there is plenty of mileage left in exploring the complexity of the brain using established ideas. As we have emphasized throughout, existing physics can generate enough exotic emergent behavior to explain many of the fundamental processes of life and the brain. "Nobody has said they are stuck in their study of the hippocampus because the fundamental laws of physics are restrictive," Arbib remarked.[132] That does not mean one should rule out the possibility of revolutions to come. Just as investigating the very small gave us quantum mechanics and investigating the very large led to general relativity, perhaps something new will come from studying the very complex. But the overwhelming majority of cognitive scientists would agree with Arbib when he says, "I would be surprised if Penrose has got the answer in quantum gravity"—and Arbib adds, "if he has, it is a pure fluke."

THE DREAM MACHINE

We are a long way from running computer simulations of the human brain in its full glory. Yet surprises continue to emerge in even highly simplified simulations. As one example, a computer model of the hippocampus, quite distinct from that of Treves and Rolls, was developed by Roger Traub at IBM in a collaboration with Columbia University to study the brain's electrical rhythms. The model connected 10,000 simulated neurons, each one described in considerable microscopic detail so that it would respond in a manner close to that of the real thing. It was a bottom-up approach to complexity similar in spirit to what we encountered in Denis Noble's work on the heart. What resulted was unexpected: Traub's network produced electrical waves similar to those generated in large populations of brain cells that can be detected by electroencephalography. One emergent behavior in the model was the "theta rhythm," which occurs during dream sleep. The origin of these waves, also called population oscillations, is not understood in the brain or the IBM 3090 computer. "It is quite a surprise," said Roger Traub of IBM.[133] "When I was starting out, we only used the model to confirm things we saw in the laboratory. Now we are beginning to do experiments on it as if it were an organism in its own right."

Working with John Jefferys of St. Mary's Hospital Medical School, London, Traub has extended his work to model the most explosive spasm of electrical activity that can occur in the brain, when a ripple of activity spreads out from a single spot during an epileptic seizure. What is particularly striking about his simulation of so-called "after-discharges" is that they compare well with experiments on slices of guinea pig hippocampus. (An after-discharge is an abnormal electrical potential that is extended in time, usually appearing as a series of oscillations.) Both the spatial and temporal properties of these electrical discharges were successfully reproduced in a computer model consisting of between 100 and 8,000 pyramidal neurons, each broken down into nineteen compartments so that their electrical properties were reasonably realistic.[134]

The most intriguing rhythm of the brain has also been simulated by the team in a network of artificial neurons. Only present during consciousness or dream sleep, the rhythm cycles forty times every second and can be detected by monitoring the electrical or magnetic activity of the brain. Some claim that it acts rather like a clock in a computer to coordinate activity in the many specialized regions. In other words, this forty Hertz rhythm could be the way the brain tackles the "binding problem" underpinning a unified consciousness, which we encountered earlier in the chapter. Various explanations have been put forward to account for this beat: perhaps individual cells have an intrinsic forty Hertz rhythm; perhaps the rhythm arises from a feedback loop between inhibitory neurons and pyramidal cells, or even between brain structures such as the cortex and the thalamus. With Jefferys and Miles Whittington, Roger Traub produced a forty Hertz rhythm in a virtual slice of hippocampus consisting of 128 inhibitory neurons, each modeled as a branching cell consisting of forty-six compartments. And indeed, in experiments where drugs are used to switch off the pyramidal cells in a hippocampal slice, the rhythm persists in the remaining active inhibitory cells. The rhythm appears to be an emergent property of networks of inhibitory neurons alone. "We have provided a tool to investigate the role of the forty Hertz rhythm in the binding problem," said Traub.[135]

The largest brain network simulated by Traub has imitated the action of only 10,000 cells. A hardware version of the brain, based on artificial neural networks, would require on the order of 10^{11} neurons or processors. The massively parallel CM-2 computer from Thinking Machines Corporation, for example, has about 65,000 processors; if each one were made to act as an individual neuron, we would need to harness the combined power of ten million such Connection Machines before we would have achieved something like parity with the brain. But, as we have seen, it is not merely numbers of neurons that are important—what matters is the way they are connected together, with complex hierarchies of networks within networks, and nets coupled to nets. This connectivity is a combined result of biological genetic programming and adaptive learning; it is not hidebound by the initial hard wiring.

Even when it eventually becomes possible to achieve simulations of hundreds of billions of neurons, the resulting networks and their emergent properties would not display intelligence and consciousness similar to that of the human brain unless they were subjected to similar sensory stimuli and experiences. This point was evident in the work of Rodney Brooks, described in the last chapter, where the computer is hooked up to a large amount of complex sensory apparatus. Conventional AI failed because it overlooked the essential importance of context-dependent knowledge and an ability to learn on the job. Instead, it was predicated on the unlikely suggestion that a programmer could design and implant something as subtle and complex as consciousness in a machine. Intelligence is an attribute that reflects brain plasticity and a direct experience of the way the world works. To be intelligent, therefore, a machine must be able to interact with the world as well as learn from it. This state of affairs is what biological evolution has wrought: it is a crucial yet often neglected ingredient of intelligence.

Throughout this book there has been one dominant theme: how we are seeking to understand complexity through a symbiosis between nature, science, and computers. In the previous chapter, we saw the remarkable progress being made in the field of artificial life. In this chapter, we have shown how computational models of neural networks offer much insight into the complexity of brain structure and function. These insights will get wider and deeper as computer power soars and important biological detail is re-created more completely within computer models.

We believe that there are good reasons to suppose that a sufficiently complex machine could one day emulate intelligence and consciousness, the most sophisticated hallmarks of the most evolved of biological species. We place our faith not in human computer programmers but rather in the complementary creative forces of self-organization and evolution. As we pointed out in Chapter 8, even the human eye, which tested Darwin's faith in his own creation and which has frequently been cited in attacks on the plausibility of biological evolution, has recently been shown to be a likely product of blind evolution.[136] Some may mourn the power of such an approach, claiming that it diminishes our existence by substituting shallow contingency and randomness for profound metaphysical meaning. However, the insights into creativity, life, and consciousness derived from an understanding of their inherent complexity in no way threaten but instead enrich the notions of chance, indeterminism, and free will so precious to us.

Chapter 10

PANORAMA

Dust as we are, the immortal spirit grows
Like harmony in music; there is a dark
Inscrutable workmanship that reconciles
Discordant elements, makes them cling together
In one society
—WILLIAM WORDSWORTH

The story of the universe is one of unfolding complexity. By emulating the processes that created the patterns and rhythms of the cosmos, science can tackle supposedly intractable problems, simulate the organization and activity of the brain, even create artificial worlds. It is now time to take stock, and attempt to draw some conclusions. Several wider issues suggest themselves. What relevance does the study of complexity have to the way science is done and to its future directions? What of the relationship between complexity and other human activities? In what ways does complexity impinge directly on our own lives?

Many people have accepted the reductionist message of contemporary science. Although sometimes powerful, reductionism can be destructively simplistic. If a mother loses her son to cancer, she desperately hunts for an explanation: was it that artificial coloring in his favorite orange juice? Was it the electric power cable outside his bedroom? Was it the cigarette smoke he inhaled? Sometimes, a simple cause can be established, but often one cannot be found. Similarly, we frequently come across surveys linking diet to health, which tell us about some particular type of food we should eat and what we should avoid. The surveys sometimes have conflicting results. However, complexity teaches us that effects can have an irreducible tangle of causes. Just as the properties of a cement slurry depend on a vast number of contributing factors, so, too, does the state of our health.

Simpleminded reductionism maintains that the whole is nothing more than the sum of its parts, each of which can be studied in isolation. But this form of reductionism is seriously limited. Take the global effort of deciphering the entire human genetic code, the human genome program. There will be many laudable benefits for medicine as the genetic errors that lead to hereditary disease are uncovered and predispositions to major killers such as heart disease, cancer, and dementia are associated with genetic markers. Yet profound dangers are also possible and none more so than in the field of behavioral genetics.[1]

Complex social behaviors such as personality, intelligence, criminality, alcoholism, schizophrenia, homosexuality, and manic depression are increasingly being touted as genetically predisposed traits, the result of "bad" genes. Studies of twins, families, and adopted children do indeed suggest that certain behaviors are at least partially hereditary: the problem is figuring out which traits are inherited and which have simply developed from a shared family environment. However, many studies that have claimed to link behavioral traits to genetics have been woefully flawed at a number of levels, and are distracting people from underlying environmental and social contributions.[2] "There are many false positives—a claim that a single gene causes schizophrenia, manic depressive psychosis, the alcoholism gene, the divorce gene and so on," says Doug Wahlsten of the University of Alberta. "If we continue along the path we are on right now, it opens this research to a number of abuses."

We may be much closer than we realize to believing that social problems are all rooted in defective genes. Garland Allen of Washington University, St. Louis, warns that the quest to link genetics with such complex behaviors could mark a new chapter in the "devastating history" of the abuse of this knowledge.[3] In the 1920s and 1930s similar claims led to the rise of the eugenics movement in the United States and to compulsory sterilization and restrictive immigration laws. "This sort of thinking also led to the Nazi eugenics movement and the Holocaust in Europe," he maintained. "Modern studies are nothing new and they're just as simplistic as nineteenth-century studies that tried to link physical features to criminality. Yet, disturbingly, these concepts are becoming increasingly prevalent in our sophisticated society. The genetic fix blames the biology of individuals rather than social circumstances for recurrent social and economic problems."

Reductionism is equally limited in the abstract worlds of mathematics. Gödel's theorem showed conclusively that the Platonic worlds of mathematics cannot be reduced to a finite alphabet of symbols and a finite set of axioms and rules of inference. "Except in trivial cases, you can decide the truth of a statement only by studying its meaning and its context in the larger world of mathematical ideas," commented Freeman Dyson of the Institute of Advanced Study in Princeton.[4] He described Gödel's theorem as a great work of art, a construction rather than a reduction: "Gödel proved that in mathematics the whole is always greater than the sum of the parts."[5] To add insult to the injury suffered by reductionists, the extension of Gödel's work by Gregory Chaitin shows that physicists will never be able to prove that a Theory of Everything— a compression of the world—is indeed the ultimate one.[6]

Real-world complex systems do not behave with clockwork regularity, and precise long-term forecasts about them are frequently moonshine. The complexity of a modern industrialized economy is such that it will never respond to the elementary manipulations of treasury secretaries. The complexity of the global climate is such that a gradual increase in levels of greenhouse gases does not always result in a gradual shift in climate: it can trigger a sudden climate flip within a single lifetime.[7] Even the behavior of some of the simplest of mechanical systems cannot be described in the complete and deterministic Newtonian manner previously thought possible. There is no simple algorithm to turn to. Instead, we must try to understand the world in more global terms, through the *interactions* between its components. Instead of attempting to take a deterministic, mechanical view of the world, we need a higher-level perspective if we are to make sense of it.

Life is also an emergent property, one that arises when physicochemical systems are organized and interact in certain ways. Similarly, a human being is an emergent property of huge numbers of cells, a company is more than the sum of its pens, papers, real estate, and personnel, while a city is an emergent property of thousands or millions of human beings. And no one should doubt that our innermost thoughts, our emotions of love and hate, are more than a rush of individual hormones, or the firing of individual neurons in the brain. The study of complexity, through its emphasis on emergent properties, goes some way to restoring a balance between the spiritual and materialistic sides of our nature.

Can we ever hope to understand such highly complex emergent properties? Some scientists already see themselves as "playing God" within the computer, for example, when they construct closed cellular automaton "universes" or carry out "directed" evolution using genetic programs, in which the rules and fitness measures are fixed externally. Might we not ourselves be dancing to some unknown tune? Perhaps. But each individual melody is too complex, too idiosyncratic, and too sensitive to history and external events to be understood in anything but broad outline.

That may sound defeatist. However, an understanding of complexity can go a long way toward helping us to make sense of the world, by providing a more global view of our role in it. Although we may not be able to precisely forecast the long-term behavior of a complex system, nonlinear dynamics shows that we can gain some insights into its global behavior—for instance, through a knowledge of the system's set of attractors. These insights may provide the bedrock of understanding for future decision-making.

Our impact on the planet has never been greater, the need to understand that impact never more urgent. Man's activities are now endangering and wiping out entire species of flora and fauna, many of which we know little or nothing about. How much more abuse can the earth take? Ecosystems—of which we are a part—are highly interconnected webs of life ruled by subtle nonlinear feedback effects. We may already be sitting on an environmental time bomb: the destruction of habitats worldwide, from rain forest to tundra, may have consequences that will not necessarily manifest themselves smoothly and gradually, but abruptly and discontinuously, possibly not until decades from now.

Even species that appear to be faring well may already be doomed, according to one nonlinear mathematical model developed by Robert May, Martin Nowak, and David Tilman, who have studied the diversity of plants that have evolved in grassland habitats, specifically the old fields and prairie at Cedar Creek Natural History Area in Minnesota.[8] These biologists recognized that to be a successful species means more than being the most abundant. To be a good survivor takes the ability to exploit a new habitat *and* adapt to a new one when under threat.

The model describes each species in the area as a "metapopulation," a set of small, local populations of plants that live in scattered sites but are linked by their ability to disperse seeds and sprout in more distant places. Observations carried out in Cedar Creek suggested that the plants that were the best competitors for resources were the poorest at dispersal. For example, bluestem grasses invest so many of their resources in their roots that they produce relatively few seeds. By building this concept of a trade-off between competitive ability and dispersal into their mathematical model, the authors could calculate the effects of loss of habitat on extinction.

The model predicts that loss of diversity (i.e., numbers of species) would be low even if half the habitat is destroyed. But any small increment in habitat loss over 60 percent causes a very sharp increase in the numbers of species that perish. Yet there is also a time lag of several generations between the loss of a habitat and species' extinction. Applied more generally, this means that it could take decades before we witness the effects of the current loss of rain forest: we are incurring an "extinction debt." Part of the reason is that destroying a habitat not only wipes out existing populations, it also removes potential colonization sites. Most surprising of all, the group's model predicts that the most successful competitors—organisms that often play a dominant role in the ecosystem—are most likely to vanish. These are the organisms that previously had no need to move from their current habitat.

The study of complexity has also shown us the importance of diversity and randomness in sustaining the capability for adaptive innovation. Our own future and that of the planet may depend on it. Lovelock's concept of Gaia is useful here: it provides a metaphor that eloquently demonstrates that destroying our environment is tantamount to destroying ourselves. That is why we must address the loss of the rain forests, the devastating decline in the number of species, depletion of the ozone layer, and global climate change. Integrative scientific studies of complex living systems should help to support the need for global action on a rational basis.

Computer simulations have led to a better understanding of the complexity of climate, of extinctions and ecosystems. In a similar fashion, computer-based models are also providing insights into the complex "ecology" of cells that protect the body from infection,[9] and what happens when they are faced with a predator such as the Human Immunodeficiency Virus.[10] Within the past five years, an ambitious effort has started that takes the approach a leap further, with major progress toward the computer simulation of evolution and the creation of artificial life. As well as "putting biological evolution into a bottle," so to speak, artificial life offers a distorted mirror for viewing ourselves. It tells us about life as it could be, rather than as it is at present on earth. Indeed, as we have seen, one can already argue that we have witnessed the birth of an alternative, digital, life-form lurking within the heart of a virtual world called *Tierra*.

The insights into the brain we outlined in the previous chapter are among the most fascinating examples of the power of the contemporary approach to complexity. What has been done should give us grounds for cautious optimism about the possibilities for the evolution of intelligent machines. But it is important not to get carried away. All that has been achieved to date falls very far short of capturing the pinnacle of complexity, a conscious thinking object. Formidable barriers remain. Basic attributes such as vision, language, and machine translation have yet to be fully reproduced, let alone united in a single device.[11] Over the next few decades we will make great strides. Genetic programming techniques will weave artificial neural networks into architectures of ever-greater complexity. A dialogue with the external world through the senses, as in the case of Cog, will hone these artificial brains. The inexorable growth in computing power will eventually match that of the brain, with or without the advent of quantum computers. And great leaps will be taken in finding the right environments, indeed cultures, in which to nurture and grow artificial brains so that they may one day match the dynamic and impressive complexity of our own.

At an increasing rate, we can expect that computer-based, goal-directed evolutionary methods will be used to evolve "intelligent" devices in ways that suit ourselves. But who decides what will "suit" us? Our moral and ethical standards of behavior, not to mention science itself, have evolved and will continue to do so in the light of political, social, and economic circumstances. These factors provide the "selection pressure" and determine in large measure what is or is not suitable.

Such "meta" processes take place in the minds of conscious individuals, as ideas compete with one another for ascendency. In this context, ideas are what Richard Dawkins has called *memes*, loosely speaking, units of cultural transmission.[12] Memes have the property of self-replication, as they propagate from brain to brain. Examples include ideas, tunes, and clothes fads. "When a craze, say for pogo sticks, paper darts, slinkies or jacks sweeps through a school it follows a history just like a measles epidemic," wrote Dawkins. "Fashions and crazes succeed each other, not because the later one is more correct or superior to earlier ones, but simply as any epidemic hits a school."[13] As the philosopher Daniel Dennett remarked, the meme concept is a good way of thinking about ideas but the perspective it provides is somewhat unsettling, even appalling. "I don't know about you, but I'm not initially attracted by the idea of my brain as a sort of dung heap in which the larvae of other people's ideas renew themselves, before sending out copies of themselves in an informational Diaspora."[14]

Dawkins does not want to apply his viral metaphor to all culture, all knowledge, and all ideas. "Not all computer programs spread because they are viruses," he believes. "Good programs—word-processors, spreadsheets and calculating programs—spread because people want them. Computer viruses spread almost entirely because their program-code says 'Spread Me.' No doubt there is a spectrum from the pure virus at one end to the useful and genuinely desirable program at the other, perhaps with addictive computer games somewhere in the middle." Dawkins' selfish gene theory recognizes a similar spectrum, from viral genes to useful genes that make animals good survivors. "The genetic instructions, 'Build a speedy, strong-boned, keen-witted, sexually attractive antelope,' are saying 'Duplicate Me' in only a very indirect sense which seems to us far less mindlessly futile than the simple and unsubtle 'Duplicate Me' programs at the virus end of the spectrum," he wrote. In the domain of culture, he believes that innovative ideas and beautiful musical works spread, not because they embody instructions that are slavishly carried out, but because they are great. "The works of Darwin and Bach are not viruses. At the other end of the spectrum, the televangelist's appeal for money to finance his appeals for yet more money is pretty directly translatable into 'Duplicate Me'."

Dawkins has taken this idea further in an attempt to draw a firm distinction between religious ideas, which he believes to be "pretty close to the virus end of the spectrum," and scientific ones. He argues that religions survive, not because of cynical manipulation by priests, and certainly not because they are true, since different religions survive equally well while contradicting each other. "Religious doctrines survive because they are told to children at a susceptible age and the children therefore see to it, when they grow up, that their own children are told the same thing." In other words, in Dawkins' opinion, religious beliefs are held for reasons of epidemiology alone.

There are, however, difficulties with Dawkins' argument. It is certainly true that while several successful strains of religion coexist and compete, ranging from Judaism, Christianity, and Islam to Buddhism and Hinduism, scientific memes tend to have an all-or-nothing feel to them, with one established theory excluding more or less all others most of the time in any given area. Religious creeds, like political ideas, are judged by every individual according to his or her own background, beliefs, and prejudices. Scientists, by this argument, are expected to believe in some things rather than others because of superior evidence in favor of them. Scientific memes will be more successful the more they can correctly account for and predict the results of experiments and observations—in short, these criteria provide the measure of the memes' "fitness." In the case of "nonscientific" concepts, such as religion, no objective yardstick or fitness measure exists for carrying out a ranking, and so which ideas win out depends on a collection of more arbitrary and subjective criteria. We must therefore expect religions to include more unprovable statements than science, while such disciplines as economics lie somewhere between these extremes.

Economics straddles the divide between science and the humanities. The world's economies possess nonlinear features characteristic of complex dynamical systems, although the marketplace is very much associated with a form of financial "survival of the fittest." There are objective measures of economic and financial success, whether of nations or companies, such as gross national product, budget deficit, market share, profits and losses, revenues, and stock prices. Yet many factors on which these quantities depend are themselves ill defined. A Wall Street catastrophe could be triggered by a financial earthquake or a whispering campaign. Beliefs and rumors generated by stockholders, analysts, and speculators can induce fluctuations in price, stock, and currency markets that in turn feed back on the objective "fitness" measures.

It is interesting to note that it has taken economists a long time to recognize the inherent complexity of their subject. For decades, the central dogma of economics revolved around stale equilibrium principles in a manner entirely analogous to the application of equilibrium thermodynamics in physics, chemistry, and even biology. For the same reasons as natural scientists, many economists have sought to shoehorn all economics into theories whose merits are their mathematical simplicity and elegance rather than their ability to say anything about the way real-world economies work. In this way, memes for classical equilibrium-based concepts have been infecting the minds of generations of science and economics students with the dogma that the behavior of a complex system can be deduced by simply summing its component parts.[15]

In more recent times, a new approach to economics has been pursued, based on evolutionary and nonlinear principles. Robert May, among others, has adapted the tools he uses for analyzing nonlinear dynamical phenomena in biology to investigate fluctuations of gilts, stock markets, and exchange rates. Part of the trick of ascertaining whether forecasts are possible is to develop methods for detecting the fingerprints of chaos in apparently random data.[16] Such work has helped to provide the all-important evidence that the complex behavior of financial markets *is* predictable to some extent. Nonlinear analysis and a range of other techniques, from neural networks and fuzzy logic to nonlinear statistics, are searching for lucrative patterns in the financial markets.[17] In 1993, it was estimated that up to a dozen firms were managing more than $100 million each on the basis of advice generated in these ways by computers.[18]

The 1994 Nobel Prize for economics was awarded for work done on game theory, fifty years after von Neumann and Morgenstern published their *Theory of Games and Economic Behaviour.* The new Nobel laureates, Americans John Harsanyi and John Nash, together with Reinhard Selten from Germany, have played a central role in establishing game theory as a powerful tool with applications ranging from the economics of industrial organization, through international trade to the theory of monetary policy. One example concerns the strategies a monopolizing company must consider in a bid to prevent a would-be competitor from encroaching on its market. One monopolist strategy might be to threaten to engage in a price war, with the intention of inflicting heavy losses on the rival. Yet this could also prove detrimental to the monopolist, unless it has substantial financial resources available. Thus, the competitor must assess how credible this threat is. Alternatively, the monopolist could pursue a more welcoming strategy. By offering mutual cooperation, a kind of cartel could develop, which would maintain profits for both companies through high prices. The benefits of a mathematical approach to such problems are now becoming more widely known. In the United States, the Federal Communications Commission is now designing auctions of the radiofrequency spectrum on the basis of game-theoretic principles.[19]

Other social "sciences" are not usually classified as scientific, despite the name. Nevertheless, social phenomena are being brought within the scope of scientific analysis as a result of the methods developed for the study of complexity. It can be maintained that the interest of the social sciences in nonlinear problems arose when Verhulst wrote down a logistic equation for the rate of human population growth in 1844, building on the ideas of Malthus.[20] Indeed, animal and human societies are replete with complex organization, from ant colonies and beehives to the dealing room of a stock exchange. We can regard such social structures as open and nonlinear, in which feedback and competition abound. In ant colonies, there have been extensive studies of strategies for food foraging and its subsequent collection and consolidation into stockpiles.[21] Attempts have been made to model the "organic" growth of an urban sprawl by drawing on the concept of dissipative structures, discussed in Chapter 6.[22] The trade-off between cheap, high-polluting transport and expensive green alternatives has been modeled in a computer to help understand our impact on the environment.[23] Others have used nonlinear models to analyze the effects of population movements and have found that they can account for the formation of uniform distributions of groups, ghettos, and restless migrations.[24]

We can draw analogies between the crisis points associated with self-organization and chaos that occur in inanimate processes like the Belousov-Zhabotinski reaction and certain phenomena that arise within human societies, such as revolutions and the breakdown of civil order. Complex human planning problems frequently involve establishing priorities among tasks with different degrees of urgency and importance: strategies in such cases have been represented by quite simple sets of rules; these rules can themselves be improved by evolutionary techniques that use genetic programming methods.[25]

The complexity of human society and organizations is immense. Business corporations and government agencies depend for their effectiveness on cooperation between their employees. However, the structure and nature of an organization can produce a conflict between the interests of the institution and those of the individuals within it. This conflict can be understood on the basis of the strategies employed by the parties concerned. In many organizations, people are moved from one position to another on short time scales, perhaps three years or less: there is thus an incentive for them to deliver immediate "results" regardless of the longer term consequences for the organization. Among other things, this means that it may not be in an individual's interests, for the sake of his or her own personal advancement, to cooperate with someone with whom in the future they are unlikely to have any further dealings; by so doing, however, the individual may cause untold difficulties for the organization as a whole. Here, game theory indicates that it would either be better for people to spend longer in any particular post, or that a record should be kept of what happens in the area of their previous responsibilities following their reassignment (or a combination of both approaches). Beyond its application to specific issues, in general terms game theory underlines how much of human behavior is irrational.

The complexity of human interactions has been fostered throughout the ages by communications technology, which facilitates the exchange of information on all levels, from individuals to governments. The more information is exchanged, the more feedback processes occur and thus, in general, the more complexity. Computer networks are now transforming the nature and speed of such communication, and the sheer volume of accessible information. To use Dawkins' terminology, these networks encourage the proliferation of memes. The Internet is the dominant superstructure linking millions of personal computers around the world. One estimate hazarded in mid-1993 put the number of networked computers at around 1.7 million, with some 17 million people using the service.[26] At the time of this writing, this figure had risen to 23 million. In several respects, the activity on the Internet is very much like the paradigms we have pursued in this book: it is evolving, highly distributed, and lacking any central control, but it has the capability to support emergent phenomena in terms of the structures that can result from information exchange. As human mobility rises and economic activity spreads throughout the world uninhibited by physical frontiers, geographically localized societies continue to fragment along many lines, of business, religion, culture, scientific, social, and leisure interests. This makes the Internet an important vehicle for maintaining an identity—albeit a faceless one— among a morass of conflicting activity. The "information superhighway," as it is now being called, has undergone an abrupt jump within the past few years with the introduction of graphical interfaces to the World Wide Web, which enables multimedia access to an almost limitless stockpile of data, including visual images.[27] These electronic webs within webs will forever change the manner in which technologically advanced societies live and operate.

More traditional and universal means of communication—through the arts—are often thought to stand separately from science, or in dogged opposition. It is commonly thought that science has little to say about or do with the aesthetics of painting, literature, or the theater, other than in the more austere sense of geometrical proportion, word processors, and stage technology. This is because, in the Newtonian tradition, science has been concerned with the abstraction from appearances of the "irreducible mathematical essence" of things; and in a reductionist manner, it has been involved not with the whole but rather with what that whole is made of. In recent times, the science of complex systems has been helping to change this perception.

We encountered the poets' disquiet with the arrogance of Newtonian physics in the verse of Alexander Pope that introduced Chapter 2. The reductionists' materialist world-view inspired deeper contempt from the likes of Blake and Keats, though this was not always the case. When Mary Shelley wrote what is without doubt the best-known novel on artificial life—*Frankenstein*—it was in the context of the debate over vitalism, which had raged in public since 1814.[28] What better than Victor Frankenstein's half-baked attempt to vivify a dead body with something "superadded," analogous to electricity, to mock the spiritualists and support the materialist world-view?

In fact, in the twentieth century, the interaction between arts and science has remained as strong as ever. Many echoes of Einstein's theory of relativity can be found in literature—for instance, in Lawrence Durrell's *Alexandria Quartet*[29] or in *Finnegan's Wake*, where Joyce plays with the finding of Einstein ("Winestain," as Joyce calls him) that light traveling in curved space time might eventually return to its source, so that we could see ourselves from behind.[30]

Complexity now influences the arts at several levels. One of the main characters of this book, Alan Turing, provided the inspiration for Hugh Whitemore's 1986 play *Breaking the Code*.[31] Complexity has offered a "cosmogenic" cocktail—the motifs of fractals, catastrophe theory, and chaos—that has caught the imagination of architects.[32] It can add an interesting twist to fiction and poetry. There was the cartoon caricature of a "chaos mathematician" who warned of the dangers of meddling with dinosaur DNA in Michael Crichton's *Jurassic Park*. Lorenz's butterfly fluttered in the poetry of Paul Muldoon.[33] Complexity also helped to mold Lemuel Falk, a Russian chaologist with "thick, callused fingers" and a "tangle of ash-dirty hair that manages to look wind-whipped even in the absence of wind." He appeared in Robert Littell's *The Visiting Professor*, where he dazzled a symposium in Prague by his quest for pure, unadulterated randomness: "Lemuel had programmed an East German mainframe computer and calculated pi out to sixty-five million, three hundred and thirty-three thousand, seven hundred and forty-four decimal places (a world record at the time) without discovering any evidence of order in the decimal expansion."[34]

Tom Stoppard's award-winning 1993 play *Arcadia* was also inspired in part by nonlinear ideas of chaos and fractals. Indeed, Stoppard even regarded the process of writing the play as one of literary self-organization.[35] The play's action flips between two centuries, hanging on a detective story spun from a web of correspondence between characters living in the past that is picked up in the present day by an unscrupulous academic determined to develop a theory about Lord Byron sensational enough to grace the pages of a tabloid newspaper. It also combines dissertations on landscape gardening and the rise of romanticism with debates about progress and perfectibility, determinism and free will.

The richest ideas emerge from the scientific discoveries glimpsed by the young heroine, Thomasina Coverly, at the feet of her tutor in 1809. She stumbles across the "geometry of irregular forms," a sly reference to the fractal geometry of Mandelbrot that to scientists is synonymous with strange attractors. Indeed, as a piece of theater, *Arcadia* displays the nested references of a fractal pattern, both in the interplay between character and prop and the resonances between dialogue and accompanying music. And Stoppard likens the structure of the play, where action zigzags between past and present, to "the crudest possible diagram of period doubling. The last scene is, as it were, chaotic, in the sense that both periods coexist."[36]

Stoppard developed the chaos themes after consulting Robert May, who attended a seminar with Stoppard, director Trevor Nunn, and the original National Theatre cast. One of May's graduate students, Alun Lloyd, developed the "Coverly set" for *Arcadia*'s Thomasina character, from a simple mathematical formula that generates the complex and "leaf-like" fractal patterns that she sought.[37] Computer-generated fractal images, ranging from ultra-abstract representations of objects like the Mandelbrot and Julia sets to realistic-looking biological and physical shapes, now adorn posters and postcards because of their popular

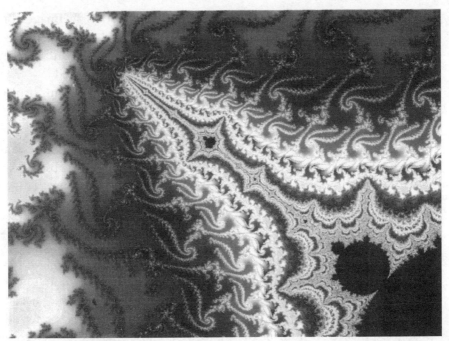

Figure 10.1   Popular chaos. Mandelbrot set produced by the Chaos Laboratory, Scarborough.

appeal (see Fig. 10.1). They have an intrinsic beauty that engenders a response in many akin to that experienced by observing nature and human works of art, whether realistic or abstract.

Digital Darwinism is also entering the picture: a new breed of computer-literate artisans are using evolutionary programming techniques to create novel forms of art. One of the leading proponents of this approach is William Latham, a British artist who has worked with IBM Research and is founder of Computer Artworks Ltd. He has employed evolutionary methods to breed alien forms in a computer. (See color plate 11.) By selecting those with eerie appeal, and then breeding from them, he generated what he calls his "garden of unearthly delights." His work not only blurs the distinctions between what is nature and what is art but further illustrates the rich contemporary symbiosis between art and science: the "Mutator" program Latham developed with IBM mathematician Stephen Todd is not only able to evolve art but also the design of a house or the parameters on a company's spreadsheets. You can then kill the variants you dislike and select the variants you want to breed improved designs. For example,

Latham has used the program to breed a rich selection of images of buildings for a computer game's virtual suburbia, and to create shampoo-bottle designs. "The Mutator comes up with extreme and subtle shapes that a human would not necessarily think of," he noted.[38]

The images shown in color plate 12 were created by Karl Sims using a population of genetic programs found by "interactive evolution." The computer is used to generate random mutations in mathematical equations known to produce colorful images, and the artist then applies an "aesthetic selection pressure" by choosing for survival and subsequent breeding those he or she finds most appealing. This is a procedure rather similar to that used by Richard Dawkins for evolving his biomorphs, as discussed in Chapter 8. After a large number of iterations of this procedure, remarkably complex and intricate pictures emerge.

In Genetic Images, an exhibition held in 1993 at the Centre Pompidou in Paris, Sims used a Connection Machine to breed a random selection of images for display on an arc of sixteen large monitors. Special step sensors that allowed visitors to select the images they preferred most were placed in front of the monitors. Those images not selected were killed off and replaced by mutants of the chosen survivors. "People in the museum were controlling the evolution," he said. "It was a case of survival of the most aesthetically interesting."[39]

Sims has taken the work a leap forward by evolving creatures with both form and function. Certain controlling "genes" determine shape, which is built up with block-shaped segments, in a similar though highly simplified version of the way our body plan is created; other genes describe a simplified program—the creature's brain—that controls its movement or reacts to sensors that respond to light, contact, or the angle of a joint.

Simulated evolution begins with a population of 300 "creatures," each randomly made of colorful blocks. Some look boring. A few are bizarre, while a handful twitch fitfully. Sims can evolve many generations of creatures in a supercomputer, selecting those with desirable characteristics, for instance, the ability to fight, swim, or move. (See color plate 13.) Intriguingly, the creatures "cheated" during his first attempts. "They did what I asked them to do but not the way that I wanted," Sims ruefully remarked. They evolved to exploit errors in the program that was intended to ensure that the creatures obeyed the laws of real-life physics. One glitch allowed the violation of Isaac Newton's law of conservation of momentum. After a few generations, creatures evolved that shuffled along by hitting themselves with a paddle. Others found an error in the "integrator"—which solved Newton's equations of motion—and this enabled them to propel themselves along with unphysical haste.

Sims uses an animated video to show, for example, the result of 100 generations of sexual reproduction achieving the evolutionary goal of becoming "good swimmers." Some digital creatures evolved into snake-like creatures that wiggle through virtual water. Others acquired a corkscrew motion or protruding paddles. "One of the interesting aspects of using simulated evolution is that you can make things more complicated than you can figure out," he says. "Luckily you don't have to." Sims speculates that he may be able to breed even stranger creatures if he could endow them with an eye for beauty.

Even "near-death" experiences are being explored using the science of complexity. One highly speculative attempt to simulate an artificial near-death experience in a neural net triggered it to regurgitate information that it had been taught, or something very similar. The net's virtual life flashed before its eyes.[40] Other artificial death studies have used computer-based nonlinear dynamical models to investigate experiences reported by numerous people, which range from swirling tunnels of light to out-of-body experiences of a singular intensity, frequently claimed to be glimpses of the after-life. At the Max Planck Institute for Molecular Physiology in Dortmund, Mario Markus has sought the formation of organized Turing structures in the visual cortex of the brain, with the expectation that these patterns may lead to geometrical visions of the kind seen near death and during hallucinations. Using the relationship between the geometry of the cortex and the way it is mapped to the retina of the eye, it is possible to infer from the simulated patterns what they would look like to an individual. Markus analyzed drawings of visions experienced under the influence of LSD that reveal strange spirals and lines converging to a point. The computer deduced that these visions would be produced by simple zebra-like stripes of activity across the cortex (see Fig. 10.2). In a similar way, Markus argued that near-death visions of tunnels with a strong—some would say heavenly—light shining forth from one end are consistent with illness or injury setting up Turing patterns in the brain.[41]

What subject other than complexity has something to say on such a diverse range of issues from emergence, life, and intelligence, through beauty and art, to death and extinction? Only philosophy. The science

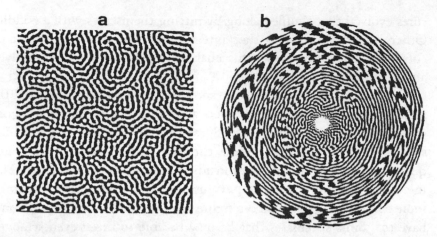

Figure 10.2   Hallucinations. The cortical pattern on the left (a) corresponds to the visual image on the right (b).

of complex systems is likely to have a major impact on many areas of philosophical inquiry of concern to our own lives.[42] As we have seen throughout, concepts such as life, intelligence, and consciousness are difficult to pin down in the abstract. In the past, a lot of somewhat nebulous theorizing has been devoted to these topics. But computer-based studies now offer a framework inside which we can study various complex models scientifically. These have the merit of being precisely defined, making explicit all the assumptions used. For example, contemporary philosophy has tended to shun the notion of emergent phenomena, but bottom-up artificial neural network models demonstrate clearly that the remarkable information-processing properties of these networks are indeed the result of their highly distributed, massively parallel *global* dynamics.

Then there is the intriguing question concerning the distinction between computer simulations and reality. In many cases, a simulation of a physical or chemical process on a computer could not be confused with the process itself. A computer simulation of a swaying bridge is not a bridge. Yet there are cases in which a simulation does produce results that are an instantiation of the process being simulated. One simple example is a computer simulation of music. We have argued that there is no reason why, under certain conditions, computers could not also display intelligent and vital attributes.

If we accept the possibility of real artificial life, we must consider important moral and ethical questions, which are of relevance to humanity at large. What if the digital organisms within an ALife simulation *were* actually alive? Do we have the right to harm or kill these organisms? Should we have the right to perform unnatural selection to engineer desired life forms, and for what purposes?

For the scientist, complexity places renewed emphasis on interdisciplinary research in the Renaissance style, and underlines the symbiosis between science and technology. We have given many examples illustrating how fruitful this approach can be, and we fully expect the trend to continue. Perhaps one of the most exciting pointers to the future direction of science has come from the first biochemical calculations, which may mark the birth of a new type of computer.[43] If they had been alive to witness it, Alan Turing and John von Neumann would surely have been delighted to see a complex computational problem being solved by molecules. Because the molecule in question was DNA, the work also marks a further blurring of the distinction between computers and living things, another aspect they would have approved of.

The idea of molecular computation dates back at least to 1959, when the Nobel laureate Richard Feynman gave a talk on the possibility of creating submicroscopic computers.[44] In previous chapters, we discussed two strands of that work, concerned with the use of BZ reactions to solve problems and the theory, design, and construction of quantum computers. Now the computer scientist Leonard Adleman of the University of Southern California has developed a new route that exploits another aspect of nature's inherent parallelism. In his experiment, massive molecular parallelism was used to crack an instance of what is called the "directed Hamiltonian path problem," an NP-hard problem that involves finding a special path through a network of vertex points.

A Hamiltonian path passes through each vertex exactly once. Consider, for example, the following set of American cities: Atlanta, Baltimore, Chicago, and Detroit. Assume that nonstop flights are scheduled only from Atlanta to Chicago, Chicago to Detroit, Chicago to Baltimore, and Baltimore to Detroit. The directed Hamiltonian path problem could then be phrased as follows: Could a traveler book exactly three flights in a sequence—starting in Atlanta and ending in Detroit—that would take him to each of the four cities? In this case it is easy to see that the answer is to fly from Atlanta to Chicago, from Chicago to Baltimore, then from Baltimore to Detroit. But in a manner characteristic of NP problems, when the number of cities and flights grows even slightly larger than this, the number of itineraries to test becomes truly enormous.

The molecular computer solved the problem by brute force and parallelism. The "computer" code consisted of a mixture of trillions of pieces of single-stranded DNA, each piece either representing a city or a route. Among the vast number of combinations that resulted from the binding together of complementary DNA strands, it was overwhelmingly probable that one combination corresponded to the solution sought. Standard molecular biological techniques were then used to fish that molecule out: the "solution" was easily distinguishable from the other molecules because of its length and details of its composition.[45]

DNA has since been shown to be capable of solving other problems that require searching a universe of solutions, a task that would defeat any conventional computer.[46] The search is now on for a practical *universal* molecular computer that could solve any computable problem. Since DNA is evidently capable of storing and retrieving data, one day a general molecular method for interacting with and interpreting data from it may be designed. Such work could throw more light on whether biological systems are indeed complicated computational devices that process the instructions held in their genes. We might uncover further details about evolution and obtain additional insights into the development of complex biological processes, such as those that turn within our brains. And, from a technological point of view, while current supercomputers are able to perform a trillion operations per second, molecular computers could conceivably run billions of times faster. As well as being orders of magnitude more energy efficient, storing information in DNA requires about one-trillionth the space needed by present-day storage media such as videotape or CD-ROM.

Who knows what astonishing symbiosis will emerge from this new conjunction of science and the computer? The future will continue to bring ever closer the two themes underpinning this book: the power of computers to simulate complexity and the problem-solving ability of complex systems found in the biological world, most notably the human brain. Imagine a superbrain, built from a global network of DNA computers. As it evolved, would such a brain lead to the emergence of a global intelligence? If so, a new era of evolution will have begun.

APPENDIX: MIND READING

The computer is not only allowing us to model the complexity of the brain but also to study it in unprecedented detail, complementing animal experiments and investigations of victims of brain damage. The ability to process vast amounts of data from a series of sensors has been part of the revolutionary development of a range of powerful *noninvasive* imaging techniques. Brain scanners have now reached a level of sophistication by which huge networks of human neurons can be captured in action, down to the last thousandth of a second. Imaging technology peers through the skull to reveal the information on the brain's mental landscape. Although such methods cannot yet tell exactly what a subject is thinking, they can see where the thinking is going on and whether it is going wrong. Neural networks will draw on these functional studies of the brain to cast light on its large-scale organization.

The oldest, most widely used method, and least reliant on the computer is electroencephalography (EEG)—the placing of electrodes on the scalp. It cannot accurately pinpoint the distribution of electrical currents in the brain because the signal is blurred by the effect of the intervening tissue. Nonetheless, EEG still yields valuable and often intriguing results, resolving patterns of neural activity within the millisecond time scale. For instance, evidence from an EEG study by Nathan Fox of the University of Maryland shows that electrical activity within the right-hand hemisphere of a baby's brain can reveal whether it is likely to grow up to be shy and retiring, marking the first time that a social behavior has been linked directly to a physiological factor.[1]

EEG is a relatively crude form of mind reading when compared with a new generation of heavily computationally based imaging techniques that are able to locate centers of activity deep within the brain. Two common methods are in current use, called Positron Emission Tomography (PET) and functional Magnetic Resonance Imaging (fMRI). Both monitor levels of brain activity through measurements of the hunger of brain cells for blood.

PET is the better established, and has provided some of the first glimpses of working brain centers by highlighting the minute changes in blood flow that occur when we think or when we take a drug like alcohol.[2] In a typical setup, subjects, whose heads are placed within a doughnut-shaped radiation detector, are injected with a small quantity of water labeled with a short-lived and harmless radioactive isotope of oxygen, oxygen-15 (see Fig. A.1). The unstable oxygen-15 nucleus decays with the emission of a positron (a positively charged antimatter electron); this positron immediately collides with an electron, leading to particle annihilation and the production of a pair of strongly correlated photons, which leave the collision site in exactly opposite directions.[3] Within a minute, this water finds its way into the brain, at a rate dependent on the local blood flow there. The greater this flow of blood, the higher the radioactivity level of the blood within the brain, and hence the more electromagnetic radiation sensed by the PET detector.

PET has highlighted the extraordinary way the brain can sift and filter information about the constant and invariant features of the external world from the shifting patterns of light cast on the retina. Such issues make this field of research as much a philosophical endeavor as a scientific one, according to one of the leading scientists investigating vision, Semir Zeki of University College, London.[4]

Wavelengths of light reflected from the surface of an object vary enormously depending on whether it is illuminated by a fifty-watt lightbulb, daylight, twilight, and so on. Yet we assign constant colors to most objects: a green tree in the morning still appears green in the evening.

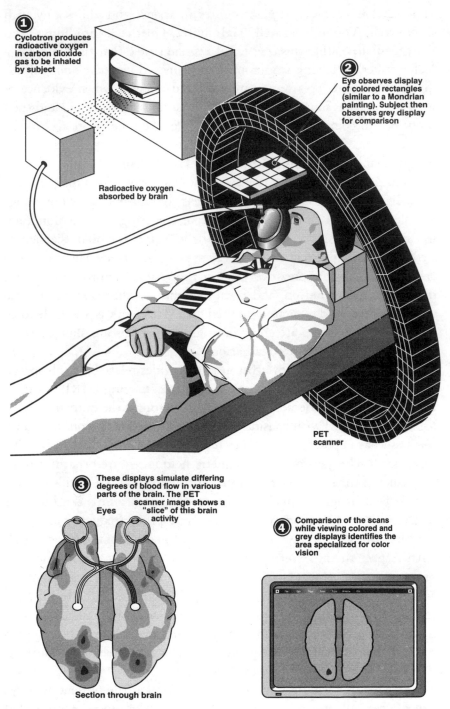

① Cyclotron produces radioactive oxygen in carbon dioxide gas to be inhaled by subject

② Eye observes display of colored rectangles (similar to a Mondrian painting). Subject then observes grey display for comparison

Radioactive oxygen absorbed by brain

PET scanner

③ These displays simulate differing degrees of blood flow in various parts of the brain. The PET scanner image shows a "slice" of this brain activity

Eyes

④ Comparison of the scans while viewing colored and grey displays identifies the area specialized for color vision

Section through brain

Figure A.1   At the Medical Research Council's cyclotron unit in Hammersmith, west London, a PET scanner is used to reveal the brain's "color processing laboratory."

This paradox has engaged giant figures in science and philosophy such as Newton, Young, Maxwell, Helmholtz, Goethe, Schrödinger, and Wittgenstein.[5] Earlier this century, it was suggested that the brain had a "color center" following studies of brain-damaged patients who had lost their color vision. This was dismissed until recently, when evidence of such a center was found by Semir Zeki in studies of macaques and then confirmed in complementary PET studies of the inner workings of the human brain. Working with Richard Frackowiak at the Medical Research Council's Cyclotron Unit at the Hammersmith Hospital, west London, Zeki compared the activity of the brain when the subject looked at an abstract Mondrian painting that was colorful but devoid of recognizable form with that when they looked at grey images of the Mondrian. In this way, the color center, a region at the back of the brain that acts as a color processing laboratory for the eyes, was discovered.[6] He went on to demonstrate separate areas for motion processing, confirming his idea that color, form, motion, and possibly other attributes of the visible world are processed separately in the brain. This work is a tour de force of scanning but there are shortcomings: PET scans are slow to carry out and require exposure to radiation.

MRI is poised to take over because it provides relatively high spatial and temporal resolution. Magnetic Resonance Imaging (MRI) has been used for more than a decade to study the structure of the human body. It has been employed with considerable success to locate tumors and study soft tissue, as opposed to the bones revealed by conventional X rays. To carry out MRI experiments, a magnetic field is required that is tens of thousands of times as powerful as that encountered at the earth's surface. In this field, magnetic nuclei at the core of many atoms can be made to align. If we direct radiowaves of an appropriate frequency at such a sample, the nuclei in these atoms can be induced to flip their orientation with respect to the external magnetic field. The hydrogen nuclei—protons—present in such abundance, both in water and organic molecules throughout the body, are among the most valuable of all such resonators. Flipping hydrogen nuclei emit radiowaves whose precise frequencies reveal the local molecular environment in which each proton sits. Moreover, the more protons present in a given environment, the more intense the emitted radiation.

The late Linus Pauling first noted in 1937 that the presence of oxygen dramatically alters the magnetic properties of blood. A group led by Seiji Ogawa at Bell Laboratories in New Jersey showed in 1990 that these changes could be mapped in the brain using MRI.[7] The strong external magnetic field used in the scanner induces an alignment of the atomic iron magnets in hemoglobin, the red molecule in blood that carries oxygen around the body. The local magnetic field in the neighborhood of hemoglobin causes a characteristic distortion of the radiowave signals emitted by nearby protons, which thus show up in the scan. Because only hemoglobin that is bereft of oxygen can be magnetized in this way, the effect is primarily seen in blood returning toward the heart, rather than in blood heading out to the brain.[8] These subtle effects can reveal changes in the blood supply to regions of the brain as small as one millimeter across, although as with PET the temporal resolution is too slow to see neural dynamics.

Using this technique, one large American team has revealed that the brains of men are organized differently from those of women when it comes to processing language.[9] In other work, Xiaoping Hu and colleagues at the University of Minnesota's Center for Magnetic Resonance Research have witnessed the imagination at work.[10] The scanner showed that centers of the brain responsible for vision were activated when subjects used their imagination, although the brain activity was approximately half as much as when they actually looked. The same team has revealed a similar phenomenon in the sound-processing centers of the brain when subjects are asked to *imagine* saying words. And they also found a part of the brain that is involved in forming a mental picture or map, situated in the fissure between the parietal and occipital lobes. This region of the brain was revealed when subjects were asked to imagine navigating their way through their own homes. In one case, a Japanese taxi driver who had lost the ability to navigate after suffering a head wound was found to have brain damage in the same region.

Activity in the brain occurs with fantastic speed. For example, from the moment an image is projected on a screen it takes around thirty milliseconds to reach the specialized vision processing regions in the cortex.[11] Until now, most scanners have been too slow to capture a thought on the wing. Not any longer. Riitta Salmelin and her colleagues[12] at the Helsinki University of Technology gave a groundbreaking demonstration of how a thought can be seen as a wave of neural activity rippling from the back to the front of the head, through both hemispheres of the brain (see Fig. A.2). The six subjects of their study were carrying out a process we are all familiar with, that of naming a picture—for instance, saying

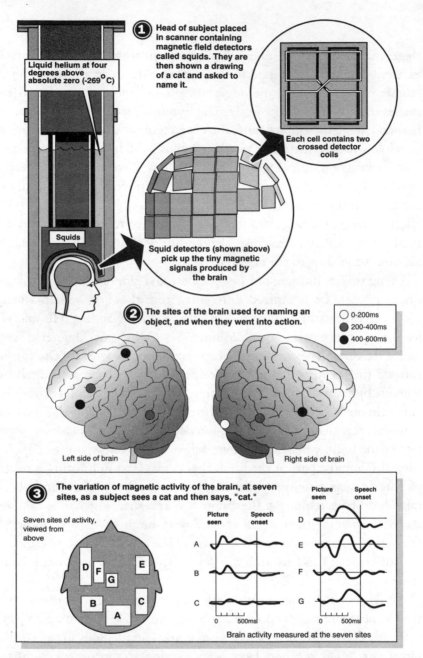

Figure A.2 Countdown to saying "cat." During the process of seeing a cat and saying "cat," the Finns witnessed the following processes at work within the brain using their scanner: 0 seconds: subject shown image of a cat; 0–200 milliseconds: visual area at the back of the head becomes active; 200–400 milliseconds: language areas of the brain, still toward the back of the head, become active; 0.5 seconds: frontal language areas become active, notably those involved in forming a name that can be spoken (though the left side of the brain is supposed to predominate in processing language, both sides were found to be active); 500–800 milliseconds: the motor areas of the brain that control the muscles of the vocal cord and mouth are activated; 830 milliseconds: subject says "cat."

"cat" on seeing a line drawing of a feline. The measurement relies on the rapidly changing magnetic field generated by the electrical activity within the conscious brain. Called magnetoencephalography, or MEG, the method was first demonstrated in principle in 1968 when David Cohen at the Massachusetts Institute of Technology showed that it was possible to record magnetic signals generated when currents flow in the brain. The technique avoids the principal problem of using EEG, which is the blurring of electrical signals due to intervening tissue. Magnetic fields pass through nonmagnetic materials such as brains without any distortion, thus enabling us to monitor rapid changes in neural activity to within a millisecond.

This technique is already helping to guide the hand of surgeons at the Department of Physiology and Biophysics at New York University Medical Center, chaired by Rodolfo Llinas. They hope that, in the near future, surgeons will be able to reveal the locations of the speech center and many other brain functions and then practice their technique with a "digital scalpel" long before they touch a hair on a patient's head. The convoluted folds of the living brain are revealed with a conventional MRI scanner. Then MEG is used to uncover the centers of the brain at work, in this case, the ones that control the fingers. MEG can show how, a few thousandths of a second before a patient starts moving his right index finger, nerve cells in the motor area of the brain send movement commands to the muscles involved. A second image, taken forty thousandths of a second later, reveals that another region processes information from the muscles and joints that the finger is moving. "It is absolutely stunning," said Llinas. "We can see the brain working in real time."

NOTES

PROLOGUE

1. His brother provides some examples of von Neumann's humor. A favorite story recounted a Berlin street scene during the First World War. A man at the corner yells repeatedly, "The Kaiser is an idiot." Two police agents appear and arrest him for high treason. "But I was referring to the Austrian Kaiser, not to our Kaiser," the man cries. The officers reply, "You can't fool us! We know who the idiot is." Taken from Nicholas A. Vonneuman, "John von Neumann, as seen by his brother" (© Nicholas Vonneuman, PO Box 3097, Meadowbrook, PA 19046 USA, 1987), p. 39, with permission.
2. A. Hodges, *Alan Turing: The Enigma* (Vintage, London, 1992), p. 95. P. Hilton, American Association for the Advancement of Science, February 1995.
3. Ibid., p. 89.
4. A. Turing, "Computing machinery and intelligence" *Mind* 59 no. 236. See also D. Hofstadter and D. Dennett (eds.), *The Mind's I* (Basic Books, New York, 1981).
5. A. Hodges, *Alan Turing: The Enigma* (Vintage, London, 1992), p. xiii.
6. Ibid., p. 488. "In a sense he had prepared for it but he did not premeditate the exact moment when it would happen. It was very significant that it happened on a long weekend, which he hated. So long as he could be at work, he could forget his circumstances." P. Hilton, interview with Roger Highfield, February 1995. Hilton joined the code-breaking effort in January 1942, where he met and worked with Turing on naval codes. Turing's first question was: "Hilton, do you play chess?" They became friends when Hilton helped him solve a chess problem that had obsessed him.
7. P. Hilton, American Association for the Advancement of Science, February 1995.

8. A. Hodges, *Alan Turing: The Enigma* (Vintage, London, 1992), p. 519.

9. Nicholas A. Vonneuman, "John von Neumann, as seen by his brother" (© Nicholas Vonneuman, PO Box 3097, Meadowbrook, PA 19046 USA, 1987), p. 4.

10. J. von Neumann, *The Computer and the Brain* (Yale University Press, New Haven, 1958), p. 82.

CHAPTER 1

1. I. Kant, *Universal Natural History and Theory of the Heavens*, S. Jaki (trans.) (Scottish Academic Press, Edinburgh, 1981), p. 87.

2. As noted, irreversibility is an essential element of macroscopic complexity, as demanded by the second law of thermodynamics. This law conflicts with other laws of microscopic physics, which describe only reversible phenomena, as we discussed in our earlier book, *The Arrow of Time* (W. H. Allen, London, 1990; Ballantine, New York, 1991).

3. O. Penrose and P. V. Coveney, *Proc R Soc London A* 447 631 (1994). Among other things, this paper shows, for a very simple model problem, that in order to make the microscopic (classical) mechanical description compatible with the macroscopic one, it is necessary to place some restriction on what can be observed in the former case.

4. M. Gell-Mann, *The Quark and the Jaguar* (Little, Brown, Boston and London, 1994), p. 352.

5. Technically speaking, these plots are of the Lyapunov exponents. A chaotic solution is one with at least one positive definite Lyapunov exponent.

6. Confirming the unification of electromagnetic and weak nuclear forces.

7. Two teams of physicists—each about 450 members strong—at the Fermi National Laboratory's particle accelerator near Chicago simultaneously transmitted their findings to *Physical Review Letters* in March 1995. However, the announcement was an anticlimax: they had narrowed the odds that they were wrong from one in 400, announced the previous year, to one in 500,000. And in the years before that, cognoscenti of particle-physics conferences had come to expect that, at an appropriate juncture in such proceedings, someone from Fermilab would talk about the top quark: the lab's detectors are the only ones capable of seeing such heavy particles. All the experimenters would usually say was that, if the particle had such and such a weight, they would have seen it, so it must be heavier than that. Hints that the top quark had indeed been found have circulated since at least 1992.

8. A. Toffler, in the foreword to I. Prigogine and I. Stengers, *Order Out of Chaos* (Heinemann, London, 1984), p. xi.

9. S. Rose, *Nature* 373 380 (1995).

10. J. Monod, *Chance and Necessity* (Fontana, London, 1972).

11. That is not to say that God has been left out of Habgood's particular picture. "You can identify the hand of God in the whole process. Putting it very crudely,

you can say it is in the mind of God that the laws of nature are written." Archbishop Habgood, interview with Roger Highfield, September 1994.

12. Archbishop Habgood, "A Theological Understanding of Life and Death" *British Association for the Advancement of Science* (September 1994).

13. But we should be clear about one thing: computational and natural phenomena will never be *identical*. The former are the result of discrete (digital) computations, whereas the latter arise from continuous (analog) processes. This distinction will be discussed further in Chapters 2, 3, and 9.

14. One example: "No definition of complexity is intrinsic. It is always context dependent." M. Gell-Mann, discussion with the authors, March 1994. Note that the term "chaos" admits a precise technical definition quite distinct from its everyday meaning. It is frequently and deliberately used ambiguously by popularizing scientists, and usually completely misunderstood by laypeople.

15. C. Emmeche, *The Garden in the Machine: The Emerging Science of Artificial Life* (Princeton University Press, 1994), p. 18. D. Johnson, Arts and Books Editor of *The Daily Telegraph*, points out that one conceptual approach to teaching fine art defines art in almost identical terms—not as something to be viewed or touched, but as a process in which the ideas are the art.

CHAPTER 2

1. A. Pope, *An Essay on Man 1732 Epistle II*. The tone of this poem is decidedly ambiguous compared with the celebrated epitaph Pope proposed for Newton's Tomb: "Nature, and Nature's Laws lay hid in Night / God said, Let Newton be! and All was Light."

2. R. Bacon, *Opus Majus* 4 III, R. B. Burke (trans.) (University of Pennsylvania Press, 1928). "This science (mathematics) is earlier than the others and naturally precedes them. Hence it is clear that it should be studied first, that through it we may advance to all the later sciences."

3. This statement is subject to the qualification that it excludes from consideration the concept of a quantum computer. Such machines are discussed further in Chapter 3.

4. As we will see later, we mean that the process must have an *algorithmic* representation.

5. *Companion Encyclopaedia of the History and Philosophy of the Mathematical Sciences*, Vol. 2, I. Grattan-Guinness (ed.) (Routledge, London, 1944), p. 1054.

6. J. Barrow, *Pi in the Sky* (Clarendon Press, Oxford, 1992), p. 296. The answer that unavoidably emerges from our book is that mathematics works because our brains have evolved it to capture aspects of the world we inhabit. This is in keeping with von Neumann's vision that we described in the Prologue.

7. Examples of proto-mathematical activity abound. A bone dated between 9000 and 6500 B.C. was dug up in the 1950s at Ishango, in what is now Zaire, that has engraved tallying marks arranged in groups. In central Czechoslovakia a 30,000-year-old wolf bone was found in 1937 that had fifty-five notches carved

in fives, perhaps a record of kills organized like the fingers on the hunter's hand. See L. Bunt, P. Jones, and J. Bedient, *The Historical Roots of Elementary Mathematics* (Prentice-Hall, Englewood Cliffs, NJ, 1976), p. 1. *Companion Encyclopaedia of the History and Philosophy of the Mathematical Sciences*, Vol. 1, I. Grattan Guinness (ed.) (Routledge, London, 1994), p. 31.

8. Aristotle, *Metaphysics*. "The mathematical arts were first set up in Egypt; for there the priestly caste were allowed to enjoy leisure." See also Proclus, *On Euclid I*, Thomas, *Greek Mathematical Works I*, I. Thomas (trans.) (Heinemann, 1939), pp. 145–47. "According to most accounts geometry was first discovered among the Egyptians, taking its origin from the measurement of areas."

9. The order used a five-pointed star (pentagram) as its symbol. His followers assigned whole (integer) numbers to basic notions like justice, soul, and opportunity. They investigated the relationship between the length of a string and the pitch of a tone, thus furnishing one of the earliest examples of a natural law found by experiment. They believed that numbers were the ultimate essence of reality. M. Kline, *Mathematical Thought from Ancient to Modern Times* (Oxford University Press, New York, 1972), p. 29. Though some ancient civilizations such as the Egyptians and Babylonians had learned to think about numbers as divorced from physical objects, there is some question as to how much they were consciously aware of the abstract nature of such thinking.

10. W. Burkert, *Lore and Science in Ancient Pythagoreanism*, E. Minar (trans.) (Harvard University Press, Cambridge, 1972), p. 112. The first historian of mathematics, Eudemus, hailed Pythagoras as the creator of pure mathematics. Adulatory reports describe him as a searching scientist-philosopher, a convincingly holy man, a consummate politician, and statesman. S. Bochner, *The Role of Mathematics in the Rise of Science* (Princeton University Press, 1966), p. 359. But his real contributions are swamped in a mire of legends. Most significant, it is doubtful that he himself was the first author of any significant mathematical discovery. *Companion Encyclopaedia of the History and Philosophy of the Mathematical Sciences*, Vol. 1, I. Grattan-Guinness (ed.) (Routledge, London, 1994), p. 46. M. Kline, *Mathematical Thought from Ancient to Modern Times* (Oxford University Press, New York, 1972), p. 28.

11. Pythagoras' theorem says that the square of the length of the hypotenuse is equal to the sum of the squares of the lengths of the other two sides. However, Pythagoras' theorem, applied to a square whose sides are of one length, says that the length of the associated diagonal would be a number expressed today as the square root of two. The Pythagorean order later proved that this number was irrational—it could not be expressed as a ratio of two whole numbers—and this caused them great consternation because its value could not be measured precisely. Their discovery of irrational numbers had a profound effect on their faith in mathematics as a secret code describing the universe, and adumbrated the philosophical problems that have arisen in the twentieth century surrounding noncomputability.

12. B. Russell, *My Philosophical Development* (George Allen and Unwin Ltd., London, 1969), p. 65.

13. N. Wiener, *Cybernetics, or Control and Communication in the Animal and the Machine*, 2nd ed. (MIT Press and John Wiley & Sons, New York, London, 1961), p. 2.

14. Despite his monumental contributions, Leibniz was an object of ridicule in his later years for his dated and overornate clothes, his huge black wig, and increasingly potty schemes. The reception to his death in 1716 stood in stark contrast to the pomp, pageantry, poetry, and statues that celebrated the life of his rival Newton. The only person who came to mourn Leibniz was his secretary. An eyewitness wrote: "He was buried more like a robber than what he really was, the ornament of his country." H. Goldstine, *The Computer: From Pascal to von Neumann* (Princeton University Press, 1972), p. 9.

15. "If we could find characters or signs appropriate for expressing all our thoughts as definitely and as exactly as arithmetic expresses numbers or geometric analysis expresses lines, we could in all subjects in so far as they are amenable to reasoning accomplish what is done in Arithmetic and Geometry.... And if someone would doubt my results, I should say to him: 'let us calculate, Sir,' and thus by taking to pen and ink we should soon settle the question." G. Leibniz, *The Philosophical Works of Leibniz*, G. Duncan (trans.) (Tuttle, Morehouse & Taylor, New Haven, CT, 1916). Though prescient, it smacks of the hubris of some twentieth-century computer scientists; even in his day, Leibniz was satirized in Swift's *Voyage to Balnibarbi*. See G. MacDonald Ross, *Leibniz* (Oxford University Press, Oxford, 1984), p. 13.

16. The date May 8, 1686, was added in the second edition, probably as a result of his bitter dispute with Leibniz on the invention of calculus. The quotation in the text omits the parentheses: "(as we are told by Pappus)." Pappus of Alexandria (A.D. 320) produced a collection of critical commentaries of previous work.

17. I. Newton, *Sir Isaac Newton's Mathematical Principles of Natural Philosophy and His System of the World*, F. Cajori (trans.) (University of California Press, Berkeley, 1934). Preface to the first edition (p. xvii).

18. J. Jeans, *Philosophy* 7 9 and 12 (1932).

19. "Arithmetic has a very great and elevating effect, compelling the soul to reason about abstract numbers, and rebelling against the introduction of visible or tangible objects into the argument." Plato, *Republic* Book VII sec. 525 in *The Dialogues of Plato*, Vol. 2, B. Jowett (ed.) (Clarendon Press, 1953).

20. Godfrey Hardy (1877–1947), an outstanding British analyst, wrote: "I believe that mathematical reality lies outside us, that our function is to discover or observe it, and that the theorems which we prove, and which we describe grandiloquently as our 'creations' are simply our notes of our observations." G. Hardy, *A Mathematician's Apology*, foreword C. P. Snow (Cambridge University Press, Cambridge, 1969), p. 123.

21. E. Wigner, *Communications in Pure and Applied Mathematics* 13 1 (1960).

22. B. Newton-Smith, interview with Roger Highfield, June 1994.

23. P. Kitcher, *The Nature of Mathematical Knowledge* (Oxford University Press, Oxford, 1983), p. 5.

24. N. Cartwright, *How the Laws of Physics Lie* (Clarendon Press, Oxford, 1983), p. 19.

25. Ibid., p. 3.

26. Recently some remarkable applications of certain theorems in abstract number theory have been found in the study of dynamical systems, particularly in explorations concerned with the distribution of their energy levels in the so-called *semiclassical* regime between classical and quantum mechanics. Among numerous examples, one due to Mike Berry of Bristol University is especially intriguing: he contends that the energy levels of a specific, yet hitherto unknown system (with a time-asymmetric Hamiltonian) would correspond with the locations of the zeros of the Riemann zeta function. He states that the validity of the Riemann hypothesis (according to which these all lie on a single line) would require that the classical limit be chaotic; M. V. Berry, *Proc R Soc London* A **400** 229 (1985). No mathematician has yet succeeded in proving this conjecture, although the numerical evidence out to exceedingly large numbers indicates its validity. It would be significant if such a proof came via physics.

27. These in turn can be accounted for completely in terms of a menagerie of subatomic particles, including electrons, protons, neutrons, and quarks—although where the regress should end is far from clear.

28. One of the most obvious difficulties for atomistic reductionism is provided by the second law of thermodynamics, which asserts the irreversibility of macroscopic phenomena in the face of microscopic equations of motion that themselves are time symmetric. This problem was the major theme developed in our earlier book, *The Arrow of Time*. Some new work that helps to resolve this problem can be found in O. Penrose and P. V. Coveney, *Proc R Soc London* A **447** 631 (1994); P. V. Coveney and A. K. Evans, *J Stat Phys* **77** 229 (1994); A. K. Evans and P. V. Coveney, *Proc R Soc London* A **448** 293 (1995).

29. Note, however, that the German Georg Cantor (1845–1918) had introduced infinite numbers in mathematics perfectly consistently on the basis of a consideration of infinite sets.

30. Coincidentally, a related bid was born when Bertrand Russell attended the 1900 International Congress of Philosophy in Paris, where he delivered a paper (B. Russell, *L'Idée d'ordre et la position absolue dans l'espace et le temps*, Congrès international de philosophie, logique, et histoire des sciences, vol. III [Paris, 1901], p. 241). There he encountered the work of Giuseppe Peano, a turning point in his life: "In discussions at the congress I observed that he was always more precise than anyone else, and that he invariably got the better of any argument upon which he embarked. As the days went by, I decided that this must be owing to mathematical logic." B. Russell, *The Autobiography of Bertrand Russell 1872–1914* (George Allen and Unwin Ltd., London 1967), p. 144. Russell joined forces with his former Cambridge mathematics tutor Alfred North Whitehead to create their *magnum opus*, *Principia Mathematica*, whose aim was to show that all pure mathematics can be reduced to primitive logic—logical atomism. Though Russell's *Principia* were as weighty as Newton's, they bear no comparison in terms of impact. "One of the great intellectual monuments of all time"

is the stock assessment of Russell's *Principia*. However, like many monuments it marks a tribute to a noble death, only this time of a great idea rather than a great individual. The book exerted little influence on contemporary mathematics. Others have referred to it as "a great and splendid failure." V. Lowe, *Alfred North Whitehead, The Man and His Work, Volume 1: 1861–1910* (The Johns Hopkins University Press, Baltimore, 1985), p. 290.

31. Constance Reid, *op cit.*, p. 73.

32. D. Hilbert, *Bulletin of the American Mathematical Society* VIII 437 (1901–2). At the congress he presented only ten problems out of the twenty-three in line with the urgings of Hermann Minkowski to shorten his speech.

33. See Note 30; Russell seemed quite out of touch with Hilbert's mission. In January 1914 he wrote that Hilbert "has grown interested in Whitehead's and my work, and that they think of asking me to lecture there next year. I hope they will. Germany (except Frege, who is not known) has hitherto been behindhand in mathematical logic, but if the mathematicians would take it up there, they would probably do wonders." *The Selected Letters of Bertrand Russell*, Volume 1: *The Private Years 1884–1914*, Nicholas Griffin (ed.) (Allen Lane, The Penguin Press, London, 1992), p. 487.

34. D. Hilbert, *Bulletin of the American Mathematical Society* VIII 445 (Macmillan, Lancaster, Pa., and New York, 1901–2). This was a rebuff to Emil duBois Reymond, a physiologist turned philosopher. When the eighteen-year-old Hilbert arrived at university in his hometown of Königsberg, the words of Reymond were much quoted, notably: *Ignoramus et ignorabimus*—"we are ignorant and we shall remain ignorant."

35. Constance Reid, *op cit.*, p. 153.

36. Ibid., p. 148; also *Bull Am Math Soc* 30 31 (1924) and 30 81 (1913) and *South African J Sci* 49 139 (1952).

37. In the twentieth century logicians such as Jan Lukasiewicz, Emil Post, and Alfred Tarski realized that they could indeed formulate logical systems different from Aristotle's through rejecting the validity of the law of the excluded middle. As Alonzo Church wrote in 1932, "There exists, undoubtedly, more than one formal system whose use as a logic is feasible, and of these systems one may be more pleasing or more convenient than another, but it cannot be said that one is right and the other is wrong." See also the discussion of fuzzy logic in Chapter 3.

38. Constance Reid, *op cit.*, p. 155. Another casualty would have been Cantor's theory of infinite sets.

39. Constance Reid, *op cit.*, p. 268.

40. One mechanical method to investigate a vast and sprawling inventory of possible proofs is by using what is jokingly called a "British Museum Algorithm." The algorithm works through all possible proofs in size order, discarding gibberish, whatever does not obey the rules, and leaving a residue of valid proofs. G. Chaitin, interview with Roger Highfield, December 1994.

41. That is, the man from Khwarizm, a Middle Eastern empire. His work survived

as a Latin translation *Algorithmi de numbro indorum*. See, for example, I. Grattan-Guinness (ed.), *Companion Encyclopaedia of the History and Philosophy of the Mathematical Sciences*, Vol. 1 (Routledge, London, 1994), p. 689.

42. Constance Reid, *op cit.*, p. 196.

43. K. Gödel, *Monatshefte für Mathematik und Physik* **38**, 173 (1931). More technically, Gödel's theorem states that, in an axiomatic formal calculus of the complexity of number theory, it is impossible to prove consistency without using methods from outside the system. Number theory involves the properties and relations of integers, including their algebraic and analytic extensions. This includes divisibility, primality and factorization, partition, irrational and transcendental numbers, algebraic independence, the description of numbers as sums of squares, and the integer solution of polynomials in several unknowns (including the Diophantine equations discussed later in this chapter). To prove his incompleteness theorem, Gödel constructed what are now called Gödel numbers, so designed to be uniquely associated with every symbol of the calculus and hence with every sequence. A valid statement in the calculus then corresponds to a particular property of the Gödel numbers. He constructed a Gödel number for the proposition that the formula associated with this number is not provable—in other words, this is akin to proving a statement that says "I am not provable."

44. The source of this surprising difficulty is related to the well-known paradoxes of self-reference studied at the turn of the century by Bertrand Russell. These arose from the work of the Russian-born mathematician Georg Cantor (1845–1918), who developed ways to handle infinities. Russell ran into a roadblock while attempting to extend the work. What resulted were logical paradoxes of an apparently trivial nature but that had devastating impact on Russell's method. A horrified Russell wrote to the logician who had provided his original inspiration, Gottlob Frege, on June 16, 1902, disclosing the contradiction he had discovered a year earlier with the understatement: "There is just one point where I have encountered a difficulty." When Russell's letter arrived, Frege was preparing the second volume of his classical work on the logical foundations of arithmetic *Grundgesetze der Arithmetik*. Frege replied on June 22, 1902: "Your discovery of the contradiction caused me the greatest surprise and, I would almost say, consternation . . . Die Arithmetik ist ins Schwanken geraten" (arithmetic is tottering).

45. J. von Neumann, J. Bulloff, T. Holyoke, and S. Hahan (eds.), *Foundations of Mathematics, Symposium Papers Commemorating the Sixtieth Birthday of Kurt Gödel* (Springer Verlag, Berlin, 1969).

46. N. Cooper (ed.), *From Cardinals to Chaos, Reflections on the Life and Legacy of Stanislaw Ulam* (Cambridge University Press, Cambridge, 1989), p. 306.

47. J. Barrow, *Pi in the Sky* (Clarendon Press, Oxford, 1992), p. 19.

48. H. Wang, *Reflections on Kurt Gödel* (MIT Press, Cambridge, MA, 1987), p. 133.

49. D. Hilbert, *Probleme der Grundlegung der Mathematik* Atti del Congresso internazionale dei matematici, Bologna 3–10 Settembre 1928 1, p. 135 (Bologna, 1929).

50. For instance, by the so-called British Museum Algorithm.

51. Hardy's 1928 Rouse Ball Lecture, quoted in A. Hodges, *Alan Turing: The Enigma* (Vintage, London, 1992), p. 93.

52. A. Hodges, *op cit.*, p. 11.

53. Turing's biographer Andrew Hodges described how the physical nature of the mind had obsessed Turing since the tragic death of his closest friend Christopher Morcom from tuberculosis in 1930. The two had discussed astronomy, ciphers, and science and Turing felt that Morcom somehow continued to help and influence him. Ironically, Turing's work on his great paper—"On computable numbers, with an application to the *Entscheidungsproblem*"—snuffed out any ideas of spiritual survival and communication. Turing became a materialist, viewing the mind as a machine. As Hodges remarked, "Christopher Morcom had died a second death, and *Computable numbers* marked his passing." A. Hodges, *op cit.*, p. 108. Note, however, that Turing was surprisingly keen on ESP—it appears as a plausible phenomenon in his *Mind* article.

54. A. Hodges, *op cit.*, p. 96.

55. A. Turing, *Proceedings of the London Mathematical Society* (ser. 2) 42 230 plus correction 43 544 (1937).

56. A. Hodges, interview with the authors, June 1994.

57. A. Hodges, *op cit.*, p. 106.

58. A contradiction arises if a universal Turing machine operates on itself. This, the crux of his discovery, turns out to be related to Gödel's work in a deep way.

59. The argument is a subtle extension of Cantor's diagonal argument used to prove the existence of irrational numbers (more accurately, to prove that the cardinality of the real numbers is greater than that of the natural numbers). The method may, at first sight, look too simple to have defeated Turing's machine. However, the problem was that there was no guarantee that the machine would, say, generate a 5,812th digit of the 5,812th number in a *finite* time: the Turing machine (or program) used to calculate it could, for instance, produce one digit, ten, one hundred, or just cycle forever without producing any output at all. There was no way to guarantee that the machine would produce each digit in the decimal expansion in a finite time, the only way to ensure that this process could be mechanized. All uncomputable numbers are irrational, but not vice versa: some irrational numbers, such as *pi* and *e*, are computable, since each digit in their decimal expansions can be computed in finite time. At root, the halting problem stems from the fact that, while there are uncountably many real numbers, there are only countably many Turing machines. (The computables have zero measure.)

60. Church made use of his elegant lambda calculus. See A. Church, *Annals of Mathematics Studies* no. 6 (Princeton University Press, 1941). The original early papers in this effort are published in M. Davis (ed.), *Computability: The Undecidable: Basic Papers on Undecidable Propositions, Unsolvable Problems and Computable Functions* (Raven Press, New York, 1965).

61. G. J. Chaitin, *The Limits of Mathematics* (1994, available by e-mail on chao-dyn@xyz.lanl.gov with "Subject: get 9407010," or via the World Wide Web on http://xyz.lanl.gov/), G. J. Chaitin, *Information-Theoretic Incompleteness*

(World Scientific, Singapore, 1992), G. J. Chaitin, *Information, Randomness and Incompleteness*, 2nd ed. (World Scientific, Singapore, 1990), G. J. Chaitin, *Algorithmic Information Theory* (Cambridge University Press, 1987), and G. J. Chaitin, in *The New Scientist Guide to Chaos*, N. Hall (ed.) (Penguin, London, 1991).

62. G. Chaitin, interview with Roger Highfield, November 1994.

63. D. Hilbert, *Bulletin of the American Mathematical Society* VIII (Macmillan, Lancaster, PA., and New York, 1901–2), p. 439. These equations include "Fermat's last theorem," one of the most famous mathematical problems (there is no cube that can be expressed as the sum of two cubes, no fourth power as the sum of two fourth powers, and so on).

64. The third-century B.C. Greek mathematician Diophantus was a somewhat mysterious character. All that is known of his life are his age when he married, when his son was born, and when he died, inferred from an arithmetical riddle found on his epitaph ("This tomb tells scientifically the measure of his life"). Only six of the supposed thirteen books of his *Arithmetica* are extant but these describe the earliest known algebraic notation and deal with a wide range of number theoretic and geometric problems.

65. M. Davis, Y. Matijasevich, and J. Robinson, *Proceedings of Symposia in Pure Mathematics XXVIII* F. Browder (ed.) (Providence, RI: American Mathematical Society, 1976), pp. 223–378.

66. A few years later, James Jones of the University of Calgary, Canada, and Yuri Matijasevich discovered that it was possible to solve Hilbert's problem without complicated number theory by drawing on a theorem that had been proved by Edouard Lucas over a hundred years ago.

67. Technically, it is an exponential Diophantine equation, which means that the powers to which various variables are raised in the equation are themselves integer variables. The proof by Jones and Matijasevich that any computation can be encoded in an exponential Diophantine equation can be found in *Journal of Symbolic Logic* 49 818 (1984).

68. G. Chaitin, interview with Roger Highfield, November 1994.

69. The details can be found in G. Chaitin, *Algorithmic Information Theory*, op cit.

70. G. Chaitin, interview with Roger Highfield, November 1994.

71. G. J. Chaitin, *Information, Randomness and Incompleteness*, op cit., p. iii.

72. A floating point number is one with a specified number of digits in which the decimal (or equivalent) point can be in any position.

73. In fact, at the present time there are many problems in P, let alone NP, that cannot be handled because the time taken to get a solution scales as a high power of the size N. This situation arises, for example, in computing turbulent fluid flows, and for *ab initio* quantum calculations on molecular clusters.

74. Strictly speaking, we need to be a little careful here: the class NP actually denotes the class of problems whose solutions can be *verified* (not actually found) in polynomial time by a *nondeterministic* algorithm. They are thus defined in terms of a decision algorithm. Actually *finding* solutions deterministically for problems in NP may only be possible in exponential time. The distinction between deterministic and nondeterministic algorithms for handling intractable—

complex—problems is very deep. Genetic algorithms, simulated annealing, and neural networks are in general nondeterministic, and this explains their effectiveness at solving "intractable" problems; see later chapters for further discussion. Smart nondeterministic algorithms are capable of checking through an exponential number of possible solutions to a problem in polynomial time, and therefore almost certainly are in general more powerful than polynomial time deterministic algorithms. This distinction makes it very likely that the two classes, P and NP, are *not* identical. No rigorous mathematical proof of this statement has yet been found.

75. See, for other examples, M. Garey and D. S. Johnson, *Computers and Intractability*, Appendix 3. Strictly, the traveling salesman problem lies in the class of so-called NP-complete problems, the hardest of all the NP problems, and for which a polynomial time algorithm for its solution can only be found if the class P is identical with NP. It follows that if just one problem in the NP-complete set could be solved in polynomial time, so, too, could all other NP problems.

76. D. Stein, *Sci Amer* **261** 40 (1989).

77. The concept of algorithmic complexity results from work done by Chaitin, Kolmogorov, Solomonoff, Martin-Löf, Willis, and Loveland. This is discussed in Greg Chaitin's collection of papers, *Information, Randomness and Incompleteness*, 2nd ed. (World Scientific, Singapore, 1990).

78. G. Chaitin, interview with Roger Highfield, December 1994. He had been inspired by the work on von Neumann and published a major paper on this work when he was nineteen.

79. G. Chaitin, *Information, Randomness and Incompleteness*, 2nd ed. (World Scientific, Singapore, 1990), p. 88.

80. See, for example, G. Chaitin, *Algorithmic Information Theory* (Cambridge University Press, 1987).

81. J. Barrow, *Pi in the Sky*, p. 163.

82. S. Weinberg, *Dreams of a Final Theory: The Search for the Fundamental Laws of Nature* (Vintage, London, 1993), p. 3.

83. S. Weinberg, *op cit.*, p. 13.

84. Ibid., p. 44.

85. For example, Pour-El and Richards proved that, depending on initial conditions, at some time after the waves ripple forth, the field represented by the wave equation is not computable: M. Pour-El and I. Richards, *Ann Math Logic* **17** 61 (1979); *Adv Math* **39** 215 (1981); *Int J Theor Phys* **21** 553 (1982); M. Pour-El and I. Richards, *Computability in Analysis and Physics* (Springer Verlag, Heidelberg, 1989). Their book provides proofs that partial differential equations associated with unbounded operators, such as the wave equation and other hyperbolic equations, may give rise to noncomputable solutions starting from computable initial conditions; while elliptic and parabolic PDEs, which include the heat equation, and are associated with bounded linear operators, cannot produce this result (their First Main Theorem). There is an interesting review of this book by D. S. Bridges, *Bulletin of the American Mathematical Society* **24** No. 1 (January 1991), pp. 216–228. He considers a question posed by Pour-El and Richards in

their "Addendum: Open Problems" concerning how their results would appear from the standpoint of nonclassical forms of mathematical reasoning, such as Brouwer's intuitionism. Bridges points out that within some rigorously constructive perspectives, which reject Church's recursive approach altogether, the problem of noncomputable solutions does not arise since such objects are then denied any existence. (The authors are grateful to Professor Pour-El for informing us about this review.)

86. The noncomputable solutions are "weak" in the sense that, although continuous, they are not twice differentiable. In his book *The Emperor's New Mind* (Oxford University Press, 1989), p. 188, Roger Penrose claims that this kind of behavior is not physically realistic, since once-differentiable solutions are not sufficiently smooth. However, many examples of nondifferentiable phenomena in nature fulfill the Pour-El-Richards criterion—one example is a shock wave.

87. P. Hilton, interview with Roger Highfield, February 1995. "Turing was fighting, in a sense, a much more primitive battle. He was fighting a battle to get people to accept that you could use words like 'think' to describe the activities of a machine. This was almost as heretical as Darwinism in its day."

88. G. Chaitin, interview with Roger Highfield, November 1994.

89. H. T. Siegelmann, *Science* **268** 545 (1995). For further discussions of related issues, see S. Wolfram, *Phys Rev Lett* **54** 735 (1985), and C. Moore, *Nonlinearity* **4** 199 (1991). Whether analog models of the kind proposed by Siegelmann and Moore possess sufficient structural stability to be capable of acting as computational devices seems open to question.

90. K. Appel and W. Haken, *Illinois Journal of Mathematics* **21** 429–567 (1977).

91. K. Appel and W. Haken, *The Mathematical Intelligencer* **8** 10 (1986).

92. C. W. H. Lam, *The Mathematical Intelligencer* **12** 1 (1990). See also C. W. H. Lam, *Canadian Journal of Mathematics* **XLI** 1117 (1989).

93. A. Wiles made the claim in a series of three lectures given at the Isaac Newton Institute in Cambridge in June 1993. See P. Swinnerton-Dyer *Nature* **364** 13 (1993). On December 6, 1993, it was retracted in a press release from Princeton University. However, a further press release from Princeton University, dated October 16, 1994, announced that a new proof was circulating, and that two papers dealing with this have been submitted to *Annals of Mathematics*.

94. G. Chaitin in *Nature's Imagination*, J. Cornwell (ed.) (Oxford University Press, Oxford, 1995).

CHAPTER 3

1. One problem was how to use shaped charges to focus the detonation energy of an antitank weapon known as the bazooka. Others arose in the development of the first nuclear weapon: for instance, the deployment of conventional explosives to form a critical mass of fissile material, such as plutonium, that would explode with nuclear intensity rather than merely fizzle. R. Rhodes, *The Making of the Atomic Bomb* (Simon and Schuster, New York, 1986), p. 480.

2. D. Swade, *Companion Encyclopaedia of the History and Philosophy of the Mathemati-*

cal Sciences, Vol. 1, I. Grattan-Guinness (ed.) (Routledge, London and New York, 1994), p. 694.

3. J. Palfreman and D. Swade, *The Dream Machine* (BBC Books, 1991), p. 13.
4. In September 1623, Schickard wrote to the great astronomer and mathematician Johannes Kepler: "I have constructed a machine consisting of eleven complete and six incomplete (actually 'mutilated') sprocket wheels which can calculate. You would burst out laughing if you were present to see how it carries by itself from one column of tens to the next or borrows from them during subtraction." From S. Augarten, *Bit by Bit* (George Allen & Unwin, London, 1985), p. 15.
5. H. Goldstine, *The Computer from Pascal to von Neumann* (Princeton University Press, 1972), p. 8.
6. G. MacDonald Ross, *Leibniz* (Oxford University Press, Oxford, 1984), p. 12.
7. "Any animal straying over the railway would be pitched into this apron, probably having its legs broken, but forming no impediment to the progress of the train." C. Babbage, *Passages from the Life of a Philosopher* (Longman, Green, Longman, Roberts, & Green, 1864), p. 318.
8. They consisted of shoes with hinged flaps that spread on downward thrust. See D. Swade, *Charles Babbage and His Calculating Engines* (Science Museum, London, 1991), p. 38, and D. Swade, *Bulletin of the Scientific Instrument Society* 28 22 (1991). For other inventions and talents, see A. Hyman, *Charles Babbage, Pioneer of the Computer* (Oxford University Press, 1982), pp. 225 and 228. He even developed a set of code-breaking dictionaries; see C. Babbage, *On the Economy of Machinery and Manufactures* (Charles Knight, London, 1832). The most accessible source that contains this and other works of Babbage is the eleven-volume set: Martin Campbell-Kelly (ed.), *The Works of Charles Babbage* (Pickering & Chatto, London, 1989).
9. C. Babbage, *Passages from the Life of a Philosopher* (Longman, Green, Longman, Roberts, & Green, 1864), p. 337. An unintentionally hilarious picture of Babbage's battle against street music in London is given. "I have obtained in my own country, unenviable celebrity, not by anything I have done, but simply by a determined resistance to the tyranny of the lowest mob, whose love, not of music, but of the most discordant noises, is so great that it insists upon enjoying it at all hours and in every street" (p. 345). "One-fourth part of my working power has been destroyed by the nuisance against which I have protested." Local children would taunt him and neighbors erected placards "in which they abused me for having put the law in force against the destroyers of my time" (p. 354).
10. C. Babbage, *Passages from the Life of a Philosopher* (Longman, Green, Longman, Roberts, & Green, 1864), pp. 34 and 37.
11. "One of these walked or rather glided along a space of about four feet, when she turned round and went back to her original place. She used an eye-glass occasionally, and bowed frequently, as if recognizing her acquaintances. The motions of her limbs were singularly graceful. The other silver figure was an admirable *danseuse*, with a bird on the forefinger of her right hand, which

wagged its tail, flapped its wings and opened its beak." C. Babbage, *Passages from the Life of a Philosopher* (Longman, Green, Longman, Roberts, & Green, 1864), p. 17. In 1834 he bought the dancing lady at an auction.

12. C. Babbage, *Passages from the Life of a Philosopher* (Longman, Green, Longman, Roberts, & Green, 1864), p. 42.

13. The inaccuracies in contemporary mathematical and navigational tables were legendary. Dionysius Lardner, one of the pioneers of popularizing science—and of Babbage's ideas—wrote that a random selection of forty volumes of mathematical tables incorporated 3,700 acknowledged errata, some of which themselves contained errors. D. Swade, *Sci Amer* **268** 62 (1993).

14. D. Swade, *Charles Babbage and his Calculating Engines* (Science Museum, London, 1991), p. 4.

15. As Babbage put it, "The first and great cause of its discontinuance was the inordinately extravagant demands of the person whom I had employed to construct it for the Government." C. Babbage, *Passages from the Life of a Philosopher* (Longman, Green, Longman, Roberts, & Green, 1864), p. 449.

16. A. Hyman, *Charles Babbage, Pioneer of the Computer* (Oxford University Press, 1982), p. 134.

17. Ibid., p. 170.

18. Ibid., p. 128.

19. In Babbage's words: "The Analytical Engine consists of two parts:
1st. The store in which all the variables to be operated upon, as well as all those quantities which have arisen from the result of other operations, are placed.
2nd. The mill into which the quantities about to be operated upon are always brought."

20. C. Babbage, *Passages from the Life of a Philosopher* (Longman, Green, Longman, Roberts, & Green, 1864), p. 137.

21. H. Goldstine, *The Computer from Pascal to von Neumann* (Princeton University Press, 1972), p. 20.

22. C. Babbage, *Passages from the Life of a Philosopher* (Longman, Green, Longman, Roberts, & Green, 1864), p. 116.

23. A. Hyman, *Charles Babbage, Pioneer of the Computer* (Oxford University Press, 1982), p. 198.

24. Ibid., p. 177.

25. L. Menabrea, *Biobliotheque Universelle de Geneva* p.xli (1842). See Martin Campbell-Kelly (ed.), *The Works of Charles Babbage* (Pickering & Chatto, London, 1989).

26. A. Hyman, *Charles Babbage, Pioneer of the Computer* (Oxford University Press, 1982), p. 196. Ada was angered by this suggestion; she had no intention of having her first publication interrupted at the eleventh hour, risking delay. On August 14, 1843, she dispatched a sixteen-page letter of protest to Babbage that contains a telling comment about their different motives: "My own uncompromising principle is to endeavor to love truth and God before fame and glory. . . . Yours is to love truth and God (yes, deeply and constantly); to love fame, glory, honors yet more." According to Babbage's biographer, Anthony Hyman, her "slightly exalted tone is attributable to the effect of opium, alcohol or both."

27. He even once cheerfully admitted that she found a "grave mistake" in his work. C. Babbage, *Passages from the Life of a Philosopher* (Longman, Green, Longman, Roberts, & Green, 1864), p. 136.

28. W. Kaufmann and L. Smarr, *Supercomputing and the Transformation of Science* (Scientific American Library, 1993), p. 27.

29. M. Wilkes, *Communications of the ACM* 35 21 (1992). Against all this, there has been the counterclaim that Ada had little mathematical prowess, translating a printer's error in Menabrea's paper to produce mathematical nonsense; see D. Stein, *Ada: Life and Legacy* (MIT Press, Cambridge, MA, 1987). However, Doron Swade feels that Stein relies more on distaste for Ada's self-opinionated and self-conceited style than on informed technical analysis. D. Swade, *Annals of the History of Computing* 11 58 (1989).

30. A. Hyman, *Charles Babbage, Pioneer of the Computer* (Oxford University Press, 1982), p. 235.

31. ADA, a high-level computing language developed at the initiative of the U.S. Department of Defense and mainly used for military programming purposes. Narain Gehani, *ADA An Advanced Introduction Including Reference Manual for the Ada Programming Language* (Prentice-Hall, Englewood Cliffs, NJ, 1984).

32. Ibid., p. vi (foreword by Arno Penzias).

33. E. Horsburgh (ed.), *Modern Instruments and Methods of Calculations* (G. Bell and Sons, 1914), pp. 19–20, and Lord Moulton Ed C. Knott, *Napier Tercentenary Memorial Volume* (London, 1915), pp. 19–21.

34. D. Swade, *Sci Amer* 268 62 (1993). See also D. Swade, *Charles Babbage and His Calculating Engines* (Science Museum, London, 1991), p. ix.

35. A. Hyman, *Charles Babbage, Pioneer of the Computer* (Oxford University Press, 1982), p. 168. Also see D. Kuck, *The Structure of Computers and Computations* (Wiley, New York, 1978), p. 55.

36. C. Babbage, *Passages from the Life of a Philosopher* (Longman, Green, Longman, Roberts, & Green, 1864), p. 142. "It soon became apparent that my progress would be seriously impeded unless I could devise more rapid means of understanding and recalling the interpretation of my own drawings" (p. 143). "I have called this system of signs the Mechanical Notation. By its application to geometrical drawing it has given us a new demonstrative science, namely, that of proving that any given machine can or cannot exist; and if it can exist, that it will achieve its desired object."

37. B. Russell, *Proceedings of the Royal Irish Academy Series B* 57 64 (1955).

38. G. Boole, *An Investigation of the Laws of Thought* (Dover, New York, 1958).

39. G. Boole, *Philosophical Transactions of the Royal Society* 134 225 (1844).

40. D. MacHale, *George Boole. His Life and Work* (Boole Press, Dublin, 1985), p. 61.

41. Ibid., p. 69.

42. G. Boole, *The Mathematical Analysis of Logic, Being an Essay Towards a Calculus of Deductive Reasoning* (Macmillan, Barclay, and Macmillan, Cambridge, 1847).

43. The "+" sign here represents the Boolean operation of union. Technically speaking, a Boolean algebra is the algebra of sets or classes, in which the operations of union, intersection, and complementation are defined.

44. Boolean algebra is isomorphic with the propositional calculus, which is concerned with the logical relations between sentences.

45. Boole died in his mathematical prime but could never have been described as robust. Shortly after his marriage, his wife even forbade him to write poetry because he was overworked and she wanted to preserve his brain from needless exertion (D. MacHale, *George Boole. His Life and Work* [Boole Press, Dublin, 1985], p. 240). On November 24, 1864, Boole walked the three miles from his house in Ballintemple to the Queen's College in the rain. He lectured in wet clothes and went down with a feverish cold. His wife gave a graphic account of his suffering, when his mind drifted to a verse "For ever, O Lord, Thy word is settled in Heaven" which he had always associated with his discoveries. "One day he told me that the whole universe seemed spread before him like a great black ocean, where there was nothing to see and nothing to hear, except that at intervals, a silver trumpet seemed to sound across the waters, 'For ever, O Lord, Thy word is settled in Heaven.' " He died on the evening of December 8, 1864, at Ballintemple, leaving his wife and family virtually destitute.

46. As with Gödel, there was a strangely irrational aspect to the death of this superb logician: Mary Boole may have hastened her husband's end. In her book *The Message of Psychic Science to the World* (C. W. Daniel, 11 Cursitor St., London 1908), p. 161, she states: "It is theoretically wrong to sleep with a hot bottle in the bed except during acute illness. It tends to increase the liability to chilliness in future. The treatment of chronic coldness of the feet may consist, if the patient is strong, in making him walk for a few minutes daily on rough gravel in a brook; if he is delicate and weakly, you should at bed time first get the feet thoroughly warm by some antipathic means, and then give a shock of cold water followed by rubbing. This is homeopathy." The youngest Boole daughter, Ethel, wrote of the Missus, as Mary Boole came to be called: "The cause of father's early death was believed to have been the Missus' belief in a certain crank doctor who advocated cold water cures for everything. Someone—I can't remember who—is reported to have come in and found Father 'shivering between wet sheets.' Now for myself I am inclined to believe that this may have happened."

47. W. Kaufmann and L. Smarr, *Supercomputing and the Transformation of Science* (Scientific American Library, 1993), p. 28.

48. D. MacHale, *George Boole. His Life and Work* (Boole Press, Dublin, 1985), p. 234.

49. Ibid., p. 235.

50. A. Hyman, *Charles Babbage, Pioneer of the Computer* (Oxford University Press, 1982), p. 244.

51. J. Dubbey, *The Mathematical Work of Charles Babbage* (Cambridge University Press, Cambridge, 1978), p. 216.

52. M. Wilkes, *Communications of the ACM* 35 15 (1992) See also D. Swade, *Charles Babbage and His Calculating Engines* (Science Museum, London, 1991), p. x.

53. A. Hyman, *Charles Babbage, Pioneer of the Computer* (Oxford University Press, 1982), p. 255. See also A. Hodges, *Alan Turing: The Enigma* (Vintage, London, 1992), p. 297.

54. N. Vonneuman, *John von Neumann, "as seen by his brother"* (© Nicholas Vonneuman, PO Box 3097, Meadowbrook, PA 19046 USA, 1987), p. 24.

55. These are serial digital computers. They are to be contrasted with analog computers and parallel digital computers.

56. The Polish mathematician Stan Ulam, a close friend of von Neumann, said that von Neumann played with Turing's descriptions of computable numbers while they toured Europe in the summer of 1938 and had praised Turing's work several times the following year. After the war, von Neumann made Turing's 1937 paper required reading for employees of his computer project. The first published reference by von Neumann to Turing's universal machine was in his lecture on "The general and logical theory of automata," delivered on September 20, 1948. It can be found in von Neumann's *Collected Works* (Pergamon Press, Oxford, 1963).

57. W. Aspray, *John von Neumann and the Origins of Modern Computing* (MIT Press, Cambridge, MA, 1990), p. 177. See also A. Hodges, *Alan Turing: The Enigma* (Vintage, London, 1992), p. 178.

58. In one letter, Ortvay first made explicit the comparison between the brain and electronic calculating equipment. In another, he went on to describe the brain in terms of a network of cells that exchanged signals. W. Aspray, *John von Neumann and the Origins of Modern Computing* (MIT Press, Cambridge, MA, 1990), p. 177. See also A. Hodges, *Alan Turing: The Enigma* (Vintage, London, 1992), p. 179.

59. In W. Aspray and A. Burks (eds.), *Papers of John Von Neumann on Computers and Computer Theory* (MIT Press, Cambridge, MA, 1987), p. 422. Warren McCulloch says: "It was not until I saw Turing's paper that I began to get going the right way around, and with Pitts' help formulated the required logical calculus. What we thought we were doing (and I think we succeeded fairly well) was treating the brain as a Turing machine."

60. W. McCulloch and W. Pitt, *Bulletin of Mathematical Biophysics* 5 115 (1943).

61. He would later remark: "It has often been claimed that the activities and functions of the human nervous system are so complicated that no ordinary mechanism could possibly perform them. . . . The McCulloch-Pitts result . . . proves that anything that can be exhaustively and unambiguously put into work is *ipso facto* realizable by a suitable finite neural network." W. Aspray and A. Burks (eds.), *Papers of John Von Neumann on Computers and Computer Theory* (MIT Press, Cambridge, MA, 1987), p. 412.

62. R. Rhodes, *The Making of the Atomic Bomb* (Simon and Schuster, New York, 1986), p. 544.

63. The Navier-Stokes equations can also be derived from microscopic physics. This derivation starts from Boltzmann's equation for the single-particle distribution function in a fluid. Solving this equation using the Chapman-Enskog procedure leads to the Navier-Stokes equations for the single-particle distribution function. However, the problem here is that the Boltzmann equation is itself only an approximation to a correct kinetic equation.

64. In reality, this is too simplistic. Asymptotic methods can often be used, and

certain nonlinear systems (such as the Korteweg–de Vries equation describing solitions) are amenable to analytical solution.

65. J. von Neumann, *Theory of Self Reproducing Automata*, A. Burks (ed.) (University of Illinois Press, Urbana and London, 1966), p. 3.

66. Ibid., p. 4.

67. H. Goldstine, *The Computer from Pascal to von Neumann* (Princeton University Press, 1972), p. 25.

68. Ibid., p. 182.

69. The ENIAC used decimal, not binary numbers. It was still reasonably successful. Its first job was actually to run a simulation of an early design for the hydrogen bomb.

70. S. Augarten, *Bit by Bit* (George Allen & Unwin, London, 1985), p. 129.

71. It can be found in W. Aspray and A. Burks (eds.), *Papers of John Von Neumann on Computers and Computer Theory* (MIT Press Cambridge, MA, 1987), p. 17.

72. The report became the focal point of a dispute over who deserved credit for the ideas. On one side were Goldstine and von Neumann, with Eckert and Mauchly on the other. At stake were the patent rights for the ideas embodied in the EDVAC. A federal court ruled that the distribution of the draft report in 1945 rendered it a public disclosure, thereby invalidating a patent claim filed by Eckert and Mauchly in 1947 (H. Goldstine, *The Computer from Pascal to von Neumann* [Princeton University Press, 1972], p. xvi). Eckert and Mauchly resigned on March 31, 1946, and the EDVAC project slowed to a halt.

73. W. Aspray and A. Burks (eds.), *Papers of John Von Neumann on Computers and Computer Theory* (MIT Press, Cambridge, MA, 1987), p. 20.

74. A. Hodges, interview with the authors, June 1994.

75. He had been put off parallel architecture because of a bad experience with an early IBM machine, according to W. Aspray, *John von Neumann and the Origins of Modern Computing* (MIT Press, Cambridge, MA, 1990), p. 30: "In March or April 1944, he [Von Neumann] spent two weeks working in the punched card machine operation, pushing cards through the various machines, learning how to wire plugboards and design card layouts, and becoming thoroughly familiar with the machine operations. He found wiring the tabulator plugboards particularly frustrating; the tabulator could perform parallel operations on separate counters, and wiring the tabulator plugboard to carry out parallel computations involved taking into account the relative timing of the parallel operations. He later told us this experience led him to reject parallel computations in electronic computers and his design of the single address instruction code where parallel handling of operands was guaranteed not to occur."

76. Ironically, however, his 1945 EDVAC report described the stored program computer in terms of the idealized neurons presented in the inspirational 1943 paper by Warren McCulloch and Walter Pitts. This step allowed von Neumann to separate the logical design from the engineering components used in implementing the logical structure. The report drew an analogy between the associative, sensory, and motor neurons of the human nervous system and the respective central processing, input, and output units of the computer. He re-

marked that simplified neurons can be imitated by telegraph relays or vacuum tubes. And he contrasted the synchronous timing of the clocked circuits with the asynchronous firing of neurons. W. Aspray, *John von Neumann and the Origins of Modern Computing* (MIT Press, Cambridge, MA, 1990), pp. 173 and 24.

77. A. Hodges, *Alan Turing: The Enigma* (Vintage, London, 1992), p. 292.

78. Turing's paper gathered dust in the NPL archives. It was eventually published in *Machine Intelligence 5*, B. Meltzer and D. Michie (eds.) (Edinburgh University Press, 1969). As Hodges explains in Chapter 6 of his book, *Alan Turing: The Enigma*, this report, like Turing's talk on similar lines to the London Mathematical Society on February 20, 1947, was met with embarrassed incredulity.

79. From Turing's talk to the London Mathematical Society, quoted in A. Hodges, *Alan Turing: The Enigma* (Vintage, London, 1992), p. 361. See Chapters 5 and 9 for further discussion of these kinds of machines.

80. Ibid., p. 363.

81. G. Tweedale, *Calculating Machines and Computers* (Shire Publications, Princes Risborough, 1990), p. 26.

82. The first popular language was Fortran, FORmula TRANslation, designed by an IBM team led by mathematician John Backus. Less technical is Cobol, COmmon Business Oriented Language, which was aimed at businesses. Many others have emerged, such as Beginner's All-purpose Symbolic Instruction Code BASIC, and ALGOL, ALGOrithmic Language. Most people who program computers use so-called high-level programming languages. Examples include Fortran, C. and Pascal. These are comparatively easy to write, consisting only of ASCII characters. However, such programs cannot be run until they are converted into machine code through the act of compiling; machine code can be interpreted directly by the computer's CPU. Machine code can be compiled from assembly language; really dedicated hackers write their programs in assembly language because it runs much faster than code written in any high-level language. One example of code written in assembly language that we discuss later is Tom Ray's Tierra (see Chapter 8). The operating system of a computer is a piece of machine code that contains all the instructions necessary for running the computer, including the monitor, keyboard, disks, and so on.

83. The compiler designed by Backus and his colleagues took three and a half years to develop, consisted of 25,000 lines of assembler, and, like any pioneering effort, contained many bugs. S. Augarten, *Bit by Bit* (George Allen & Unwin, London, 1985), p. 216.

84. W. Kaufmann and L. Smarr, *Supercomputing and the Transformation of Science* (Scientific American Library 1993), p. 233.

85. S. Augarten, *Bit by Bit* (George Allen & Unwin, London, 1985), p. 239.

86. One of the most important techniques is molecular-beam epitaxy. In an ultrahigh vacuum chamber, atomic layers a mere two angstroms thick can be deposited on semiconductor substrates via effusion from adjacent vaporization cells.

87. We are not overlooking the rise of the personal computer (PC). These computers are used for general-purpose computing tasks, but being "low level" devices they are not part of the subject of the present book, which is concerned with

high performance scientific computing. Nevertheless, in today's era, one can use such PCs as "front ends" to link into national and international networks, and thereby connect a PC to very powerful but remote supercomputing facilities.

88. Computer companies see the world as divided into two parts. The larger of these is the business sector, and mainframe is the term usually used for its most powerful computers. The smaller sector is concerned with scientific computing and, although people have referred to larger computers as mainframes, today these are called supercomputers.

89. The (dimensionless) Reynolds number is defined as the ratio of the inertial to the viscous forces: the weaker the viscous forces, the greater the tendency to turbulence. At values of the Reynolds number of order 100, flows are usually laminar; at values of order 1,000,000, flows possess fully developed turbulence; intermediate values indicate the transition regime between the two states—the onset of turbulent motion.

90. For these reasons, people are at work on so-called subgrid modeling, feverishly hoping to use basic physics to short circuit the numerical brick wall. Alternative approaches use spectral methods, based on Fourier series with a very large number of terms.

91. J. Harrington, R. LeBeau, K. Backes, and T. Dowling, *Nature* **368** 525 (1994). Tim Dowling, interview with Roger Highfield, February 1995.

92. "Grand challenges 1993: High Performance Computing and Communications," report by the Committee on Physical, Mathematical and Engineering Sciences, Federal Coordinating Council for Science, Engineering and Technology, Washington, D.C. (1992).

93. G. Karniadakis and S. A. Orszag, *Physics Today*, March 1993, p. 42.

94. Ibid., p. 36.

95. *From Desktop to Teraflop: Exploiting the U.S. Lead in High Performance Computing*, NSF Blue Ribbon Panel on High Performance Computing, August 1993.

96. This is for Reynolds-number-averaged computations. Estimates based on Kolmogorov's scaling arguments indicate that a direct numerical simulation of air flow past a complete aircraft will require at least an exaflop (10^{18}) computer. Thus, even with teraflop computing capabilities, there is still a long way to go, so that the interplay between theory, experiment, and coding will remain of vital importance over the next decade or more.

97. At least for problems easily vectorized, such as fluid dynamics calculations. Not all problems can be vectorized naturally; however, by enhancing his machine's scalar speed, Cray ensured that it would run *all* codes substantially faster.

98. As John Mauchly told attendees of his 1946 Moore School Lecture: "There are . . . two extremes in machine design. At one extreme we have a completely parallel machine. . . . At the other extreme is the completely serial machine. . . . There are of course many possible intermediate variations which are to some extent serial and to some extent parallel." J. Mauchly, Digital and Analog computing machines, Lecture 3 in *The Moore School Lectures: Theory and Techniques for Design of Electronic Digital Computers*, Vol. 1, M. Campbell-Kelly and M. Williams (eds.) (MIT Press, Cambridge, MA, July 8, 1946), p. 37.

99. *The Boston Globe*, Business Section, September 6, 1994, p. 21. PVC is grateful to Bruce Boghosian for providing this reference.

100. D. Hillis, *Sci Amer* **256** 108 (1987). See also B. Boghosian, *Computers in Physics* **4** 14 (1990).

101. Parallel architectures can be further divided into two categories, depending on whether processing power and memory are connected within the computer by a route called a high-speed "bus" or an alternative, such as the method shared by the nCUBE and the Connection Machine. Both are related to the Cosmic Cube experiments conducted in the early 1980s by Charles Seitz and Geoffrey Fox at the California Institute of Technology. C. Seitz, *Communications of the ACM* **28** 22 (1985).

102. The last version of the Connection Machine built by Thinking Machines Corporation, the CM-5, achieved convergence between SIMD and MIMD architectures, and used a special "data parallel" programming language, so-called CM-Fortran, which was highly optimized to accelerate processing on this machine. It provided a very natural environment for running cellular automaton simulations (see Chapter 4).

103. Danny Hillis, interview with Roger Highfield, October 1993.

104. J. L. Gustafson, *Communications of the ACM* **31** 5 (1988).

105. W. Kaufmann and L. Smarr, *Supercomputing and the Transformation of Science* (Scientific American Library, 1993), p. 47.

106. D. Patterson and D. Ditzel, *Computer Architecture News* **8** 25–33 (1980).

107. R. Colwell, C. Hitchcock, and E. Jensen, *Computer Architecture News* **11** 44 (1983).

108. At present the fastest clock speeds of the IBM RISC machines are about one-quarter that of the Cray Y-MP. But the IBM machines can execute several instructions per CPU cycle. This superscalar feature means that they can match the Y-MP supercomputer on some types of problems. Moreover, by connecting a cluster of such RISC machines, one can today achieve parallel supercomputing capabilities.

109. L. Zadeh, *Information and Control* **8** 338 (1965).

110. L. Zadeh, foreword to *Fuzzy Computing, Theory, Hardware and Applications*, M. Gupta and T. Yamakawa (eds.) (Elsevier Science Publishers, Amsterdam, 1988).

111. A horizontally challenged individual.

112. L. Zadeh, e-mail to Roger Highfield, February 1995.

113. Statistical inference and probability theory.

114. A recent survey was provided in T. Munakata and Y. Jani, *Communications of the ACM* **37** 69 (1994). The patent estimate was provided by L. Zadeh to Roger Highfield in February 1995.

115. T. Munakata and Y. Jani, *Communications of the ACM* **37** 70 (1994).

116. L. Zadeh, *Communications of the ACM* **37** 77 (1994).

117. B. Kosko and S. Isaka, *Sci Amer* **269** 62 (1993). There are also combinations of neural nets and fuzzy logic called neuro-fuzzy learning systems.

118. See, for example, the definition of "digital computer" provided in the *Collins Reference Dictionary of Computing* (Collins, Glasgow, 1988).

119. It can also fundamentally reduce power for interconnection, provide voltage isolation and immunity to electromagnetic interference. David Miller (head of

the Advanced Photonics Research Department of AT&T Bell Laboratories) personal communication with Roger Highfield, July 1994.

120. J. L. Jewell, J. P. Harbison, and A. Scherer, *Sci Amer* 265 86 (1991).

121. Applying an electric field to some materials can alter their optical properties. The field may originate from the laser light wave itself or from an externally applied voltage. One optical property that varies as a result of a changing field is the absorption of light. Another is the refractive index, the ratio of the velocity of light in a vacuum to that in the more optically dense medium. For light of fixed frequency (or equivalently of fixed wavelength), the refractive index is a constant for a given substance at a given temperature and pressure. But under the influence of an electric field the refractive index may vary strongly in what are called optically "nonlinear" materials. These show a whole range of curious optical effects, and provide a very sensitive way of manipulating beams of light, some at speeds of one-trillionth of a second (1/1,000,000,000,000) or faster. Lasers complement this work by providing intense beams of light and associated intense electric fields, usually at a fixed wavelength, allowing engineers to selectively exploit nonlinear effects. By combining lasers with nonlinear effects, light can be used for communication, computation, and storing information.

122. F. McCormick, F. Tooley, T. Cloonan, J. Brubaker, A. Lentine, R. Morrison, S. Hinterlong, M. Herron, S. Walker, and J. Sasian, *Applied Optics* 31 5431 (1992).

123. D. Miller, D. Chemla, and S. Schmitt-Rink, *Optical Nonlinearities and Instabilities in Semiconductors*, H. Haug (ed.) (Academic Press, Orlando, 1988), p. 325.

124. D. Miller, letter to Roger Highfield, July 1994. FET-SEED technology integrates field effect transistors with photodiodes and quantum well modulators to produce systems operating at 155 Mb/s.

125. J. L. Jewell, J. P. Harbison, A. Scherer, Y. H. Lee, and L. T. Florez, *IEEE Journal of Quantum Electronics* 27 1332 (1991).

126. G. Fox and P. Messina, *Sci Amer*, October 1987.

127. D. Miller, "Optics in Digital Computing," draft preprint, 1994.

128. For example, the FET-SEED relies on the quantum confined Stark effect.

129. See P. Coveney and R. Highfield, *The Arrow of Time*, Chapters 4 and 8, for more details of the controversies surrounding the interpretation of quantum mechanics and the associated measurement problem. For example, whether this collapse is a real physical event, or is due to parallel copies of an observer being aware of different observed values of the object measured, is a matter of dispute.

130. Put somewhat more abstractly, two questions drive this work: (1) are there other axiomatic systems that can be used to compute, in which the fundamental notions differ substantially from those of Turing? (2) how easy is it to embody these different axiomatic systems in real-world hardware? Quantum mechanics is associated with its own brand of fuzzy logic, since coherent superpositions represent an "included," as opposed to excluded, middle.

131. P. Benioff, *J Stat Phys* 22 563 (1980); *Phys Rev Lett* 48 1581 (1982).

132. R. Feynman, *Int J Theor Phys* 21 467 (1982); *Opt News* 11 11 (1985); *Found Phys* 16 507 (1986).

133. For an overview, see D. Deutsch, *Physics World* 5 57 (1992).

134. D. Deutsch, interview with Roger Highfield, June 1994.

135. D. Deutsch, *Proc R Soc A* **400** 97 (1985); D. Deutsch, *Proc R Soc A* **425** 97 (1989).

136. One example is the "square root of NOT" gate. Two in series give the classical NOT gate.

137. This has been developed by Charles Bennett from IBM and colleagues. It is based on a technique called quantum public key distribution, which is described in C. Bennett, G. Brassard, and A. Ekert, *Sci Amer* **267** 26 (1992).

138. P. Shor, "Algorithms for Quantum Computation: Discrete Log and Factoring," AT&T Bell Labs, New Jersey (1994). This paper appeared in the *Proceedings of the 35th Annual Symposium on Foundations of Computer Science*, held in late November 1994 and published by IEEE Press.

139. D. Deutsch, interview with Roger Highfield, May 1994.

140. J. S. Bell, *Speakable and Unspeakable in Quantum Mechanics* (Cambridge University Press, 1987), p. 117.

141. As we discussed in Chapter 8 of *The Arrow of Time*, our view is that irreversibility is intricately bound up with wave function collapse, a view shared by Ilya Prigogine, Oliver Penrose, Roger Penrose, and numerous others (not all for the same technical reasons). The fact that the quantum equations of motion are symmetrical in time is the reason why the theory cannot account for collapse. Thus, it seems quite likely that "coherence" is a very special property of some forms of dynamical evolution, but it is not true in general.

142. The reason many cosmologists embrace the many-worlds interpretation is that they erroneously believe that without it one must face up to the collapse of wave functions owing to the role of observers and, in the case of the universe, who is the observer if not God? Irreversibility is not taken seriously by most cosmologists, since they are wedded to time-symmetric mechanical equations, be they classical, relativistic, or quantum mechanical, for the description of the universe.

143. B. de Witt and N. Graham, *The Many Worlds Interpretation of Quantum Mechanics* (Princeton University Press, Princeton, 1973).

144. F. Tipler, *The Physics of Immortality* (Macmillan, London, 1995), p. 169.

145. Deutsch views the computation as one going on within an enormous cellular automaton (see Chapter 4) whose sites each correspond to different universes. D. Deutsch, interview with Roger Highfield, May 1994.

146. Skeptics would argue that all these references to unobservable universes are unnecessary metaphysical baggage, though Lloyd and Deutsch are adamant that this is the only way to *explain* quantum computing processes.

147. This limitation is, however, a boon to cryptographers who can ensure that there is no eavesdropping using quantum methods.

148. S. Lloyd, *Science* **261** 1569 (1993), and interview with Roger Highfield, September 1993.

149. D. Deutsch, interview with Roger Highfield, May 1994.

150. M. Reed, *Sci Amer* **268** 98 (1993).

151. The quotations are from D. Deutsch, interview with Roger Highfield, May 1994.

152. B. Boghosian, interview with Roger Highfield, October 1993.

153. Greg Chaitin has also remarked: "If mathematics can be made out of Darwin, then we will have added something basic to mathematics; while if it cannot, then Darwin must be wrong, and life remains a miracle which has not been explained by science." G. Chaitin, *Association for Computing Machinery SIGACT News* 4 12 (1970).

154. C. Babbage, *Passages from the Life of a Philosopher* (Longman, Green, Longman, Roberts & Green, 1864), p. 389.

CHAPTER 4

1. T. H. Huxley, *Lay Sermons, &c., iii, A Liberal Education*.

2. See W. Aspray and A. Burks (eds.), *Papers of John von Neumann on Computers and Computer Theory* (MIT Press, Cambridge, MA, 1987), p. 391. Von Neumann's conception of automata theory was close to Weiner's conception of cybernetics, and each influenced the other. But von Neumann's automata theory placed more emphasis on logic and digital computers, while Wiener's cybernetics was oriented more around physiology and control engineering. Another figure in the story is the German engineer Konrad Zuse. Hiding from the Nazis on an isolated Austrian mountain near the end of the Second World War, Zuse conceived of digitized mechanical systems as evolving in discrete time steps, rather than continuously. This led him to consider artificial rules controling the updating of such a system in time; in this way his "computing spaces" were born. A good general introduction to the field is T. Toffoli and N. Margolus, *Cellular Automata Machines* (MIT Press, Cambridge, MA, 1987).

3. W. Aspray and A. Burks (eds.), *Papers of John von Neumann on Computers and Computer Theory* (MIT Press, Cambridge, MA, 1987), p. 421.

4. Ibid., p. 538.

5. S. Ulam, *Proc Int Cong Mathem* 2 264 (1952).

6. This is not such a strange venue for inspiration as it sounds. Shortly after the seventeen-year-old Ulam enrolled at Lwów Polytechnic Institute, he discovered that the mathematics that really mattered was not taught in the classroom but was instead to be found in an environment akin to Babbage's Analytical Society. In one of the large cafés in town, the Scottish Café, the Lwów mathematicians would gossip and argue daily.

7. The term is due to Arthur Burks.

8. W. Aspray and A. Burks (eds.), *Papers of John von Neumann on Computers and Computer Theory* (MIT Press, Cambridge, MA, 1987), p. 495. According to Kendall Preston and Michael Duff, *Modern Cellular Automata, Theory and Applications* (Plenum Press, New York, 1984), p. 1, "Von Neumann is now strongly associated with the old-fashioned single CPU computer architecture. Many of today's computer scientists and engineers have completely forgotten that John

von Neumann was also the major pioneer in parallel computing via his research on arrays of computers, or cellular automata."

9. The lectures were to be best remembered because of the popular account written by John Kemeny, who covered them for *Scientific American* 192 58 (1955) and von Neumann's book *The Computer and the Brain* (New Haven, CT, 1958) published posthumously, which drew on a considerable portion of his first three lectures.

10. J. Kemeny, *Sci Amer* 192 67 (1955).

11. J. von Neumann, *Theory of Self Reproducing Automata*, A. Burks (ed.) (University of Illinois Press, Urbana and London, 1966).

12. Iterative circuit computers exhibit two principal characteristics: (1) the entire processing part of the computer can be made up of identical modules uniformly interconnected; (2) within limits of overall storage capacity, the resulting computer can execute an arbitrary number of different computer programs simultaneously.

13. A. Burks (ed.), *Essays on Cellular Automata* (University of Illinois Press, Urbana and Chicago, 1970), p. xxii.

14. J. Holland, *Journal of the Association for Computing Machinery* 9 297 (1962).

15. J. Holland, "Universal Spaces: A Basis for Studies in Adaptation" in *Automata Theory* (Academic Press, New York, 1966), pp. 218–230.

16. A. Burks (ed.), *Essays on Cellular Automata* (University of Illinois Press, Urbana and Chicago, 1970), p. 219.

17. Ibid., p. 233.

18. J. Holland, *Automata Theory* (Academic Press, New York, 1966), pp. 218–30.

19. M. Gardner, *Sci Amer* 223 120 (1970).

20. M. Gardner, *Sci Amer* 224 112 (1971). Achim Flammenkamp has compiled a massive, though still not comprehensive, list of Game-of-Life objects.

21. E. Berlekamp, J. Conway, and R. Guy, *Winning Ways for Your Mathematical Plays* (Academic Press, New York, 1982). In short, gliders are analogous to bits, the glider gun to a clock, and other structures are able to model various logical gates.

22. From the Introduction, *Cellular Automata Machines*, T. Toffoli and N. Margolus (MIT Press, Cambridge, MA, 1987).

23. Lattice-gas models of such systems offer a significant advantage over molecular dynamics (MD) simulations. The prime reason is that the particles need not then represent individual molecules. Also, lattice-gas time steps are on the order of the mean free time, whereas MD time steps are typically two orders of magnitude smaller than the mean free time. However, a very recent technique called "dissipative particle dynamics" claims to enable the best features of LGA and MD to be combined; see P. J. Hoogerbrugge and J. M. V. A. Koelman, *Europhysics Letters* 19 155 (1992).

24. U. Frisch, B. Hasslacher, and Y. Pomeau, *Phys Rev Lett* 56 1505–8 (1986). They demonstrated that a class of deterministic lattice gases with discrete Boolean elements simulates the Navier-Stokes equations, and can be used to design simple, massively parallel computing machines. More or less simultane-

ously, Stephen Wolfram, then at the Institute for Advanced Study in Princeton, published a paper in which he derived continuum equations for lattice gas cellular automata models; S. Wolfram, *J Stat Phys* 45 471–526 (1986).

25. Note that the numerical solution of the Navier-Stokes equations in any case requires us to discretize time and space, so that the continuum model begins to take on the aspects of a cellular automata.

26. One example are periodic flows. On the other hand, simple Poiseuille flow at low Reynolds number is a case where the lattice gas (LGA) method is much better than the vortex or spectral methods (although it is not better than the spectral element, finite element, lattice Boltzmann, or even finite difference). On the other hand, it is sometimes argued that the LGA method is itself a way of solving the Navier-Stokes equations. Intrinsic drawbacks are associated with the lack of Galilean invariance of lattice models.

27. M. Sahimi, *Rev Mod Phys* 65 1393 (1993). See also B. Boghosian, P. Coveney, and A. Emerton, *A Lattice-gas Model of Microemulsions*; preprint (1995).

28. S. Wolfram, *Rev Mod Phys* 55 601 (1983). Class I cellular automata (CA) possess fixed points; Class II evolve to limit cycles; Class III are associated with strange or chaotic attractors. There is no analog of Class IV CA in continuous time dynamical systems (cf. Chapter 6): class IV CA are associated with the existence of very long-lived transients. However, in the context of artificial life studies, a great deal has been made of the importance of complex dynamics corresponding to the Class IV behavior. The so-called "edge of chaos" we later discuss refers to Class IV dynamical behavior sandwiched between Class II and Class III (Chapter 8).

29. C. Langton, Ph.D. thesis, University of Michigan (1991). Langton tuned what he called the Lambda parameter to show that the behaviors moved from Class I to II to IV to III.

30. K. Preston and M. Duff, *Modern Cellular Automata, Theory and Applications* (Plenum Press, New York, 1984), p. 3.

31. N. Margolus, "CAM-8. A Computer Architecture Based on Cellular Automata" in *Pattern Formation and Lattice Gas Automata*, E. Lawniczak and R. Kapral (eds.) (American Mathematical Society, Washington, D.C., 1995).

32. From the Introduction, *Cellular Automata Machines*, T. Toffoli and N. Margolus (MIT Press, Cambridge, MA, 1987).

33. They were interested in reversible computation. See, for example, C. H. Bennett, *IBM Res Dev* 17 525 (1973); E. Fredkin and T. Toffoli, *Int J Theor Phys* 21 219–53 (1982); N. Margolus, *Physics* 10D 81–95 (1984); T. Toffoli, *Automata, Languages and Programming*, De Bakker and van Leeuwen (eds.) (Springer-Verlag, New York, 1980), pp. 632–44. Actually defining information is a difficult problem.

34. N. Margolus, T. Toffoli, and G. Vichniac, *Phys Rev L* 56 1694 (1986).

35. N. Margolus, interview with Roger Highfield, February 1995.

36. CA-based neural net modules are grown in a two-phase process. Three-cell-wide CA trails are grown by sending a sequence of growth signals (e.g., extend and turn right) down the middle of the trail. When an instruction hits the end of the trail, it executes its function. This sequence of growth instructions is treated

as a chromosome in a genetic algorithm and is evolved: the sequence maps to a CA network. When trails collide, they form "synapses." Once the CA network is grown, it is used as a neural network in a second neural signaling phase. Some neural signals can be tapped to control certain processes and we can measure the quality or fitness of the control. This fitness measure is used to drive the evolution.

37. H. de Garis and T. Kaloudis, *The "CAM-Brain" Project*, and H. de Garis, *Brain Building: The Evolutionary Engineering of Artificial Nervous Systems* (Wiley, in press), 1995.

38. For full details, see D. Bentz, P. Coveney, E. Garboczi, M. Kleyn, and P. Stutzman, *Modelling and Simulation in Materials Science and Engineering* 2 783 (1994).

39. The algorithm used was derived from previous work concerned with modeling porous carbonate rocks, where both the solid matrix and pore space are percolating over a very wide range of porosities. It was found that a 3D image of striking visual resemblance and similar statistical properties to real carbonate rocks could be generated by thresholding suitable Gaussian convolutions of 3D images in which white noise was initially made to populate all the sites. See P. A. Crossley, L. M. Schwartz, and J. R. Banavar, *Appl Phys Lett* 59 3553 (1991).

40. R. D'Angelo, T. Plona, L. Schwartz, and P. Coveney, *Advanced Cement Based Materials* 2 8 (1995).

41. This is particularly true in oilfield operations, where cement is used to line wellbores. One needs to know that the cement is not setting prematurely while still being pumped, and also that it has set once it is in place down the hole.

42. E. Garboczi and D. Bentz, *Journal of Materials Science* 27 2083 (1992); D. Bentz and E. Garboczi, NISTIR 5125 (1993).

43. In other words, it should introduce some randomness into the search for a solution. An analogous idea occurred to Stanislaw Ulam in 1946, when wrestling with how to predict the explosive power of an atom bomb from the development of neutron chain reactions. He was convalescing from an illness and playing solitaire. After fruitlessly attempting to calculate the odds that he would win, he hit on the idea of experimenting: the more games he played, the closer the win/lose ratio was to the value that would be predicted from complex analysis. A game of chance could also trace the history of a given neutron. Called the Monte Carlo method, after the famous gambling haunt, it was tested on ENIAC in 1947 and has since taken hold in many areas of research.

44. Technically, we mean the least *free* energy, as defined in thermodynamics.

45. D. Stein, *Sci Amer* 261 36 (1989).

46. The statistical approach was pioneered by S. F. Edwards and P. W. Anderson, *Journal of Physics* F 5 965 (1975), who produced an idealized model of spin glasses whose properties were amenable to calculation. David Sherrington then tried to simplify their model, and Scott Kirkpatrick became involved in computer simulations, experimenting with various ways to prevent the spin glass from settling in a local minimum.

47. In fact there are exponentially many. Note, too, that spin glasses are not intrinsically metastable, unlike the silicate (window) glasses. There may be a single

ground state for a spin glass that will be its strict equilibrium state, but many other metastable states are also associated with low-lying local minima that are separated from it by very high energy barriers (compared to thermal energy). These states differ from one another macroscopically. In practice, it makes little sense to focus on only one of these—all are just about as stable.

48. D. Sherrington and S. Kirkpatrick, *Phys Rev Lett* 35 1972 (1975). A longer version appeared in S. Kirkpatrick and D. Sherrington, *Phys Rev B* 17 4384 (1978). Sherrington first had the idea and spoke about it while at Imperial College. It is worth bearing in mind the tremendous difference that exists between the physicists' approach to understanding the world and that of biologists. Whereas physicists are quite happy to make idealized models that they can analyze mathematically and then attempt to draw general conclusions from, biologists tend to insist on retaining all the detail at all costs.

49. David Sherrington, interview with Roger Highfield, July 1994. The solution is exact in the mean field approximation.

50. S. Kirkpatrick, C. Gelatt, Jr., and M. Vecchi, *Science* 220 671 (1983). In this paper, the authors gave results of optimizations for the traveling salesman problem involving thousands of cities. The technicalities involve the choice of algorithm and cooling schedule.

51. This is the internal, configurational energy, not the free energy.

52. Simulated annealing is in fact very closely intertwined with the Monte Carlo algorithm. The Monte Carlo method, first conceived of by Stanislaw Ulam, and usually implemented today according to the so-called Metropolis algorithm, is effectively a stochastic search technique in which moves to lower energies are accepted with unit probability, but moves to higher energies are allowed only with a temperature-dependent Boltzmann weighting. It is commonly used for modeling the equilibrium behavior of large assemblies of ions, atoms, and/or molecules; variations on the theme include the numerical evaluation of complicated integrals and other mathematical expressions.

53. A. Brünger, J. Kuriyan, and M. Karplus, *Science* 235 458 (1987).

54. This is the "protein folding problem." See, for example, S. Schulze-Kremer in *Parallel Problem Solving from Nature 2*, R. Männer and B. Manderick (eds.) (North-Holland, Amsterdam, 1992), pp. 391–400; T. Dandekar and P. Argos, *Protein Engineering* 5 637 (1992).

55. M. Waldrop, *Complexity* (Penguin, 1994), p. 163.

CHAPTER 5

1. "In every case that I have studied, one can easily think of improvements," says John Holland. "For example, in the human eye, marvellous as it is, the neural soma are in front of the retina!" John Holland, letter to Roger Highfield, July 6, 1994.

2. M. Waldrop, *Complexity* (Penguin, London, 1994), p. 167.

3. Holland was awarded the first Ph.D. in Computer Science in the United States

(S. Levy, *Artificial Life* [Pantheon New York, 1992], p. 60). For an early reference to genetic algorithms, see J. Holland, *Journal of the Association for Computing Machinery* 3 297 (1962).

4. J. Holland, *Sci Amer* 267 66 (1992).

5. J. Holland, *Adaption in Natural and Artificial Systems: An Introductory Analysis with Applications to Biology, Control, and Artificial Intelligence* (MIT Press, Cambridge, MA, 1992). The 1975 edition was published by the University of Michigan. The first mention of the words "genetic algorithm" came in 1967, when a dissertation described a game-playing program based on a highly simplified version of chess called hexapawn. See J. Bagley, *The behavior of adaptive systems which employ genetic and correlation algorithms*, Ph.D. dissertation, University of Michigan (1967).

6. J. Holland, *Sci Amer* 267 66 (1992). Note that genetic algorithms are nondeterministic, owing to the random nature of crossover and mutation.

7. An important point emphasized by Holland is that the genetic algorithm's—and nature's—primary source of effectiveness comes from the crossover operation, coupled with survival of the fittest. Random mutations are relatively unimportant. This is an observation that disagrees with other attempts to solve problems adaptively using only asexual mutation and survival of the fittest. See, for example, R. Dawkins, *The Blind Watchmaker* (Longman, Harlow, 1986).

8. J. Hadamard, *The Psychology of Invention in the Mathematical Field* (Princeton University Press, Princeton, 1949), p. 29.

9. The algorithm was seeded with designs from an expert system, one that uses inference rules based on experience to predict the effects of design changes.

10. D. Goldberg, *Genetic Algorithms in Search, Optimization and Machine Learning* (Addision-Wesley, Reading, MA, 1989), Chapters 4 and 7. See Foreword for the quoted comments by Holland.

11. Genetic programming has certain advantages over the use of genetic algorithms. The data input to a genetic programming task is usually taken directly from the problem at hand and is therefore its "natural representation." Genetic algorithms (as well as neural networks) generally require that their input data be preprocessed—transformed in some specific manner—before it can be handled.

12. J. Koza, *Genetic Programming* (MIT Press, London, 1992), p. 4. It is to be hoped and expected that mathematicians will in the future find ways of providing such proofs.

13. D. Goldberg, *Genetic Algorithms in Search, Optimization and Machine Learning* (Addision-Wesley, Reading, MA, 1989), p. 309.

14. See, for example, *Artificial Intelligence at MIT: Expanding Frontier*, P. Winston and S. A. Shellard (eds.) (MIT Press, Cambridge, MA, 1990). The foreword admits: "Unfortunately a definition of intelligence seems impossible at the moment because intelligence appears to be an amalgam of so many information-processing and information-representation abilities."

15. J. Haugeland, *Artificial Intelligence: The Very Idea* (MIT Press, Cambridge, MA, 1985).

16. I. Aleksander and P. Burnett, *Thinking Machines: The Search for Artificial Intelligence* (Oxford University Press, Oxford, 1987), p. 195.

17. R. Penrose, *The Emperor's New Mind* (Vintage, London, 1990), p. 578.

18. A. Newell and H. A. Simon, *Computer Simulation of Human Thinking*, The RAND Corporation P-2276 (April 20, 1961). Both Alan Turing and Ada Lovelace had realized long before this that computers could do much more than manipulate numbers.

19. E. S. Shapiro, *Encyclopaedia of Artificial Intelligence*, Vol. 1 (Wiley Interscience, John Wiley, New York, 1987), p. xi.

20. B. Buchanan and E. Feigenbaum, *Journal of Artificial Intelligence* 11 5–24 (1978).

21. E. Feigenbaum, IJCAI-77 Proceedings, p. 1014.

22. F. Hayes-Roth and N. Jacobstein, *Communications of the ACM* 37 36 (1994).

23. J. Weizenbaum, *Communications of the ACM* 9 36 (1966).

24. H. Dreyfus, *What Computers Still Can't Do* (MIT Press, Cambridge, MA, 1992), pp. 57–62.

25. T. Winograd, "A Procedural Model of Language Understanding," in *Computer Models of Thought and Language*, R. Schank and K. Colby (eds.) (W. H. Freeman, San Francisco, 1973). SHRDLU is an acronym that does not stand for anything at all—an idea taken by Winograd from *Mad Magazine*. Minksy said of the model: "It talks about a fairyland in which things are so simplified that almost every statement about them would be literally false if asserted about the real world." H. Dreyfus, *What Computers Still Can't Do* (MIT Press, Cambridge, MA, 1992), p. 9.

26. Ibid., p. xi.

27. W. S. McCulloch and W. Pitts, *Bull Math Biophysics* 5 115 (1943).

28. Their work proposed formal binary neurons that could only take one or two inputs; Rosenblatt extended this to arbitrarily many.

29. N. Wiener, *Cybernetics* (MIT Press, Cambridge, MA, 1986).

30. N. Wiener, *Cybernetics, or Control and Communication in the Animal and the Machine*, 2nd ed. (MIT Press and John Wiley & Sons, New York, London, 1961), p. 116.

31. F. Rosenblatt, *Psychological Review* 65 386 (1958). In fact, Rosenblatt emphasized the statistical or random nature of networks, and placed far less emphasis on the logical analysis of McCulloch and Pitts.

32. I. Aleksander and P. Burnett, *Thinking Machines: The Search for Artificial Intelligence* (Oxford University Press, 1987), p. 157.

33. Technically, it couldn't solve the XOR problem, which is "linearly inseparable." M. Minsky and S. Papert, *Perceptrons: An Introduction to Computational Geometry* (MIT Press, Cambridge, MA, 1969), p. 227: "Many of the theorems show that perceptrons cannot recognize certain kinds of patterns."

34. M. Minksy and S. Papert, *Perceptrons: An Introduction to Computational Geometry*, *Expanded Edition* (MIT Press, Cambridge, MA, 1988), Prologue.

35. Ibid.

36. It has now been shown that one layer of sigmoid hidden units is sufficient for uni-

versal approximation. The question left unanswered is how the number of hidden units scales with the size of the problem. See K. Hornick, M. Stinchcombe, and H. White, *Neural Networks* 2 359 (1989), and K. Hornick, M. Stinchcombe, H. White, and P. Auer, *Neural Computation* 6 1261 (1994).

37. The basic requirement for backpropagation to work at all is that the input/output mapping be once differentiable. In fact, it was the Boltzmann machine (discussed later) that provided the first learning algorithm for multilayer networks with hidden units, coming in just ahead of backpropagation.

38. P. Werbos, *Beyond Regression: New Tools for Prediction and Analysis in the Behavioral Sciences* (Ph.D. thesis, Harvard University, 1974). However, his idea made no impact since he did not demonstrate that the backpropagation algorithm would work. Geoffrey Hinton is credited with establishing its capabilities.

39. Other remedies for the learning difficulties of MLPs include lowering the "gain" term, that is, slowing down the rate at which connection weights are altered, adding more nodes in the hidden layer, and adding noise (cf. the Boltzmann machine, discussed later). A more drastic solution is to invoke radial basis function representations of artificial neural networks, also described later.

40. J. Hopfield, *Proc Natl Acad Sci USA* 79 2554 (1982).

41. See, for example, M. Lea in *Advanced Digital Information Systems*, I. Aleksander (ed.) (Prentice-Hall, Englewood Cliffs, NJ, 1984).

42. Note that associative memories and neural networks have a fuzzy interface in this area: one does not imply the other. Other examples include Willshaw's associative net, and ADAM (advanced distributed associative memory network), an improvement on the former.

43. An important issue concerns the nature of the attracting states themselves. Just as in nonlinear dynamics where one encounters fixed points, limit cycles, tori, and strange attractors (Chapter 6), the same type of states can occur in principle in attractor neural networks. The simplest type are fixed points, meaning that the network settles down to a fixed repeating pattern of activity. These are the most commonly studied to date, because of the simplifying assumption usually made about detailed balance, namely that the connection strength between neuron A and neuron B is identical to that between B and A, for which there is no physiological basis. In fact, for biological applications, this symmetry condition is wrong, but it turns out to be not badly so. Limit cycle attractors imply the memorization of dynamic behavior—period patterns—while strange attractors would correspond to aperiodic information storage and memory. Now that we understand the properties of these symmetrical neural connections, it is easier to deal with more complex, asymmetric networks.

44. Hopfield—and others—frequently refer to the "energy surface" and not the error surface, because of the analogy between the network error and the free energy in the spin glass model.

45. D. Hebb, *Organization of Behavior* (Wiley, New York, 1949).

46. J. J. Hopfield and D. W. Tank, *Biological Cybernetics* 52 141 (1985). Actually this network is an embarrassment to the field since it performs so poorly, and

even today is not competitive with other techniques. The real significance of this paper was that it showed how constrained optimization problems could be formulated in terms of neural nets.

47. G. Hinton, T. Sejnowski, and D. Ackley, *Boltzmann Machines: Constraint Satisfaction Networks That Learn*, Tech. Rep. CMU CS 84, 111 (Carnegie-Mellon University, Pittsburgh, 1984); also G. Hinton and T. Sejnowski in *Parallel Distributed Processing*, Vol. 1, D. E. Rumelhart and J. L. McClelland (eds.) (MIT Press, Cambridge, MA, 1986), p. 282.

48. False or spurious states (minima) typically arise when a network self-organizes into a set of attractors where memories (particular neural firing patterns) are meant to reside, through, for example, Hebbian learning. One almost inevitably then induces additional spurious states (local minima) that must be eliminated. Information is stored intentionally by the adjustment of neural connection strengths, when these additional states also appear, and could be construed as corresponding to patterns not intentionally learned. They can often be removed by the addition of noise as discussed here, and sometimes by introducing synaptic asymmetries. Instead of thinking of these spurious minima as the result of creative thinking, the physicist Daniel Amit suggests that it may be preferable to consider the phenomenon as "the psychiatric metaphor of schizophrenia" (D. Amit, *Modelling Brain Function*, Chapter 2 [Cambridge University Press, 1989]).

49. This probability is the well-known Boltzmann factor, suitably normalized, so that the higher the energy—the greater the error—the lower the probability of ending up in such a state, and by an exponentially decreasing amount.

50. Faster annealing algorithms are now available. See, for example, H. Szu in *Neural Networks for Computing*, J. S. Denker (ed.) (American Institute of Physics, New York, 1986), p. 420.

51. G. Carpenter and S. Grossberg in *Computer Vision, Graphics and Image Processing* No. 37 (Academic Press, New York, 1987), pp. 54–115.

52. ART is completely described by a set of nonlinear differential equations, thus providing an example of self-organization of the kind discussed in Chapter 6.

53. G. A. Carpenter and S. Grossberg, *Computer* 21 77 (1988). In fact, ART is an extension of the unsupervised, competitive learning algorithms pioneered by Kohonen and others, whose motivation is greater biological realism. Kohonen's self-organizing nets are discussed in Chapter 9.

54. T. Caudell, "Genetic Algorithms as a Tool for the Analysis of Adaptive Resonance Theory Network Training Sets," in *COGANN-92* (IEEE Computer Society Press, Los Alamitos, 1992), pp. 184–200.

55. P. Fletcher and P. Coveney, *Advanced Cement Based Materials* 2 21 (1995); P. Coveney and P. Fletcher in *Information Technology Awareness in Engineering*, J. A. Powell (ed.) (EPSRC/DRAL, 1994). Also P. Fletcher, P. Coveney, T. Hughes, and C. Methven, *Journal of Petroleum Technology* 47 129 (February 1995).

56. A "sigmoidal" firing profile is often assumed.

57. These figures will vary depending on the details of the way in which training is carried out, and on the nature of the computer used. The figures quoted refer to

use of a fast workstation, together with some preprocessing of the raw spectral data. See P. Fletcher and P. Coveney, *op cit*. and forthcoming publications.

58. There is in fact a great deal more information that can be obtained from the infrared spectrum of a cement powder, including chemical composition and particle size distribution. See the papers by P. Fletcher et al., *op cit.*, and T. Hughes, C. Methven, T. Jones, P. Fletcher, S. Pelham, and C. Hall, *Advanced Cement Based Materials* 2 91 (1995).

59. J. Moody and C. J. Darken, "Learning with Localized Receptive Fields" in Proceedings of the 1988 connectionist models summer school, pp. 133–45; *Neural Computation* 1, 281–93 (1989); D. Broomhead and D. Lowe, HMSO, RSRE Report, April 1988. The neural nets used for predicting cement setting routinely are in fact based on radial basis functions. These effectively transform the data to be learned into a higher dimensional space in which the problem becomes linearly separable, and the training process amounts to linear optimization, thereby providing a guarantee of finding the globally optimal solution. There are some drawbacks, however. One is that there is a tendency to overtrain, and then generalization is considerably weakened.

60. This is done using standard statistical techniques, such as "K-means" clustering.

61. I. Aleksander and P. Burnett, *Thinking Machines. The Search for Artificial Intelligence* (Oxford University Press, 1987), p. 158. Details of computer-assisted cervical cancer screening using neural networks are given in L. Mango, *Cancer Letters* 77 155 (1994).

62. H. Thodberg, interview with Roger Highfield, January 1994.

63. The work is based on the Kohonen self-organizing feature map (see Chapter 9) and concepts from Darwinian evolution. S. Amin and J-L Fernández, interview with Roger Highfield, May 1994. See also S. Amin, *Neural Computing and Applications* 2 129 (1994); F. Favata and R. Walker, *Biological Cybernetics* 64 516 (1991); S. Lin and B. Kernigham, *Operations Research* 21 516 (1971); and E. L. Lawler et al. *The Traveling Salesman Problem: A Guided Tour of Combinatorial Optimization* (John Wiley and Sons, New York, 1985).

CHAPTER 6

1. P. Dirac, *Proceedings of the Royal Society*, A 123 714 (1929).

2. Indeed, Dirac was well aware of this shortcoming. His triumphalist claim for quantum mechanics is often quoted but rarely in its entirety. He also warned that "the exact application of these laws leads to equations much too complicated to be soluble."

3. One method that gives essentially "exact" results and does not demand the use of any approximation schemes is the so-called quantum Monte Carlo technique. This method has certain similarities with the conventional Monte Carlo technique, mentioned in Chapter 4, although now a solution to the quantum mechanical Schrödinger equation is sought by statistical sampling of the wave function itself. This approach has been used to calculate the ground state energy of the helium dimer on a CM-5 supercomputer. The calculation predicts

the existence of a stable, bound dimer, which is also observed experimentally. It is interesting to note that again a stochastic computational technique has proved to be the most effective method for solving what might at first sight be regarded as a simple problem. See J. B. Anderson, C. A. Trainor, and B. Boghosian, *J Chem Phys* 95 7418 (1991); J. B. Anderson, C. A. Trainor, and B. Boghosian, *J Chem Phys* 95 345 (1993). However, as noted repeatedly, the fundamental problem remains that the large-scale, macroscopic behavior of complex processes is time asymmetric, whereas the microscopic equations are based on time-symmetric equations.

4. As we discussed in Chapter 4, another field, called statistical mechanics, attempts to link these macroscopic properties to the behavior of individual units considered en masse.

5. This is Prigogine's theorem of minimal entropy production. A more detailed discussion of nonequilibrium thermodynamics in general can be found in Chapter 5 of *The Arrow of Time*.

6. We are here assuming that time is continuous, not discrete. Were time taken to be discrete, we would be dealing with difference equations. In population dynamics, for example, people frequently model time as if it advances discontinuously from one year to the next. The logistic equation (discussed briefly in Chapter 1) is a discrete time difference equation for the time dependence of a population.

7. The reason is the complexity of the solutions of these equations. It is common to encounter highly oscillatory solutions, for example, and these demand a tremendous amount of precision and therefore available memory to cope with the large quantities of numerical data generated by the integrators as they step through time.

8. J. K. Platten and J. C. Legros, *Convection in Liquids* (Springer Verlag, Berlin, Heidelberg, 1984), p. 318; M. Velarde and C. Normande, *Sci Amer* 243 92 (1980).

9. G. Nicolis in *The New Physics*, P. C. W. Davies (ed.) (Cambridge University Press, 1989), p. 319.

10. E. Bodenschatz, J. de Bruyn, G. Ahlers, and D. Cannel, *Phys Rev Lett* 67 3078 (1991).

11. M. Assenheimer and V. Steinberg, *Phys Rev Lett* 70 3888 (1993). See also M. Assenheimer and V. Steinberg, *Nature* 367 345 (1994).

12. H. Xi, J. Gunton, and J. Vinals, *Physical Review E* 47 2987 (1993).

13. At first sight, it may seem surprising that the father of the modern computer should pop up in such an improbable role as one of the first theoretical biologists. However, his interest in morphogenesis was part of his lifelong fascination with establishing connections between the logical and physical structure of the brain. In a letter written to the neurophysiologist J. Z. Young on February 8, 1951, he discussed the storage capacity of a human brain network composed of 10^{10} neurons, adumbrating as he had done before in his *Intelligent Machinery* paper of 1947 (see Chapter 3) the connectionist approach to learning. Stating that he was as yet very far from "asking any anatomical questions [about the brain],"

he revealed that he was working on a mathematical theory of embryology that he believed gave "satisfactory explanations of (i) gastrulation (ii) polygonal symmetrical structures, e.g., starfish, flowers (iii) leaf arrangement, in particular the way the Fibonacci series (0,1,1,2,3,5,8,13, . . .) comes to be involved (iv) color patterns on animals, e.g., stripes, spots and dappling (v) patterns on nearly spherical structures such as some Radiolaria, but this is more difficult and doubtful." He said that he was doing this work because it was more tractable than directly attacking similar questions concerning the brain. But, he told Young, "The brain structure has to be one which can be achieved by the genetical embryological mechanism, and I hope that this theory that I am now working on may make clearer what restrictions this really implies." Quoted in A. Hodges, *Turing: The Enigma* (Vintage, London, 1983), pp. 436–37.

14. A. Turing, *Phil Trans R Soc London* **B237** 37 (1952).
15. See, for example, J. Murray, *Mathematical Biology* (Springer-Verlag, Berlin, 1989).
16. B. Chance, A. K. Ghosh, E. K. Pye, and B. Hess, *Biological and Biochemical Oscillators* (Academic Press, New York, 1973); also C. Vidal and P. Hanusse, *Int Rev Phys Chem* **5** 1–55 (1986).
17. I. Prigogine and R. Lefever, *J Chem Phys* **48** 1695 (1968).
18. A. Hodges, *Alan Turing: The Enigma* (Vintage, London, 1992), p. 466.
19. In the technical literature, the term "chemical clock" is usually reserved for reactions in which one or more species start out present in very low or even vanishing concentrations but at a well-defined time their concentrations grow extremely rapidly—as happens in thermal explosions. See, for example, J. Billingham and P. Coveney, *J Chem Soc Faraday Trans* **89** 3021 (1993).
20. The term "crosscatalator" was introduced by A. Chaudry, P. Coveney, and J. Billingham, *J Chem Phys* **100** 1921 (1994). The crosscatalator is in fact *not* a direct development of the Brusselator but rather of a similar but simpler system called the "autocatalator." The basic model underlying the crosscatalator was originally introduced in a variety of guises by Peter Gray, Stephen Scott, and their coworkers; see, for example, P. Gray and S. K. Kay, *J Phys Chem* **94** 3005 (1990); S. K. Scott and A. S. Tomlin, *Phil Trans R Soc London A* **332** 51 (1990); S. K. Scott, B. Peng, A. S. Tomlin, and K. Showalter, *J Phys Chem* **94** 1134 (1990). V. Petrov et al. performed related studies in *J Chem Phys* **97** 1921 (1992). Today there are a vast number of model nonlinear chemical reaction schemes in the scientific literature.
21. Specifically, if we allow for the precursor chemical species to be rapidly consumed during the crosscatalator reaction, then the exquisite details of the complex dynamics are suppressed.
22. Biologically realistic neural networks are discussed in Chapter 9. See also T. R. Chay and J. R. Rinzel, *J Biophys* **47** 357 (1985). Mixed modes are also to be found in the Belousov-Zhabotinski reaction mentioned later in this chapter; J. Maselko and H. L. Swinney, *J Chem. Phys* **85** 6430 (1986).
23. E. Lorenz, *The Essence of Chaos* (UCL Press, London, 1993), p. ix.
24. D. Ruelle and F. Takens, *Comm in Math Phys* **20** 167–92 (1971). In his book,

Chance and Chaos, Ruelle points out that this work was rejected by two academic journals; eventually he used his prerogative as an editor of this journal to ensure publication of this work.

25. To be correct, it seems that strange attractors are useful for describing the *onset* of turbulence, but not necessarily for fully developed turbulence, which remains largely a mystery. (The same distinction holds good in cardiology, between fibrillation and its onset in tachycardia caused by rotating waves.) Essentially, Ruelle and Takens showed that any system passing through three or more successive Hopf (limit-cycle) bifurcations (regardless of the initial conditions) will necessarily end up in a chaotic state.

26. J. Swift, *On Poetry* (1733).

27. M. Markus and J. Tamames, "Fat Fractals in Lyapunov Space," to be published in *Fractals in the Future*, C. A. Pickover (ed.) (St. Martin's Press, New York). Fat fractals have an integer dimension. However, the coarse-grained measure of fat fractals varies exponentially with the scale length, allowing quantification via a "fatness exponent."

28. B. Mandelbrot, *Science* 156 636 (1967). The linear distance ("as the crow flies") between adjacent seaside towns would give one crude estimate. But if you strolled around the coast, you would find that the country's coastline had grown, having traversed every cove and inlet. To an ant, mere pebbles could add considerably to the journey, while for a wriggling bacterium the length of Britain's coast would soar again. Clearly, the answer depends on the scale of measurement used because there is structure on essentially *all* scales of length. Indeed, if we could shrink the length scale to the infinitesimal, then the coastline would have an infinite length. The apparently paradoxical result is that the coast is in fact a "line" of infinite length contained quite happily within a finite area (draw a circle around Britain).

29. X. Shi, M. Brenner, and S. Nagel, *Science* 265 219 (1994).

30. The term "chaology" was introduced by Mike Berry of Bristol University.

31. D. Ruelle, *Trans N Y Acad Sci Ser II*, 35 66 (1973). Ruelle's short paper was initially rejected but afterwards accepted by another journal. Only later were chaotic oscillations discovered in chemistry.

32. The discovery of an oscillating chemical reaction in the conversion of hydrogen peroxide to water by William Bray of the University of California at Berkeley in 1921 was similarly dismissed. He was told that the oscillations were probably an artifact caused by poor experimental procedure.

33. R. Field and R. Noyes, *J Chem Phys* 60 1877 (1974).

34. J. Roux, R. Simoyi, and H. Swinney, *Physica D* 8 257 (1983). The first experimental reconstruction of a strange attractor was carried out by J. C. Roux, A. Rossi, S. Bachelart, and C. Vidal, *Phys Lett* 77A 391 (1980).

35. L. Gyorgyi and R. Field, *Nature* 355 808 (1992).

36. V. Petrov, V. Gáspár, J. Masere, and K. Showalter, *Nature* 361 240 (1993).

37. Other examples include spreading depressions on the retina, waves on fertilized eggs, disease propagation in living communities, and shock-driven star

formation in spiral galaxies. The important common feature is that waves in these media propagate without attenuation.

38. See, for example, A. Winfree, *When Time Breaks Down* (Princeton University Press, 1987), and M. Markus and G. Kloss, "An Analogue to Cardiac Fibrillation in a Chemical System" preprint submitted for publication (1994).

39. V. Castets, E. Dulos, J. Boissonade, and P. De Kepper, *Phys Rev Lett* 64 2953–65 (1990).

40. J. Boissonade, *J Phys (Paris)* 49 541 (1988).

41. Q. Ouyang and H. Swinney, *Nature* 352 610 (1991).

42. S. Turing, *Alan M Turing* (Heffer, Cambridge, 1959), p. 105.

43. K-J Lee, W. McCormick, J. Pearson, and H. Swinney, *Nature* 369 215 (1994).

44. J. Boissonade, *Nature* 369 188 (1994).

45. O. Steinbock, A. Tóth, and K. Showalter, *Science* 267 868 (1995).

46. Indeed, the concept of determining optimal paths in this manner was first reported at a 1991 conference on neural networks. J. Sepulchre, A. Babloyantz, and L. Steels, *Proceedings of the International Conference on Artificial Neural Networks* T. Kohonen, K. Makisara, O. Simula, and J. Kangas (eds.) (Elsevier, Amsterdam, 1991), p. 1265.

47. The eight-tank system, interconnected with sets of tubes, valves, and pumps, was developed by Ross and Hjelmfelt with Jean-Pierre Laplante and Maria Payer of the Military College of Canada, Ontario. It relies on a slow bistable chemical reaction, the iodate-arsenous reaction, that fluctuates between a high iodine state (blue) and low (colorless). It can store up to three different patterns, hard wired into the system by the way the tanks are connected (the Hebbian weights). An initial pattern is put into the system by filling each chamber with either high or low concentrations of iodine. The pumps are turned on and, if the pattern does not contain too many errors, it stabilizes on one of the stored patterns after an hour. It is also possible to teach a chemical neural net. J. Ross, American Association for the Advancement of Science, February 1995. See also A. Hjelmfelt, J. Ross, *Proc Natl Acad Sci* 91 63 (1993); A. Hjelmfelt, E. Winberger, and J. Ross, *Proc Natl Acad Sci* 88 10983 (1991) and 89 383 (1992).

48. B. Madore and W. Freedman, *Science* 222 615 (1983).

49. M. Markus and B. Hess, *Nature* 347 56 (1990).

50. This was achieved not by altering the geometry of the lattice itself but by parking one point randomly within each square and then defining the neighborhood of a square to be all those squares whose points lie within a circle of fixed radius from the point within that square.

51. F. Mertens and R. Imbihl, *Nature* 370 124 (1994).

52. M. Markus and H. Schepers, "Turing Structures in a Semi-random Cellular Automaton" in *Mathematics Applied to Biology and Medicine* (Wuerz Publishing Ltd., Winnipeg, 1993), p. 473. Markus and Stavridis have also studied experimentally the deformation of chemical wavefronts in a light-sensitive version of the Belousov-Zhabotinski reaction, and successfully reproduced the observed effects using their semi-random cellular automaton; *Phil Trans R Soc Lond A*

347 601 (1994). In other work on the same system, these authors report on observations of chemical "turbulence"—spatiotemporal aperiodicities reminiscent of hydrodynamic turbulence; *Journal of Bifurcations and Chaos* 4 1233 (1994) and M. Markus, G. Kloss, and I. Kusch, *Nature* 371 402 (1994). We are grateful to Mario Markus for providing these references.

53. M. Markus and I. Kusch, *Proceedings of the 2nd European Conference on Mathematics Applied to Biology and Medicine* (Lyon, in press, 1995). See also Y. Gunji, *BioSystems* 23 317 (1990); G. B. Ermentrout and L. Edelstein-Keshet, *J Theor Biol* 160 97 (1993). The fall 1983 issue of *Los Alamos Science*, p. 21, has a picture of Steven Wolfram holding one of these shells, and p. 6 has an enlargement of the shell itself. Mario Markus maintains that the similarity between shells and Wolfram's automata is accidental. However, the CA approach has several advantages over continuous Turing-style reaction-diffusion models in the description of complex biophysical processes, of which little is often known in sufficient detail for the latter to be applied effectively.

54. William Blake, "Auguries of Innocence," 1, *The Poetical Works of William Blake*, J. Simpson (ed.) (1913).

55. W. Press, *Communications in Modern Physics* C 7 103 (1978).

56. P. Bak and K. Chen, *Sci Amer* 264 26 (1991).

57. P. Bak, C. Tang, and K. Wiesenfeld, *Phys Rev Lett* 59 381 (1987). Another influential model has been that from M. Kardar, G. Parisi, and Y. Zhang, *Phys Rev Lett* 56 889 (1986).

58. In the case of dry sand, it is around $34°$ with respect to the horizontal.

59. The need to incorporate the physics of interactions on *all* length scales in order to describe critical phenomena correctly was originally recognized by Kenneth Wilson at Cornell University, who introduced the so-called renormalization group to handle this mathematically. K. Wilson, *Phys Rev Lett* 28 548 (1972), and K. Wilson and J. Kogut, *Physics Reports* 12c 77 (1974). Wilson was later (1982) awarded the Nobel Prize in physics for this work.

60. Another, less elegant way of describing this self-organized state is to say that it is a generic, scale-invariant, steady state.

61. G. Held, D. Solina, D. Keane, W. Haag, P. Horn, and G. Grinstein, *Phys Rev Lett* 65 1120 (1990). This research was repeated and extended in work that purports to show that while one may obtain critical fluctuations of the avalanche size for piles that are small enough, there is a crossover to classical fluctuations; see P. Evesque, D. Fargeix, P. Habib, M. Luong, and P. Porion, *Physical Review E* 47 2326 (1993). However, other recent experiments have found no crossover out of self-organized criticality: M. Bretz et al., *Phys Rev Lett* 69 2431 (1992), and S. Grumbacher et al., *Amer J Phys* 61 329 (1993).

62. P. Bak and K. Chen, *Sci Amer* 264 28 (1991).

63. P. Bak, K. Chen, and M. Creutz, *Nature* 342 780 (1989).

64. P. Bak and C. Tang, *J Geophys Res* 95 15635 (1989), and Z. Olami, H. Feder, and K. Christensen, *Phys Rev Lett* 68 1244 (1992).

65. K. Chen, P. Bak, and M. Jensen, *Phys Lett* A 149 207 (1990); H. Feder and J. Feder, *Phys Rev Lett* 66 2669 (1991); H. Rosu and H. Canessa, *Physical Re-*

view E 47 3818 (1993); B. Suki, A. Barabási, Z. Hantos, F. Peták, and H. Stanley, *Nature* 368 615 (1994); J. Machta, D. Candela, and R. Halloc, *Physical Review E* 47 4581 (1993).

66. K. Chen and P. Bak, *Physics Letters A* 140 299 (1989).

67. P. Anderson, *Bulletin of the Santa Fe Institute* 4 13 (1989).

68. P. Bak and K. Chen, *Sci Amer* 264 32 (1991).

69. H. Jaeger, C. Liu, and S. Nagel, *Phys Rev Lett* 62 40 (1989), and S. Nagel, *Reviews of Modern Physics* 64 321 (1992). "We found no evidence of critical behavior. Instead we found what is more similar to a first order phase transition where either a large avalanche occurs or no flow occurs at all," said Nagel in a letter to Roger Highfield, November 1994. "If you doubt this it is easy to try the experiment yourself in a sugar bowl. Tilt the bowl very slowly until it starts to flow. If you keep tilting the bowl at a constant rate you will see that once the first avalanche stops the angle must be increased by a finite amount (a few degrees) before the next big avalanche occurs. There are no small avalanches of size 1, 2, 3, or 4 occurring but only the large system spanning events." In addition, a cellular automaton model of a sandpile failed to show the effects: A. Mehta and G. Barker, *Europhysics Letters* 27 501 (1994). Similar criticisms can be leveled at the KPZ equation. One theoretical model that captures this behavior correctly is from J-P Bouchaud et al., *J Physique (France) I* 4 1383 (1994).

70. A. Mehta and G. Barker, *Rep Prog Phys* 383 (1994).

71. M. Cross and P. Hohenberg, *Reviews of Modern Physics* 65 1078 (1993).

72. A. Winfree, *When Time Breaks Down* (Princeton University Press, Princeton, 1987), p. 216.

CHAPTER 7

1. J. Hudson, *The History of Chemistry* (Macmillan, London, 1992), p. 104. Materials obtained from animal and vegetable sources had been classifed and studied separately from those obtained from mineral sources ever since the days of the Arab alchemists. Bergman in 1790 first referred to "inorganic and organic bodies" and Berzelius in 1806 first used the term "organic chemistry."

2. One of the enduring myths in chemistry is that vitalism was disproved in 1828 when Fridrich Wöhler made urea (its importance was that it provided an early example of isomerism). Vitalism did receive a serious setback in 1844 when Hermann Kolbe synthesized acetic acid from nonorganic materials.

3. As we saw in Chapter 6, the second law of thermodynamics says that any change in an *isolated* system will make it more disordered. Living organisms are not isolated systems. They take in energy and matter from their environment and use it to maintain order within their boundaries; the disorder inexorably created in the process is then eliminated. Animals use oxygen and food as their inputs; plants harness carbon dioxide, oxygen, water, and sunlight. The importance of this open system aspect of life was emphasized to physicists by Erwin Schrödinger in his book *What Is Life?* (Cambridge University Press, 1944).

4. H. Dietz et al., *Nature* 352 337 (1991).

5. V. McKusick, *Nature* 352 279 (1991).

6. R. Acharya, E. Fry, D. Stuart, G. Fox, D. Rowlands, and F. Brown, *Nature* 337 709 (1989).

7. G. Binnig, W. Haeberle, F. Ohnesorge, D. Smith, H. Hörber, and C. Czerny, *Scanning Tunnelling Microscopy "STM '91,"* Interlaken, Switzerland, August 1991, and W. Haeberle, J. Hörber, F. Ohnesorge, D. Smith, and G. Binnig, *Ultramicroscopy* 42–44 1161 (1992).

8. R. C. Lewontin, *The Doctrine of DNA* (Penguin, London, 1993), p. 13. Richard Lewontin is scathing in his comments on the practical benefits of science for medicine: "Most cures for cancer involve either removing the growing tumor or destroying it with powerful radiation or chemicals. Virtually none of this progress in cancer therapy has occurred because of a deep understanding of the elementary processes of cell growth and development, although nearly all cancer research, above the purely clinical level, is devoted precisely to understanding the most intimate details of cell biology. Medicine remains, despite all the talk of scientific medicine, essentially an empirical process in which one does what works," p. 5.

9. R. Dawkins, *The Selfish Gene*, 2nd ed. (Oxford University Press, 1989).

10. E. Schrödinger, *What Is Life? The Physical Aspect of the Living Cell* and *Mind and Matter* (Cambridge University Press, Cambridge, 1967).

11. E. Schrödinger, *What Is Life? The Physical Aspect of the Living Cell* and *Mind and Matter* (Cambridge University Press, Cambridge, 1967), p. 79.

12. M. Eigen, *Steps Towards Life* (Oxford University Press, 1992), p. 39. See also J. M. Smith, *The Problems of Biology* (Oxford University Press, 1986), p. 7.

13. H. Hartman, J. Lawless, and P. Morrison (eds.), *Search for the Universal Ancestors* (NASA, Washington, SP-477, 1985), p. 1.

14. R. Dawkins, *The Blind Watchmaker* (Longman, Harlow, 1986), p. 116.

15. The molecular optimization that has been achieved today through evolution in DNA must have reached its natural limit at a sequence length of only 100 to 1,000 repeating units.

16. This approach has been used to suggest that the genetic code cannot be more than about 3.8 billion years old. M. Eigen, B. Lindemann, M. Tietze, R. Winkler-Oswatitsch, A. Dress, and A. von Haeseler, *Science* 244 673 (1989). However, the result depends on the dating of the last common ancestor (2.5 billion years for this study) when other estimates range from one billion to (most likely) something approaching 3.8 billion years, introducing a large error margin into the age of the genetic code.

17. Darwin to J. D. Hooker quoted in *Origin of Life*, Wolman (ed.) (D. Reidel, Dordrecht, 1981), p. 1.

18. Conversation between James Lake and Roger Highfield, January 1988, and see J. Lake, *Nature* 331 184 (1988). Lake performed a detailed analysis of the genetic material in the "factories" used by cells to make proteins. Called *ribosomes*, these factories are common to all classes of cell found in nature (Eubacteria, Halobacteria, Methanogens, Eocytes, Eukaryotes). Traditionally, the classes of

cells have been divided into two groups. One consists of prokaryotes, which do not contain a nucleus. They include the common bacteria or eubacteria and the "archaebacteria," which survive within extreme habitats such as places saturated with salt (halobacteria), the hot water found in springs (eocytes or "dawn cells"), and the oxygen-deprived sediments where methane-belching bugs graze (methanogens). The second group consists of more complex cells that contain a nucleus. They are called eukaryotes and include the cells from which man and plants are made.

Lake's genetic analysis suggested splitting off one group of archaebacteria, the eocytes, and placing them closer to the eukaryotes. As a result, he suggested that human cells and eocytes should be grouped together into the "Karyotes," while the eubacteria, halobacteria, and methanogens should be grouped into "Parkaryotes."

19. With Maria Rivera, James Lake has also compared the sequences of amino acids in different forms of a naturally occurring molecule called EF-tu, which cells use to make proteins. In eukaryotes and eocytes, the researchers found a distinctive segment in EF-tu's structure. In eubacteria and other types of archaebacteria, that segment is replaced. This fits Lake's theory that eukaryotes and eocytes share a special relationship. James Lake, interview with Roger Highfield, September 1994.

20. L. Orgel, *Sci Amer* 271 53 (1994).

21. A. Oparin, *The Origin of Life*. (Proiskhozhdenie zhizny Moscow Izd Moskovshii Rabochii, 1924) English trans. in J. D. Bernal, *The Origin of Life* (Weidenfield and Nicolson, London, 1967), pp. 199–234. This and many other key papers in the field can be found in D. Deamer and G. Fleischaker, *Origins of Life: The Central Concepts* (Jones and Barrtlett, Boston, 1994).

22. Scientists later reasoned that because hydrogen is common in the solar system, the early atmosphere must have been reducing rather than oxidizing—that is, the most common biogenic elements carbon, oxygen, and nitrogen would have been in their hydrogenated or reduced forms.

23. S. Miller, *Science*, 117 528 (1953); see also S. Miller and L. Orgel, *The Origins of Life on the Earth* (Prentice-Hall, Englewood Cliffs, NJ, 1974), p. 55.

24. Years later, a meteorite that struck near Murchison, Australia, was found to contain a number of the same amino acids in roughly the same relative amounts, lending credence to the idea that Miller's protocol approximated the chemistry of the prebiotic earth. More recent findings suggest the prebiotic atmosphere was less reducing than Miller and Urey thought. Repetitions of the spark discharge experiments with the initial gas phase made up of carbon monoxide or carbon dioxide in place of methane, and molecular nitrogen instead of ammonia, were found to give essentially the same range of amino acids, providing that the initial gas mixture was reducing overall.

25. W. Groth and H. Suess, *Naturwissenschaften* 26 77 (1938); J. Ferris, C-H Huang, and W. Hagan, *Origins of Life and Evolution of the Biosphere* 18 121 (1988).

26. J. D. Bernal, *Origin of Life* (Weidenfield and Nicholson, London, 1967), p. 8.

See also K. Kvenvolden, J. Lawless, K. Pering, E. Peterson, J. Flores, C. Ponnamperuma, I. Kaplan, and C. Moore, *Nature* 228 923 (1970), and C. Chyba, P. Thomas, L. Brookshaw, and C. Sagan, *Science* 249 366 (1990).

27. Although other possibilities can occur, such as the more complex crosscatalysis that provides feedback through a more indirect sequence of interlocking reactions.

28. R. Driscoll, M. Youngquist, and J. Baldeschwieler, *Nature* 346 294 (1990).

29. L. Orgel, "Evolution of the Genetic Apparatus: A Review," *Cold Spring Harbor Symposium Quant Biol* 52 9–16 (1987).

30. The central dogma of molecular biology, laid down by Francis Crick in 1958, specified a unidirectional and irreversible flow of chemical instruction from DNA to RNA and from RNA to protein: "The transfer of information from nucleic acid to nucleic acid, or from nucleic acid to protein, may be possible, but transfer from protein to protein, or from protein to nucleic acid is impossible." F. Crick, *Symp Soc Expl Biol* 12 138 (1958).

31. H. Temin and S. Mizutani, *Nature* 226 1211 (1970).

32. A term coined by W. Gilbert, *Nature* 319 618 (1986).

33. C. Guerrier-Takada, K. Gardiner, T. Marsh, N. Pace, and S. Altman, *Cell* 35 849 (1983), and F. Westheimer, *Nature* 319 534 (1986).

34. L. Orgel, interview with Roger Highfield, October 1994.

35. Nobel citation, Royal Swedish Academy of Sciences Information Department, October 12, 1989. See also J. Rajagopal, J. Doudna, and J. Szostak, *Science* 244 692 (1989); J. McSwiggen and T. Cech, *Science* 244 679 (1989).

36. L. Orgel, interview with Roger Highfield, October 1994.

37. J. Nowick, Q. Feng, T. Tjivikua, P. Ballester, and J. Rebek, *J Am Chem Soc* 113 8831–39 (1991).

38. G. von Kiedrowski, *Angew Chem* 98 932 (1986), and G. von Kiedrowski, *Angew Chem Int Ed Engl* 25 932 (1986).

39. J. Rebek, *Chemistry in Britain* 30 286 (1994).

40. T. Tjivikua, P. Ballester, and J. Rebek, *J Am Chem Soc* 112 1249 (1990).

41. V. Rotello, interview with Roger Highfield, July 1992.

42. J. Hong, Q. Feng, V. Rotello, and J. Rebek, *Science* 255 848 (1992).

43. R. Wyler, J. de Mendoza, and J. Rebek, *Angew Chem Int Ed Engl* 32 1699 (1993). J. Rebek, interview with Roger Highfield, April 1994.

44. This is ultimately due to the manner of Watson-Crick hydrogen bond basepairing between the four nucleotide base pairs found in DNA and RNA. It means that in general the process is crosscatalytic rather than autocatalytic.

45. D. Sievers and G. von Kiedrowski, *Nature* 369 221 (1994). Broadly similar conclusions were arrived at by T. Li and K. C. Nicolaou, *Nature* 369 218 (1994) using much longer oligonucleotide chains.

46. D. R. Mills, R. L. Peterson, and S. Spiegelman, *Proc Natl Acad Sci USA* 58 217 (1967), and D. R. Mills, F. R. Kramer, and S. Spiegelman, *Science* 180 916 (1973). This work made use of Q-beta replicase, an enzyme that catalyzes the synthesis of the nucleotide sequence in the RNA based Q-beta virus (whose host is the bacterium *Escherichia coli*). The original experiments took the RNA genome from the virus and stewed it with monomeric triphosphates. New

strands of RNA were produced, which had the same infectiousness as the original RNA. However, when small portions of the product were taken and incubated with more monomeric triphosphates, and this procedure repeated many times over, it was found that the RNA so produced lost its infectiousness. A selection pressure was applied in some experiments by, for example, introducing an inhibitor into the incubation phase. Then the RNA template for replication was found to evolve a mutant that could resist the inhibitor.

47. C. K. Biebricher, *Cold Spring Harbor Symp Quant Biol.* **52** 299 (1987).

48. A. Beaudry and G. Joyce, *Science* **257** 635 (1992).

49. N. Lehman and G. Joyce, *Current Biology* **3** 11 (1993).

50. Louis Bock and his colleagues at Gilead Sciences in Foster City, California, reported that they had selected out DNA molecules with the ability to anchor themselves to and inhibit thrombin, a glycoprotein that helps clot blood: L. Bock, L. Griffin, J. Latham, E. Vermaas, and J. Toole, *Nature* **355** 564 (1992). They call the molecules "aptamers" and hope the research could lead to a compound that helps dissolve blood clots in people with cardiovascular problems. As with Joyce's protocol, Bock and his colleagues generated an initial population of 10 trillion or so variants by mutation, and deployed the polymerase chain reaction to amplify aptamers for the next generation of molecules. They point out, for example, that while only one-hundredth of a percent of the DNA bound to the thrombin in the first selection cycle, almost 40 percent stuck to the glycoprotein in the fifth cycle, a significant increase in efficiency. "It may be that the affinity of an aptamer for its ligand (binding site on the thrombin) is comparable to that of an antibody for its antigen," they say.

51. S. Brenner and R. Lerner, *Proc Natl Acad Sci USA* **89** 5381 (1992).

52. G. Joyce, *The New Biologist* **3** 399 (1991).

53. S. Kauffman, *The Origins of Order* (Oxford University Press, 1993), p. 340.

54. Stuart Kauffman, naturally enough, disputes this. S. Kauffman, interview with Roger Highfield, February 1995.

55. L. Orgel, interview with Roger Highfield, October 1994.

56. L. Orgel, *Sci Amer* **271** 55 (1994).

57. J. Maynard Smith, *Nature* **280** 445 (1979).

58. M. Eigen, *Naturwissenschaften* **58** 465 (1971), and M. Eigen and P. Schuster, *The Hypercycle—A Principle of Natural Self-Organization* (Springer, Heidelberg, 1979).

59. M. Boerlijst and P. Hogeweg, "Self-structuring and Selection: Spiral Waves as a Substrate for Prebiotic Evolution" in *Artificial Life II*, C. Langton et al. (eds.) pp. 255–76 (1992); also *Physica D* **48**, 17 (1991).

60. M. Eigen, *Sci Amer* **269** 36 (1993).

61. Ibid., p. 32.

62. M. Eigen, interview with Roger Highfield, at the *Our Place in Nature* symposium celebrating the 125th anniversary of *Nature*, Royal Institution, November 1994.

63. M. Ho, P. Saunders, and S. Fox, *New Scientist* 27 (February 1986). See also S. Fox and K. Harada, *Science* **128** 1214 (1958). They synthesized protein-like polymers (proteinoids) by heating dry mixtures of amino acids. Fox has argued that

the amino acid composition of the polymers differs from that of the original mixture, from which he concludes that such polymers contain intrinsic information.

64. C. Avers, *Molecular Cell Biology* (Addison-Wesley, Reading, MA, 1986). See also W. Hargreaves, S. Mulvihill, and D. Deamer, *Nature* **266** 78 (1977).

65. F. J. Varela, H. R. Maturana, and R. Uribe, *Bio Systems* **5** 187 (1974); also F. J. Varela, *Autopoiesis: A Theory of Living Organization* (North Holland, New York, 1981). The definition of autopoiesis proposed by these authors actually goes much wider than its application to prebiotic systems alone. The concept has found application in philosophy, sociology, and economics.

66. P. A. Bachmann, P. Walde, P. L. Luisi, and J. Lang, *Journal of the American Chemical Society* **112** 8200 (1990); P. A. Bachmann, P. L. Luisi, and J. Lang, *Nature* **357** 57 (1992).

67. J. Billingham and P. V. Coveney, *J Chem Soc: Faraday Transactions* **90** 1953 (1994); P. V. Coveney and J. A. D. Wattis, "Analysis of a Generalized Becker-Döring Model of Self-reproducing Micelles," preprint (1995).

68. P. Walde, A. Goto, P.-A. Monnard, M. Wessicken, and P. L. Luisi, *J Amer Chem Soc* **116** 2541 (1994); P. L. Luisi, P. Walde, and T. Oberholzer, *Ber Bunsenges Phys Chem* **98** 1160 (1994).

69. M. Markus and B. Hess, *Proc Natl Acad Sci USA* **81** 4394 (1984); B. Hess and M. Markus, *Trends in Biochemical Sciences* **12** 45 (1987).

70. B. Hess, *Trends in Biochemical Sciences* **2** 37 (1985).

71. J. Lechleiter, S. Girard, E. Peralta, and D. Clapham, *Science* **252** 123 (1991).

72. B. Hess and A. Mikhailov, *Science* **264** 223 (1994). In fact, at normal temperatures the random motion of molecules due to heat is such that, within a micrometer, any two will bump into one other roughly once a second. Because of this, the molecules within cells should be regarded as a network of agents in constant communication.

73. J. Tabony, *Science* **264** 245 (1994).

74. J. L. Martiel and A. Goldbeter, *Biophys J* **52** 807 (1987); J. J. Tyson and J. D. Murray, *Development* **106** 421 (1989).

75. A. Hodges, *Turing: The Enigma* (Vintage, London, 1992), p. 435.

76. The reader may find it interesting to reflect on the widespread occurrence of such antagonistic elements in complex systems. In Chapter 4, we encountered the equivalent of activation and inhibition in frustrated spin systems, which have ferromagnetic and antiferromagnetic interactions simultaneously present. In neural networks (Chapters 5 and 9), we find that their complexity—and ensuing richness—is caused by the presence of both excitatory and inhibitory synaptic connections.

77. J. Murray, American Association for the Advancement of Science, 1995, J. Murray, *Sci Amer* **256** 80 (1988).

78. In the United States alone, some 400,000 people die each year when the heart develops a fatal arrhythmia without warning. It is interesting to note that the human heart is the only one whose function can be destroyed in this way.

79. Denis Noble, conversation with the authors, June 1994. See also D. Noble and G. Bett, *Cardiovascular Research* **27** 1701 (1993).

80. This sensitivity to damaged tissue makes the heart sound delicate, when we know it to be a robust organ that beats for decades and has about ten times more tissue than required for everyday use. Heart attacks are much more common in postreproductive people, suggesting that this phenomenon is still with us owing to a lack of evolutionary pressure to eradicate it. See R. Winslow, A. Varghese, D. Noble, C. Adlakha, and A. Hoythya, *Proc R Soc Lond B* 254 55–61 (1993).

81. A nonlinear demographic model has been used to predict shifts in flour beetle behavior, from stable equilibria to cycles, aperiodic, and even chaotic fluctuations as a result of changes in beetle survival rate. What makes the work significant is that elegant experiments confirmed these predictions, marking a milestone in research on nonlinear demographic models. R. Cosantino, J. Cushing, B. Dennis, and R. Desharnais, *Nature* 375 227 (1995).

82. The interested reader should consult R. Dawkins, *The Selfish Gene* (Oxford University Press, 2nd ed., 1989) or H. Cronin, *The Ant and the Peacock* (Cambridge University Press, 1991) for a lucid discussion of kinship-based cooperation. Reciprocity cooperation is concerned with symbiosis or cooperation, often between very different species, for example, the alga and fungus that make up a lichen; the fig-wasp and fig tree, wherein the wasps, which are parasites of fig flowers, serve as the only means of pollination for the tree; and a hermit crab and a sea anemone. Sometimes in such symbioses, the participants also show signs of antagonism. The application of reciprocity theory is probably the most recent addition to the biological understanding of group behavior.

83. R. Axelrod, *The Evolution of Co-operation* (Penguin Books, London, 1990), back cover.

84. One of us *(PVC)* has attended more than one course in which the work and approach advocated by Axelrod et al. has been cited as a model for understanding personal relationships inside commercial organizations. It was also remarkable to find out through role-play and in the workplace how few of those participating explicitly recognized the importance of cooperation in their daily affairs.

85. R. Axelrod, *The Evolution of Cooperation* (Penguin, London, 1990).

86. Ibid., p. xi. The results were published in R. Axelrod, *Journal of Conflict Resolution* 24 3 (1980), and R. Axelrod, *Journal of Conflict Resolution* 24 379 (1980).

87. R. Axelrod and W. Hamilton, *Science* 211 1390 (1981).

88. R. Trivers, *Quarterly Review of Biology* 46 35 (1971).

89. Ibid., p. 54.

90. M. Lombardo, *Science* 227 1363 (1985).

91. Work done by Eric Fischer and Egbert Leigh of the Smithsonian Institution.

92. T. Hobbes, *Leviathan* (Collier Books, New York, 1962), p. 100.

93. In this case, the number is 2^{70}, the exponent being the number of genes required to represent all possible combinations of the last three moves, where each move has four possible combinations, $(4 \times 4 \times 4 = 64)$ plus 3! for the hypothetical three rounds prior to the start. The factor 2 occurs because there is a choice of two moves for every one of the seventy possible sequences.

94. R. Axelrod in *Genetic Algorithms and Simulated Annealing*, L. Davis (ed.) (Pitman, London, 1987), pp. 32–41.

95. R. Axelrod, *op cit.*, pp. 37–38.

96. Note that fitness landscapes were first introduced by Sewall Wright many years ago; see S. Wright, *Evolution* 36 427 (1982).

97. W. Hamilton, *Oikos* 35 282 (1980); *Science* 218 384 (1982).

98. R. May, *Nature* 327 15 (1987).

99. M. Nowak and K. Sigmund, *Nature* 355 250 (1992).

100. M. Nowak, interview with Roger Highfield, September 1994.

101. A. Rapoport and A. Chammah, *Prisoner's Dilemma* (University of Michigan Press, Ann Arbor, 1965).

102. M. Nowak and K. Sigmund, *Nature* 364 56 (1993).

103. This sounds hardhearted but is necessary. Tit-for-tat and generous tit-for-tat can be invaded after having established themselves by any exploiters that emerge after a mutation. By contrast, softies cannot subvert a Pavlov population.

104. M. Milinski, *Nature* 325 433 (1987).

105. M. Nowak, R. May, and K. Sigmund, *Sci Amer* 272 76 (1995).

106. Blue represents a cooperating site that was cooperating in a previous generation; red a site that has remained a defector; yellow, a cooperator turned defector; and green, a defector turned cooperator. The amount of yellow and green highlights the amount of change from one generation to the next.

107. M. Nowak and R. May, *International Journal of Bifurcation and Chaos* 3 35 (1993).

108. M. Nowak and R. May, *Nature* 359 826 (1992). See also M. Nowak, S. Bonhoeffer, and R. May, *International Journal of Bifurcation and Chaos* 4 33 (1994).

109. The punctuated equilibrium concept is a controversial one, largely due to the "overblown rhetoric" used by its advocates, which frequently appears to imply that it cannot be accommodated within conventional Darwinism. See, for example, R. Dawkins, *The Blind Watchmaker* (Penguin, London, 1986), and J. Maynard Smith, *Did Darwin Get It Right?* (Penguin, London, 1993), Part 3.

110. Catastrophe theory was devised by the French Fields Medallist René Thom to describe situations in which continuous causes—in this case, changes in the evolutionary landscape—produce discontinuous effects.

111. E. C. Zeeman, *Colloque des systèmes dynamiques* (Fondation Louis de Broglie, September 1984).

112. P. Bak, H. Flyvsbjerg, and K. Sneppen, *New Scientist* 141 1916 (1994).

113. J. Lovelock, *Gaia: A New Look at Life on Earth* (Oxford University Press, Oxford, 1979). The theologian William Paley published a book called *Natural Theology* in 1802. In it, he passionately argued that the machinery of life is so intricate that it must have had a designer—God.

114. G. Ayers, J. Ivey, and R. Gillett, *Nature* 349 404 (1991).

115. R. Charlson, J. Lovelock, M. Andrea, and S. Warren, *Nature* 326 655 (1987).

116. See, for example, J. Lovelock and L. Kump, *Nature* 369 732 (1994), which analyzes the effects of temperature change on the feedbacks induced by changes in surface distribution of marine algae and land plants. In the model, they assume that algae affect climate primarily though DMS.

117. J. Lovelock, *Phil Trans R Soc Lond B* 338 383 (1992).

118. J. Lovelock, letter to Roger Highfield, January 1995.

119. H. Cronin, *The Ant and the Peacock* (Cambridge University Press, 1991), p. 279.

120. For a discussion of evolution, information, and complexity, see E. Szathmáry and J. Maynard Smith, *Nature* 374 227 (1995). Some, like Stuart Kauffman, argue that the Darwinian view that survival of the fittest is the only mechanism driving evolution does not account for all the complexity of the living world. "Darwin did not know about self-organization. And we are only just beginning to," he claims. (Interview with Roger Highfield, February 1994.) Similar points are made by Brian Goodwin of the Open University: "Darwinism sees the living process in terms that emphasize competition, inheritance, selfishness, and survival as the driving forces of evolution. These are certainly aspects of the remarkable drama that includes our own history as a species. But it is a very incomplete and limited story." B. Goodwin, *How the Leopard Changed Its Spots* (Weidenfield & Nicholson, London, 1994), p. xiv.

CHAPTER 8

1. J. Bernal, *The World, The Flesh, and The Devil* (E. P. Dutton, New York, 1929).

2. Hixon Symposium lecture, "The General and Logical Theory of Automata," Princeton University, March 2–5, 1953.

3. S. Levy, *Artificial Life* (Jonathan Cape, London, 1992), p. 28.

4. E. Moore, *Sci Amer*, October 1956, p. 118.

5. F. Dyson, "The Twenty-first Century," Vanuxem Lecture, February 1970.

6. Cf. the Game of Life (Chapter 4); see also C. Langton, "Artifical Life" in *Artificial Life: Proceedings of an Interdisciplinary Workshop on the Synthesis and Simulation of Living Systems*, C. Langton (ed.) (Addison-Wesley, Redwood City, CA, 1989), pp. 1–47. But recall from Chapter 2 that analog computers are capable of "super-Turing" computation.

7. E. Sober, *Artificial Life II*, Santa Fe Institute Studies in the Sciences of Complexity, Vol. X, C. Langton, C. Taylor, J. Farmer, and S. Rasmussen (eds.) (Addison-Wesley, Redwood City, CA, 1991), p. 750.

8. Ibid., p. 749.

9. D. Farmer and A. Belin in ibid., p. 815.

10. See, for example, E. Sober's article in ibid., p. 749.

11. A. K. Dewdney, *Sci Amer* 250 14–22 (1984); *Sci Amer* 252 14–23 (1985); *Sci Amer* 256 14–20 (1987); *Sci Amer* 260, 110 (1989); E. H. Spafford, K. A. Heaphy, and D. J. Ferbrache, "Computer Viruses, Dealing with Electronic Vandalism and programmed threats," ADAPSO 1300 N. 17th Street, Suite 300, Arlington, VA 22209 (1989).

12. A. Solomon, *PC Viruses* (Springer-Verlag, Berlin, 1991), p. 19.

13. E. Spafford, *Artificial Life II*, SFI Studies in the Sciences of Complexity, Vol. X, C. Langton, C. Taylor, J. Farmer, and S. Rasmussen (eds.) (Addison-Wesley, Redwood City, CA, 1991), p. 730.

14. B. Blumberg, *The Croonian Lecture of the Royal College of Physicians*, May 1994, and interview with Roger Highfield, May 1994.

15. A. Lindenmayer, *Journal of Theoretical Biology* 18 280 (1968).

16. C. Langton, C. Taylor, J. Farmer, and S. Rasmussen (eds.), *Artificial Life II*, SFI Studies in the Sciences of Complexity, Vol. X (Addison-Wesley, Redwood City, CA, 1991).

17. His work has since been extended to represent the development of cellular structures such as the spiral cleavage pattern of cells in a developing limpet embryo. See M. de Boer, F. Fracchia, and P. Prusinkiewicz, *Artificial Life II*, SFI Studies in the Sciences of Complexity, Vol. X, C. Langton, C. Taylor, J. Farmer, and S. Rasmussen (eds.) (Addison-Wesley, Redwood City, CA, 1991), p. 465.

18. C. Langton in *Artificial Life*, C. Langton (ed.) (Addison-Wesley, Redwood City, CA, 1988), p. xvi.

19. R. Dawkins in ibid., p. 201.

20. D. Nilsson and S. Pelger, *Proc R Soc London B* 256 53 (1994).

21. R. Dawkins, Cheltenham Festival of Literature, 1994. See also R. Dawkins, *River Out of Eden* (Weidenfeld & Nicolson, London, 1995), p. 78.

22. S. Levy, *Artificial Life* (Jonathan Cape, London, 1992), p. 160.

23. J. Holland, *Sci Amer* 267 72 (1992).

24. D. Jefferson, R. Collins, C. Cooper, M. Dyer, M. Flowers, R. Korf, C. Taylor, and A. Wang in *Artificial Life II*, SFI Studies in the Sciences of Complexity, Vol. X, C. Langton, C. Taylor, J. Farmer, and S. Rasmussen (eds.) (Addison-Wesley, Redwood City, CA, 1991), p. 549.

25. The ANNs used were recurrent, that is, feedback, nets; they do not perform any learning during their lifetimes in the simulation. It is the job of the GA—which encodes them as bit strings representing network connection weights—to select the best attuned ANNs for the tracking task. Similar results were found regardless of whether using finite state automata or ANN, and it is not clear whether one or the other should be regarded as generally preferable; however, the finite-state automata do not scale well with the size of the problem, making them impractical for representing more sophisticated sensory apparatus in ants or other organisms.

26. R. Collins and D. Jefferson in *Artificial Life II*, SFI Studies in the Sciences of Complexity, Vol. X, C. Langton, C. Taylor, J. Farmer, and S. Rasmussen (eds.) (Addison-Wesley, Redwood City, CA, 1991), p. 579. The colonies are composed of genetically identical ants, whose behavior is determined by an ANN. The ants have the ability to detect and transport food, but they can also deposit and sense pheromones for purposes of communication.

27. S. Appleby and S. Steward, *British Telecom Technology Journal* 12 No. 2 (1994).

28. S. Appleby, interview with Roger Highfield, May 1994.

29. J. Holland, *Sci Amer* 267 72 (1992).

30. Steven F. Smith introduced variable-length character strings (Ph.D. dissertation, University of Pittsburgh, 1980), cited in J. Koza, *Artificial Life II*, p. 605. Holland's classifier system has further extended the generality of the structures that can adapt via the GA method; see, for example, J. H. Holland in *Machine Learning: An Artificial Intelligence Approach*, R. S. Michalski et al. (eds.), Vol. II, (Los Altos, Morgan Kaufman, 1986), p. 593. We have also described Koza's own

genetic programming approach in Chapter 5, wherein whole programs are mutated and combined.

31. A rather widely subscribed to explanation for why sex evolved is to thwart parasites. See P. Coveney and R. Highfield, *The Arrow of Time*, pp. 257–58; also M. Ridley, *The Red Queen* (Viking, London, 1993).

32. C. Winter, P. McIlroy, and J. L. Fernández-Villacanas, *British Telecom Technology Journal* 12 No. 2 (1994).

33. This kind of thing happens in the papers by A. K. Dewdney, *Sci Amer* 253 21 (1985); R. Dawkins, "The Evolution of Evolvability," in *Artificial Life: Proceedings of an Interdisciplinary Workshop on the Synthesis and Simulation of Living Systems*, C. Langton (ed.) (Addison-Wesley, Redwood City, CA, 1989), pp. 201–20; N. Packard, ibid., pp. 141–55. Open-ended evolution—the holy grail of ALife—cannot happen with a fixed, a priori fitness function. However, a fixed, finite genome size can still allow open-ended evolution if it is governed by a fitness function that is changing a posteriori. How can it be changing if the genome size is fixed? By continual coevolution. If the effective a posteriori fitness function for creatures of type X is determined primarily by the ecology of other types of creatures Y, Z, and so on that form X's effective environment, and if the same is true for Y, Z, and so on, then the evolution of arms races, cooperation, and so on means that everybody's fitness function is continually changing in an unpredictable manner. Though finite, the size of the genome space can well be to all intents and purposes infinite. To sum up: open-ended evolution and the a posteriori fitness functions this entails are central in life, and variable-sized genomes are one possible source of this, but they are not required.

34. T. Ray, interview with Roger Highfield, June 1993.

35. T. Ray, *Artificial Life* 1 179 (1994).

36. J. D. Farmer and A. Belin, in *Artificial Life II*, *loc. cit.*, pp. 815–40.

37. T. Ray, interview with Roger Highfield, June 1993.

38. One other special feature of the *Tierran* language was introduced by biological analogy. This is that the targeting or addressing of data is carried out in *Tierran* by means of "addressing by template," a computerized analogy of template recognition between biomolecules (such as proteins), in which interactions are mediated by complementary chemical structural features. Template addressing has a dramatic effect on the size of the instruction set by removing numeric operands.

39. T. Ray, interview with Roger Highfield, June 1993.

40. C. Langton, "Artifical Life," in *Artificial Life III*, C. Langton (ed.) (Addison-Wesley, Redwood City, CA, 1989).

41. Technically speaking, *Tierra* is a "virtual" computer, in the sense that the entire computer is emulated in software within another piece of hardware. The main reason for doing this was as a security measure: the entire evolving digital community cannot then escape to wreak havoc in the computer system and, even more worryingly, spread across a local and/or wide area network.

42. Note that constant developments, refinements, and improvements are being

made to Tierra since the original release. Readers who are interested in following developments more closely (as well as obtaining source code) should use anonymous ftp to tierra.slhs.udel.edu (128.175.41.34).

43. Note that genetic algorithms, which use "crossover," as discussed in Chapter 2, are actually predicated on sexual, rather than vegetative, reproduction and selection, which is what we start off with in Tierra. It is surely no coincidence that, as noted in Chapter 5, crossover is a much more significant evolutionary component than mutations in conventional GAs.

44. T. Ray, interview with Roger Highfield, June 1994.

45. S. M. Stanley, *Proc Natl Acad Sci USA* 70 1486 (1973).

46. *Artificial Life II*, SFI Studies in the Sciences of Complexity, Vol. X, C. Langton, C. Taylor, J. Farmer, and S. Rasmussen (eds.) (Addison-Wesley, Redwood City, CA, 1991), p. 371.

47. T. Ray, in *Artificial Life II*, p. 393.

48. R. May, interview with Roger Highfield, June 1993.

49. T. Ray, interview with Roger Highfield, June 1993.

50. J. Skipper, in *Proceedings of the First European Conference on Artificial Life* (1992), p. 355.

51. P. R. Montague, P. Dayan, and T. Sejnowski, in *Advances in Neural Information Processing Systems 6*, J. Cowan, G. Tesauro, and J. Alspector (eds.) (Morgan Kaufman Publishers, San Mateo, California, in press).

52. T. Sejnowski, interview with Roger Highfield, October 1994.

53. D. Terzopoulos, X. Tu, and R. Grzeszczuk, *Proceedings of the Artificial Life IV Workshop* (MIT Press, Cambridge, MA, in press).

54. D. Terzopoulos, interview with Roger Highfield, August 1994.

55. For example, if the left front leg is raised, the right front leg stays put. R. Brooks, interview with the authors, August 1994. See also R. Brooks, "A Robust Layered Control System for a Mobile Robot," *AI Memo 864* (MIT AI Lab, Cambridge, MA, 1985). Genghis uses finite state automata hooked up in a "subsumption" architecture, a hierarchical control strategy in which one behavior can inhibit others depending on the priority of the action.

56. Takashi Gomi, president of Applied AI Systems, interview with Roger Highfield, August 1994.

57. R. Brooks, interview with the authors, August 1994.

58. D. Cliff, I. Harvey, and P. Husbands, *Adaptive Behaviour* 2 1 (1993).

59. P. Husbands, I. Harvey, and D. Cliff, "Circle in the Round: State Space Attractors for Evolved Sight Robots," submitted to *Robotics and Autonomous Systems*.

60. They are developing a gantry robot that can distinguish two types of target. I. Harvey, P. Husbands, and D. Cliff, *From Animals to Animats 3*, D. Cliff, P. Husbands, J.-A. Meyer, and S. Wilson (eds.) (MIT Press, Cambridge, MA, 1994), p. 392.

61. R. Brooks, interview with the authors, August 1994.

62. Ibid.

63. D. Hillis, in *Artificial Life II* (1992), p. 313.

64. K. Thearling and T. Ray, in *Artificial Life IV* (MIT Press, Cambridge, MA, 1994). The authors have a massively parallel (CM-5) version of Tierra now running.

65. C. Adami and C. Titus Brown to appear in *Artificial Life IV*, Proceedings of the Workshop on the Synthesis and Simulation of Living Systems, Cambridge, MA (July 6–8, 1994).

66. A recent proof was given by K. Lingren and M. G. Nordahl, *Complex Systems* 4 299 (1990), using reduced site alphabet and local rule complexity. One should note, however, that there are universal CA arbitrarily far from this "edge of chaos," so there is no uniqueness to Class IV as far as computation is concerned.

67. The whole issue of finding complex, highly adapted behavior at some transitional region between extreme order and extreme disorder has been illuminated by Eigen and Schuster's work on the so-called error threshold (see P. Schuster, *Artificial Life* 1 1994). The work of the American philosopher Mark Bedau seems to shed further significant light on this issue, too. He has found that there is a precisely quantifiable distinction between two qualitatively different kinds of evolving systems—roughly, those genetically very similar, and those genetically quite dissimilar. This distinction applies across complex adaptive systems generally, whether or not natural selection is building adaptive strategies. According to his work, the ability of natural selection to build highly adaptive strategies is maximized at the transition between these two kinds of "order" and "disorder"; furthermore, higher-order evolution can tune the parameters of first-order evolution to keep it poised at this transition. See M. Bedau and A. Bahm, in *Artificial Life IV*, R. Brooks and P. Maes (eds.) (MIT Press, Cambridge, MA, 1994), and M. Bedau and R. Seymour, *Complex Systems— Mechanisms of Adaptation* (IOS Press, Amsterdam, 1994).

68. R. Solé, O. Miramontes, and B. Goodwin, *Journal of Theoretical Biology* 161 343 (1993), and D. Gordon, B. Goodwin, and L. Trainor, *Journal of Theoretical Biology* 156 293 (1992).

69. B. Goodwin, *How the Leopard Changed Its Spots* (Weidenfeld & Nicolson, London, 1994), p. 175.

70. C. Langton, in *Artificial Life II*, p. 41.

71. C. Langton, "Computation at the Edge of Chaos: Phase-Transitions and Emergent Computation" (Ph.D. dissertation, University of Michigan, 1991); see also Chapter 4.

72. C. G. Langton, *Physica D* 42 12 (1990).

73. The system must, however, be organized in the right kind of way, says Langton: "A capacity for maximal information processing is not the same thing as the actual utilization of that capacity." C. Langton, e-mail to Peter Coveney, February 1995.

74. S. Kauffman, *The Origins of Order* (Oxford University Press, Oxford, 1993), p. 261.

75. P. Bak, K. Chen, and M. Creutz, *Nature* 342, 780–82 (1989).

76. P. Bak, C. Tang, and K. Wiesenfeld, *Phys Rev A* 38, 364–74 (1988).

77. The onset of chaos was proved to be a phase transition by Mitchell Feigenbaum: M. Feigenbaum, *J Stat Phys* 21 669 (1979).

78. In his doctoral thesis, Chris Langton explored the computational capabilities of automata, tuning their behavior by adjusting his so-called lambda parameter. However, he says that the parameter does not uniquely locate all potentially critical rules. C. Langton, e-mail to Peter Coveney, February 1995.

79. J. Crutchfield and K. Young, *Phys Rev Lett* 63 105 (1989); J. Crutchfield and K. Young, *Entropy, Complexity and the Physics of Information*, W. Zurek (ed.), Santa Fe Institute Studies in the Sciences of Complexity VIII (Addison-Wesley, 1990), pp. 223–69.

80. N. H. Packard, *Complexity in Biological Modelling*, J. A. S. Kelso, A. J. Mandell, and M. F. Schlesinger (eds.) (World Scientific, Singapore, 1988).

81. J. Crutchfield, e-mail to Roger Highfield, February 1995.

82. J. Crutchfield, letter to Roger Highfield, February 1994.

83. Letter from J. Crutchfield to Roger Highfield, March 7, 1994. Today, Langton still maintains that his measure is good enough to reveal and explore the association between complex dynamics and the edge of chaos, though he admits that it cannot find all complex rules.

84. M. Mitchell, P. T. Hraber, and J. P. Crutchfield, *Complex Systems* 7 89 (1993). C. Langton says these studies were "poorly done." C. Langton, e-mail to Peter Coveney, February 1995.

85. M. Mitchell, interview with Roger Highfield, February 1994.

86. M. Mitchell, P. T. Hraber, and J. P. Crutchfield, *op cit.*

87. Langton continues to defend his claim of a deep link between complexity, critical phase transitions, transitions between order and chaos, and computational capacity. And his critics continue to criticize the vagueness of the concepts used in his CA work. He agrees that the proposed four classes of CAs, "although a useful qualitative first approximation, are far too crude." Equally, he says his measure of complexity "is too crude a measure to be of much practical use for the CAs we typically want to work with."

CHAPTER 9

1. *Spiritus intus alit, totamque infusa per artus. Mens agitat molem et magno se corpore miscet.*

2. J. Searle, letter to Roger Highfield, December 1993.

3. T. Sorell, *Descartes* (Oxford University Press, Oxford, 1987), p. 8.

4. Though prepared in 1632, *The World* was suppressed because it supported the heresy that the earth moved around the sun.

5. J. Rée, *Descartes* (Allen Lane, London, 1974), p. 62.

6. C. Sherrington, *Man on His Nature* (Cambridge University Press, Cambridge, 1951).

7. S. Rose, *The Making of Memory* (Bantam Press, London, 1992), p. 172. Even our own bodies contain several ganglia or "computing" centers, though they have considerably diminished autonomy compared with those of a wasp, which can continue eating even when its head has been severed from its abdomen. Our

"head ganglion," the brain, may be located in the skull but the nerves that pass inside the backbone down the spinal cord to the rest of the body communicate with our other ganglia and senses.

8. What follows is one *possible* scenario.

9. C. Blakemore, interview with the authors, June 1994.

10. Ibid.

11. S. Jones, R. Martin, and D. Pilbeam, *The Cambridge Encyclopaedia of Human Evolution* (Cambridge University Press, Cambridge, 1992), p. 115. Note that the Neanderthals of Europe and the Middle East also had a modern brain size.

12. M. Nedergaard, *Science* 263 1768 (1994), and V. Parpura, T. Basarsky, F. Liu, K. Jeftinija, S. Jeftinija, and P. Haydon, *Nature* 369 744 (1994).

13. G. Fishbach, *Sci Amer* 267 25 (1992).

14. C. von der Malsburg, *Internal Report 93-06* (Ruhr Universität, Bochum, 1993).

15. F. Crick, interview with Roger Highfield, October 1994.

16. J. Carey (ed.), *Brain Facts* (Society for Neuroscience, Washington, D.C., 1993), p. 8.

17. Hearing, balance, smell, and taste neurons originate outside the neural crest and tube.

18. J. Carey (ed.) *Brain Facts* (Society for Neuroscience, Washington, D.C., 1993), p. 8.

19. B. Alberts, D. Bray, J. Lewis, M. Raff, K. Roberts, and J. Watson, *The Molecular Biology of the Cell*, 2nd ed. (Garland, New York, 1989), p. 1123.

20. R. Campenot, *Proc Natl Acad Sci USA* 74 4516 (1977).

21. C. Shatz, *Sci Amer* 267 60 (1992).

22. G. Tononi, interview with Roger Highfield, October 1994.

23. G. Tononi, O. Sporns, and G. Edelman, *Proc Natl Acad Sci USA* 91 5033 (1994). Their approach does, however, neglect many complicating factors, such as the heterogeneity of the components of a real neural system. Indeed, the complexity measure C_N can only be applied to *static* neural connectivity; it does not capture anything to do with neural dynamics, the seat of all intelligent behavior.

24. Differences are often associated with such things as sensory functions, motor functions, and long-distance transmission.

25. S. Rose, *The Making of Memory* (Bantam Press, London, 1992), p. 259.

26. S. Ramón y Cajal, *Recollections of My Life* (Garland, New York and London, 1988).

27. P. S. Churchland and T. J. Sejnowski, *The Computational Brain* (MIT Press, Cambridge, MA, 1992), p. 43.

28. There is a high internal concentration of potassium and a low internal concentration of sodium.

29. In digital computers, the typical time scale over which events take place are nanoseconds (10^{-9} seconds), while biological neurons act over periods of milliseconds (10^{-3} seconds).

30. R. Douglas and K. Martin, *Trends in Neurosciences* 14 286 (1991).

31. R. Douglas, interview with Roger Highfield, June 1994. The images in Figure 9.3 (a) and (b) were drawn from O. Bernander, R. Douglas, K. Martin, and C. Koch, *Proc Natl Acad Sci USA* 88 11569 (1991), and J. Anderson, K. Martin, and D. Whitteridge, *Cerebral Cortex* 412 1047 (1993).

32. R. Lyon and C. Mead, *IEEE Trans Acoustics Speech and Signal Processing* 36 1119 (1988).

33. M. Mahowald and C. Mead, *Neural Networks* 1 91 (1988).

34. The artificial retina consisted of a 64 × 64 array of light-sensitive receptor cells—handfuls of transistors—that generated in real time the sort of output one would expect to see from the real thing. Gaze at an image of yourself relayed from the retina and you find that you disappear when standing still: like the real thing, the receptors only extract information salient to subsequent processing. Overall, the similarities between the simulation and reality were remarkable; its success has not only helped to emphasize more generally the success of biological computing, but also demonstrated that the principles of neural information processing offer a powerful new engineering paradigm.

35. This work was preceded by the use of simplified networks to demonstrate how self-organization within an analog system can deal with degraded information. M. Mahowald, *Analog VLSI Chip That Computes Stereocorrespondence Using a Cooperative Multiscale Algorithm*, American Association for the Advancement of Science, February 1994. The network can take information from two artificial retinas and calculate how far away an object is on the basis of the difference in perspective. This network has successfully tackled one of the classical problems in machine vision—stereocorrespondence, the determination of corresponding points in a pair of images taken from two vantage points.

36. A review of neuromorphic analog VLSI can be found in R. Douglas, M. Mahowald, and C. Mead, *Annual Review of Neuroscience* 18 (1995), in press.

37. M. Mahowald, interview with Roger Highfield, June 1994.

38. M. Mahowald and R. Douglas, *Nature* 354 515 (1991).

39. The problem of how to store synaptic weights is the main stumbling block facing analog neural network design.

40. A similar construction-kit approach is being attempted in an extraordinary experiment to grow living networks of brain cells by scientists at the Naval Research Laboratory's Center for Biomolecular Science and Engineering in Washington D.C., in collaboration with groups at the Science Applications International Corporation, Virginia, the National Institutes of Health, Maryland, and the University of California at Irvine. The original inspiration came from a project at AT&T Bell Laboratories in New Jersey, where rat neurons were grown on lines that were created by using conventional chip-fabrication technology. See C. Robinson, *Signal*, February 1994, p. 15. Some of the most impressive work in this field has been carried out by Peter Fromherz of the Max Planck Institute of Biochemistry in Martinsried, near Munich. P. Fromherz, A. Offenhäusser, T. Vetter, and J. Weis, *Science* 252 1290 (1991), and P. Fromherz and H. Schaden, *Eur J Neurosci* 6 1500 (1994). Fromherz has succeeded in growing neurons on chips, setting up a dialogue between neuron and chip, and has persuaded neurons to grow tendrils in desired geometries. At the time of going to press, forming synapses between the cultured neurons still presented difficulties. P. Fromherz, interview with Roger Highfield, May 1995.

41. The disease could be alleviated by administering a drug called L-dopa, which is

transformed to dopamine within the brain. This remedy is still used today, though its effectiveness declines in the long term. New approaches rely on the use of fetal neurons, growth factors to stimulate the repopulation of the substantia nigra, or even genetic engineering to "program" body cells to make dopamine. T. A. Larson and D. B. Calne, *Trends in Neurosciences* 5 10 (1982).

42. P. Seeman, H-C Guan, and H. Van Tol, *Nature* 365 441 (1993). Schizophrenia affects around one person in 100 and, though it cannot be precisely defined, it is characterized by disturbances in thinking, emotional reaction, and behavior. Delusions, hallucinations, muddled thinking, and speaking can all follow after a sufferer initially becomes withdrawn and introspective. It had been suspected for some time that there is a link between schizophrenia and dopamine. Doctors found that amphetamines can cause schizophrenia-like psychosis and it turns out that amphetamines stimulate the release of dopamine in the brain. This "dopamine hypothesis" was furthered when it was discovered that antischizophrenic drugs, which prevent amphetamines exerting this psychostimulant effect, act at dopamine receptors. The efficacy of the drugs used to treat the illness was known to vary according to how well they bound to the receptors. L. Iversen, *Nature* 365 393 (1993).

43. P. Seeman, H.-C. Guan, and H. Van Tol, *Nature* 365 441 (1993).

44. P. Seeman, interview with Roger Highfield, September 1993.

45. E. Tanzi, *A Text-Book of Mental Diseases* (Rebman, 1909).

46. D. Hebb, *Organization of Behavior* (Wiley, New York, 1949), p. 62.

47. S. Rose, *The Making of Memory* (Bantam Press, London, 1992), p. 91.

48. T. V. Bliss and T. Lømo, *Journal of Physiology* 232 331 (1973).

49. T. Brown, E. Kairiss, and C. Keenan, *Annual Review of Neuroscience* 13 475 (1990).

50. R. Morris, E. Anderson, G. Lynch, and M. Baudry, *Nature* 319 774 (1986).

51. S. Rose, *Time dependent biochemical and cellular processes in memory formation*, Mexican Physiology Society Symposium, August 1993.

52. A. R. Luria, *The Mind of a Mnemonist: A Little Book About a Vast Memory* (Basic Books, New York, 1968).

53. N. L. Desmond and W. B. Levy, *Brain Research* 265 21 (1983); A. S. Artola, S. Brocher, and W. Singer, *Nature* 347 69 (1990).

54. An elegant study of dozing rats by Matthew Wilson and Bruce McNaughton of the University of Arizona shows how activity of the hippocampus during sleep helps us to consolidate our memories. Using ultrafine electrodes, they studied the electrical activity of large numbers of nerve cells in the hippocampus, as the rats explored a box and then a maze. When the rats slumbered, they found that the same patterns of electrical activities were repeated. This suggested that the rats relived their experiences during dreams, transferring memories from short-term storage in the hippocampus to create long-term memories in the cortex. M. Wilson, interview with Roger Highfield, August 1994; see also M. Wilson and B. McNaughton, *Science* 265 676 (1994).

55. D. Marr, *Phil Trans R Soc London B* 262 23 (1971).

56. E. Rolls, "Learning and Memory," p. 4 of a draft, Chapter 28 in *Physiology*, C. Blakemore, C. Ellory, J. Morris, and P. Nye (eds.) (Gower, London, 1995).

57. E. T. Rolls, in *Parallel Distributed Processing: Implications for Psychology and Neuroscience*, R. G. M. Morris (ed.) (Oxford University Press, 1989), p. 286.

58. E. Rolls, interview with Roger Highfield, June 1994.

59. E. Rolls and S. O'Mara, in *Brain Mechanisms of Perception: From Neuron to Behavior*. T. Ono, L. Squire, M. Raichle, D. Perrett, and M. Fukuda (eds.) (Oxford University Press, Oxford, 1993), p. 282.

60. A. Treves and E. Rolls, *Hippocampus* (in press, 1995).

61. For example, there are several inputs to CA3: perforant path, associational and commissural fibers (between them forming the great majority of synapses), and mossy fibers. Rolls and Treves predict the mossy fibers are important for learning, but not recall.

62. Some, like Gerald Edelman, believe neural nets offer a "metaphor, not a theory" and are not able to categorize an unlabeled world without supervision or a program. His colleague Paul Verschure attacks models such as NETtalk as optimization techniques that say more about the data set than the neural net approach to the brain. (See P. Verschure and G. Dorffner [eds.], *Neural Networks and a New AI* [Chapman Hall, London, in press, 1995].) Others, like Francis Crick, believe that the connectionist approach may prove valuable so long as the resulting neural networks contain sufficient biological realism. Interviews with Roger Highfield, October 1994.

63. S. Zeki, *A Vision of the Brain* (Blackwell Scientific Publications, Oxford, 1993), p. 19.

64. B. Milner, in *The Neurosciences: Third Study Program*, F. O. Schmitt and F. G. Worden (eds.) (MIT Press, Cambridge, MA, 1973), p. 75.

65. R. Adolphs, D. Tranel, H. Damasio, and A. Damasio, *Nature* 372 669 (1994).

66. C. Von der Malsburg, *Kybernetik* 14 85 (1973).

67. D. Willshaw and C. von der Malsburg, *Proc R Soc London* B194 431 (1976).

68. M. Merzenich et al., *Neuroscience* 8 33 (1983), and M. Merzenich et al., *Neuroscience* 10 639–65 (1983).

69. G. Hinton, interview with Roger Highfield, November 1993.

70. C. von der Malsburg and W. Singer, *Neurobiology of Neocortex* (John Wiley and Sons, London, 1988), p. 70.

71. T. Kohonen, *Self-Organization and Associative Memory*, 3rd ed. (Springer, Berlin, 1989), pp. 122–23.

72. J. Maddox, *Nature* 369 517 (1994). Maddox cites one explanation and a neural net to go with it, made by L. Molgedey and H. Schuster, *Phys Rev Lett* 72 3634 (1994).

73. Strictly, we should mention that this is for letter accuracy on the written output, for an unlimited vocabulary. It also involves the use of a grammar rule base, primarily to deal with the problems caused by co-articulation, in which phoneme pronunciation is altered because of the presence of neighboring phonemes.

74. The technique is called "learning vector quantization."

75. C. Blakemore, interview with authors, June 1994.

76. G. Hinton, *Sci Amer* 267 145 (1992).

77. Brains have a limited type of supervision in the form of "teacher" cells. This is the nervous system's way of avoiding the combinatorial explosion that would otherwise occur if the network had to learn *all* possible functions that map to the input data when the amount of data to be handled grows large.

78. G. Hinton, interview with Roger Highfield, November 1993; see R. Linsker, *Proc Natl Acad Sci USA* **83** 7508 (1986), **83** 8390 (1986), and **83** 8779 (1986). To find out more about the dream work, see G. Hinton, P. Dayan, B. Frey, and R. Neal, *Science* **268** 1158 (1995). If the "recognition" Helmholtz network works well at converting pictures into representations, it can be used to train the "generative" Helmholtz network, and vice versa. It seems that dreams are essential, if we are to recognize objects and make sense of the world. G. Hinton, interview with Roger Highfield, May 1995.

79. Visual acuity in human infants improves over the first few months of life. The visual cortex in an adult has six layers of neurons; it takes the first year of a baby's life for these layers to all be put in place, with the interior ones developing first, the outer layer last. In recent years, it has proved possible to evaluate the effect of this neural development on an infant's visual capabilities through studies of eye movement. At an age of one month, the lowest-lying layer of visual cortex enables control of fixation—the voluntary holding of the eyes on a stimulus. So-called obligatory looking—fixated staring at objects for long periods—occurs from one to four months. After this time, babies become progressively more able to control their own orientation toward visual events.

80. The expression of the NMDA receptor around twelve weeks after birth. C. Blakemore, interview with the authors, June 1994.

81. A very interesting neural network model simulating the development of kitten visual cortex has been performed by K. D. Miller and M. P. Stryker, in *Connectionist Modeling and Brain Function: The Developing Interface*, S. J. Hanson and C. R. Olson (eds.) (MIT Press, Cambridge, MA, 1990), p. 255.

82. C. Blakemore, interview with the authors, June 1994.

83. C. Blakemore, interview with Roger Highfield, March 1995.

84. T. Bonhoeffer, V. Staiger, and A. Aertsen, *Proc Natl Acad Sci USA* **86** 8113 (1989). There is also an example of a non-Hebbian synapse in *Aplysia*; see E. R. Kandel et al., in *Synaptic Function*, G. M. Edelman, W. E. Gall, and W. M. Cowan (eds.) (Wiley, New York, 1987), p. 471.

85. Within his neural net, he found that the first layer of cells, so-called spatial-opponent cells, amplified any contrast between signals received by closely spaced cells on the model's "retina." A few layers deeper into the net, another category of cells emerged that closely modeled the "orientation selective" cells in the real visual cortex that respond to edges, the V1 region of the cortex first discovered by Hubel and Wiesel. The patterns of cells were more complex than had been seen in the laboratory. However, experimental work conducted around the same time came up with a qualitatively similar finding.

86. M. Enquist and A. Arak, *Nature* **361** 446 (1993).

87. C. Darwin, *The Descent of Man and Selection in Relation to Sex* (Murray, London, 1871).

88. R. Johnstone, *Nature* 372 172 (1994); also R. Johnstone, interview with Roger Highfield, November 1994.

89. A. Arak, interview with Roger Highfield, November 1994. M. Enquist and A. Arak, *Nature* 372 169 (1994).

90. There is a breakdown in communication between the front and back of the brain in sufferers of this common psychiatric disorder, according to Karl Friston of the Institute of Neurology in London. Previous studies have suggested that a breakdown occurs in the dialogue between the left and right halves of the brain. However, "more remarkable and more exciting" in terms of understanding the illness is the discovery that there is a lack of communication (see Appendix) between frontal regions of the brain, which deal with intentions, and the temporal lobes toward the back of the brain responsible for language, word processing, and representations. This breakdown was found by Chris Frith using a technique called Positron Emission Tomography (PET) (see Appendix) at the Hammersmith Hospital in London, and by processing the data using analytical techniques developed by Friston. "If you like, it is a failure to integrate what we are doing to the world with our perception of the consequences of what we have done that leads to many of the symptoms experienced by schizophrenics," said Friston. This failure to integrate the intention to speak with the consequent recognition of the speech is the reason that schizophrenics claim to hear voices. K. Friston, interview with Roger Highfield, February 1995.

91. W. Singer, in *International Review of Neurobiology*, O. Sporns and G. Tononi (eds.), 37 153 (1994), and C. Gray, A. Engel, P. König, and W. Singer, *Visual Neuroscience* 8 337 (1992).

92. L. Finkel and G. Edelman, *The Journal of Neuroscience* 9 3188 (1989); G. Tononi, O. Sporns, and G. Edelman, *Cerebral Cortex* 2 310 (1992).

93. O. Sporns, G. Tononi, and G. Edelman, *Proc Natl Acad Sci USA* 88 129 (1991); G. Edelman, *Neuron* 10 115 (1993); and K. Friston, G. Tononi, G. Reeke, O. Sporns, G. Edelman, *Neuroscience* 59 229 (1994).

94. J. Taylor, *When the Clock Struck Zero* (Picador, Basingstoke, 1993), p. 165; J. Taylor and F. Alavi (eds.), *Mathematical Approaches to Neural Networks* (Elsevier Science Publishers, 1993), p. 341; J. Taylor and F. Alavi, *Neural Network World* 5 477 (1993); J. Taylor, S. Hameroff et al. (eds.), *Toward a Scientific Basis for Consciousness* (MIT Press, Cambridge, MA, in press, 1995).

95. J. Taylor, interview with Roger Highfield, July 1993.

96. W. Calvin, *Sci Amer* 271 84 (1994).

97. T. Shallice, interview with Roger Highfield, British Association for the Advancement of Science, August 1993; see also G. Hinton, D. Plaut, and T. Shallice, *Sci Amer*, October 1993, p. 58.

98. G. Hinton, D. Plaut, and T. Shallice, *Sci Amer*, October 1993, p. 65.

99. This is an example of "vector completion"—the filling in of visual information even when input vectors are essentially auditory; see A. Damasio, D. Tranel, and H. Damasio, *Annual Review of Neuroscience* 13 89 (1990).

100. M. Virasoro, *Europhysics Letters* 7 293 (1988); see also D. Sherrington, *Speculations in Science and Technology* 14 319 (1991).

101. Christopher Longuet-Higgins of Sussex University provides an amusing view of the problem. For him, there are at least five different problems of consciousness: "First: What is its biological function? Answer: It's safer to be wide awake sometimes than sound asleep all the time. Second: Are any parts of the brain specially active when one is conscious? Answer: Undoubtedly. Third: Are worms, or plants, or robots conscious? Answer: It is impossible to tell, because consciousness is an essentially private affair. (I like to think my friends are conscious, but one can never be sure.) Fourth: Isn't consciousness just another natural phenomenon, like gravity, now ripe for scientific explanation? Answer: No, the boot is on the other foot. Science, the story we tell about nature, could not arise in a world devoid of conscious beings. Fifth: Then why is consciousness so much in the news these days? Answer: Because neuroscientists and neurophilosophers have recently discovered what fun it is thrashing around on one another's territory." C. Longuet-Higgins, letter to Roger Highfield, December 1993.

102. G. Fishbach, *Sci Amer* **267** 24 (1992).

103. F. Crick, letter to Roger Highfield, December 1993.

104. F. Crick, interview with Roger Highfield, October 1994.

105. Of course, "the yellowness of yellow" is itself subjective! It is not necessary that everything science discusses need be observable; for instance, note that, as discussed in Chapter 3, the wave function in quantum mechanics is not an observable property, yet it is the source from which everything we can measure is supposed to be derived.

106. *Chambers 20th Century Dictionary*, New Edition (1983).

107. G. Edelman, interview with Roger Highfield, October 1994.

108. Edelman has started to test his ideas about neuronal group selection in adaptive automata. One, called NOMAD (Neurally Organized Mobile Adaptive Device) starts out with a basic set of reflexes and learns how to sort red and blue blocks in a playpen on the basis of differences in color associated with their appearance and inbuilt values on how to react to their conductivity or "taste." The "nervous system" of NOMAD is Darwin IV. See G. Edelman, *Neuron* **10** 115 (1993), and G. Reeke, O. Sporns, and G. Edelman, *Proceedings of the IEEE* **78** 1498 (1990). The aim is to show in a controlled environment how a given behavior emerges in real time when visual and taste maps associate with a motor map, on the basis of Edelman's selectionist principles. Paul Verschure of the Neurosciences Institute argues that the visual-motor map will adapt to distinguish the blocks, without explicit instructions about their color, once NOMAD has learned to associate a given color with a preset reaction to a taste. Although this sounds very similar to Kohonen's concept of a self-organizing feature map, Edelman argues that labels, such as "blue block," are written into all forms of artificial neural networks by their programmers and that therefore true learning does not occur in the nets. Perhaps surprisingly, in view of his automaton modeling, Edelman believes that it is not possible to simulate brain function or evolution in a Turing machine. He maintains that both elude Turing's definition of computation because of the essential role played by (noncomputable)

randomness. In the light of our discussions in Chapter 2, this leaves the possibility that brains are carrying out either quantum computable or noncomputable ("super-Turing") processes. Since Edelman disagrees with Roger Penrose's position on consciousness, we would infer that he supports a quantum computational view of the brain. However, Edelman strongly disagrees with this conclusion. Interview with Roger Highfield and demonstration of NOMAD, October 1994 and February 1995.

109. G. Edelman, G. Reeke, E. Gall, G. Tononi, D. Williams, and O. Sporns, *Proc Natl Acad Sci USA* **89** 7267 (1992).

110. G. Edelman, interview with Roger Highfield, October 1994.

111. D. Dennett, letter to Roger Highfield, December 1993. See also D. Dennett, *Consciousness Explained* (Penguin, London, 1991).

112. C. von der Malsburg, IIAS Symposium on Cognition, Computation and Consciousness, Kyoto, August/September 1994.

113. See, by way of illustration, the critical essays in *Minds and Machines*, Alan Anderson (ed.) (Prentice-Hall, Englewood Cliffs, NJ, 1964), which includes Turing's *Mind* article.

114. R. Penrose, *Shadows of the Mind* (Oxford University Press, 1994). Part I attempts to demonstrate that consciousness is inherently noncomputable; Part II asserts that as yet unknown advances in physics will yield a scientific (but noncomputable) account of consciousness. It is a development of the ideas in *The Emperor's New Mind*, although the basic point of view is changed only in certain (important) details. It contains a response to criticisms of the earlier book, and includes a discussion of microtubules as a bridge from an as yet unknown "quantum gravity" theory to neurons.

115. Roger Penrose, letter to Peter Coveney, 22 February 1995.

116. Note, however, that the arguments made by Penrose in Part I of *Shadows of the Mind* do not require one to adopt a Platonic position.

117. Penrose maintains that the collapse of wave functions must be noncomputable.

118. R. Penrose, *Shadows of the Mind* (Oxford University Press, 1994), p. 358. Hameroff and his colleagues argue that microtubules may play a role as cellular automata that transmit and process complex signals.

119. R. Penrose, *Shadows of the Mind* (Oxford University Press, 1994), sections 7.6, 7.7 and 8.6; and letter to P. Coveney, 22 February 1995.

120. R. Penrose, *Shadows of the Mind* (Oxford University Press, 1994), p. 376.

121. See R. Penrose, "Précis of *The Emperor's New Mind: Concerning Computers, Minds and the Laws of Physics*," responses by critics and Penrose's replies, *Behavioural and Brain Sciences* **13** 643 (1990). Note that, unlike the later *Shadows*, there is no mention of cytoskeletons and related ideas to be found in *Emperor*.

122. F. Crick, interview with Roger Highfield, October 1994.

123. A. Klug, interview with Roger Highfield, December 1994. "Apparently the hole in the middle of the microtubule is the right size for certain quantum gravity effects. But there are many biological objects with holes in the middle. I am totally unconvinced that microtubules are there for the purpose of quantum gravity calculations rather than for providing structural elements both in

neurons and elsewhere. The reason there is a hole is a simple one: it is a consequence of the structural design. Why is a drinking straw rigid? It is because its mass is concentrated on the outside, thereby providing resistance to bending." Sir Roger is unclear as to what Sir Aaron is referring to, but points out that he has never claimed that the hole in the middle of the microtubule is the right size for quantum gravity effects. (Letter to Peter Coveney, 22 February 1995.)

124. G. Edelman, interview with Roger Highfield, October 1994. Sir Roger states that it would be very surprising if the drug dissolved all the brain's microtubules: "There are many *admitted* roles for microtubules in the actions of neurons and it would be very dangerous to interfere substantially with these." (Letter to Peter Coveney, 22 February 1995.) In further correspondence, Sir Roger wrote: "Evidently the drug that Edelman is referring to is colchicine, but when given by mouth or injection it does *not* get into the brain. In experiments on animals in which colchicine is directly introduced into the cerebrospinal fluid, however, it causes dementia. Thus, one might indeed conclude that dissolving the brain's microtubules also dissolves the 'soul'! See G. Bensimon and R. Chernat, *Pharmacol Biochem Behavior* 38 141 (1991). The effects seem to be similar to those of Alzheimer's disease." Letters to P. Coveney, 8 and 17 March 1995.

125. M. Arbib, interview with the authors, August 1994. See also, M. Arbib, *The Daily Telegraph*, 2 November 1994, p. 16. In discussing Gödel's theorem, his book *Brains, Machines, and Mathematics* (Springer-Verlag, Berlin, 1987) includes John Myhill's result that a computer can be designed such that, whatever formal system L it embodies, it can reproduce itself in a form that embodies the old L with the consistency claim for L appended as a new axiom. It needs no consciousness to add the new axiom. By contrast, Roger Penrose rests his claim that human consciousness is noncomputable on its ability to detect a "missing truth" that, according to Arbib, can be computed by an unconscious machine.

126. M. Hesse and M. Arbib argued in their Gifford Lectures, *The Construction of Reality* (Cambridge, 1986), that much of human knowledge is fallible and metaphorical. But for these authors, fallibility does not imply noncomputability. Arbib and Hesse offered "schema theory" as a *computational* account of a fallible knower. It is not based on a fixed logical system, but changes constantly as the knower makes mistakes or encounters new aspects of the world or society. Roger Penrose discusses these kinds of issues at length in Chapter 3 of *Shadows of the Mind* (Oxford University Press, 1994) and comes to opposite conclusions.

127. R. Penrose, *Shadows of the Mind* (Oxford University Press, 1994), p. 201. He argues that "human mathematical understanding cannot be reduced to (knowable) computational mechanisms, where such mechanisms can include any combination of top down, bottom up or random procedures. We appear to be driven to the firm conclusion that there is something essential in human understanding that is not possible to simulate by any computational means."

128. A. Turing, "Intelligent machinery," in *Machine Intelligence 5*, B. Meltzer and D. Michie (eds.) (Edinburgh University Press, 1969). Note, however, that

learning systems are discussed extensively in Chapter 3 of *Shadows of the Mind*, where Penrose argues that Gödel's theorem still applies.

129. M. Arbib, interview with the authors, August 1994.

130. I. J. Good, *Brit J Phil Sci* 18 359 (1969).

131. See, for example, H. T. Siegelmann, *Science* 268 545 (1995), and our discussion in Chapter 2 near footnote 89. There is no need to introduce "quantum gravity" to achieve noncomputability.

132. M. Arbib, interview with the authors, August 1994.

133. R. Traub and R. Miles, *Seminars in the Neurosciences* 4 27 (1992).

134. R. Traub, R. Miles, and J. Jefferys, *Journal of Physiology* 461 525 (1993); R. Traub, J. Jefferys, and R. Miles, *Journal of Physiology* 472 267 (1993); R. Traub and J. Jefferys, *Journal of Physiology* 472 267 (1993); R. Traub, J. Jefferys, and M. Whittington, *Journal of Physiology* 478 379 (1994).

135. M. Whittington, R. Traub, and J. Jefferys, *Nature* 373 612 (1995).

136. D. Nilsson and S. Pelger, *Proc R Soc London B* 256 53 (1994). The authors quote Darwin's words: the thought "that the eye . . . could have been formed by natural selection seems, I freely confess, absurd in the highest possible degree."

CHAPTER 10

1. S. Rose, *Nature* 373 380 (1995).

2. D. Wahlsten, American Association for the Advancement of Science, February 1995. A recurring theme in such studies is the lack of a well-defined trait. Intelligence, alcoholism, or criminality are not single entities and no one has ever come up with an accurate definition of them.

3. G. Allen, American Association for the Advancement of Science, February 1995.

4. F. Dyson, in *Nature's Imagination*, J. Cornwell (ed.) (Oxford University Press, Oxford, 1995), p. 6.

5. Ibid., p. 8.

6. J. Barrow, ibid, p. 47.

7. Because abrupt climate changes have not occurred during the past 10,000 years, human civilization has flourished with a false sense of security, according to Wallace Broecker of Columbia University's Lamont Doherty Earth Observatory. "The Earth's climate system is endowed with a disturbing characteristic: it is able to jump from one mode of operation to another," he said. Evidence from seafloor and lake sediments, ice cores, fossilized pollen, corals, Andean glacial movements, and other records demonstrate unequivocally that during the last ice age from 70,000 to 10,000 years ago, the Earth's climate system changed frequently, and often within the span of a lifetime. The shifts had worldwide consequences: in the North Atlantic, temperatures rose and fell by 10°F or more; tropical rainfall changed radically; the world's oceans circulated in a different way; icebergs and sea ice proliferated, and atmospheric concentrations of gases varied." American Association for the Advancement of Science, February 1995. See also W. Broecker, *Nature* 372 421 (1994).

8. D. Tilman, R. M. May, C. L. Lehman, and M. Nowak, *Nature* 371 65 (1994).

9. At the Santa Fe Institute in New Mexico, Alan Perelson, Gerard Weisbuch, Lee Segel, Rob De Boer, Avidan Neumann, and others have been attempting to create an immune system in a computer. When we are infected with a bacterium, B cells in the blood start to manufacture antibodies, proteins that are able to recognize the bacterium by latching on to it. However, the process does not stop there: an antibody is made that recognizes the antibody and so on and so forth, weaving vast interconnected loops of activity. One important application of these computer models is to understand how the immune system can distinguish its tissue from an invader so that it does not turn on itself, as occurs in arthritis and multiple sclerosis.

10. Another nonlinear model, similar to the one due to Tilman et al. for species extinction, provides an explanation for why the time lag between infection with the virus and the manifestation of AIDS (acquired immunodeficiency syndrome) can range from a year to a decade or more. The strains of the virus that are best able to compete for white blood cells tend to be worst at spreading. One implication of this is that a common medical dogma—that fast-spreading diseases always become less virulent—is wrong. Competitively superior viruses will tend to produce many strains, none of which are very abundant. Similarly, abundant viruses that are worse at spreading will have fewer strains. It turns out that the ones that mutate faster are the more deadly. The model suggests that for AIDS to develop, the diversity of different strains of HIV in the body must reach a certain threshold level. Beyond that point, the immune system can no longer keep the virus under control. R. M. May and R. M. Anderson, *Sci Amer* 266 58 (1992), and British Association for the Advancement of Science, Keele, 1993.

11. See, for example, J. L. McClelland and D. E. Rumelhart, *Parallel Distributed Processing*, Vol. 2 (MIT Press, Cambridge, MA, 1986), and S. Pinker, *The Language Instinct* (Morrow, New York, 1994).

12. R. Dawkins, *The Selfish Gene* (Oxford University Press, 1976), p. 206.

13. R. Dawkins, "Is Religion Just a Disease?," *The Daily Telegraph*, December 15, 1993. To his annoyance, Dawkins received a copy of the St. Jude chain letter, which is said to have been around since 1903, and promises good luck within four days to those who distribute twenty copies and bad luck to those who do not. He said that it could cause "mental distress as real as, in its own way, the physical distress caused by the common cold virus." "By inducing guilt, fear, greed and piety, it causes susceptible hosts to multiply it 20-fold and transmit the 20 copies to potential hosts," he wrote to the journal *Nature*. Fortunately many people were "immune," or within eight generations, everyone in the world would receive an average of 4.5 copies. O. Goodenough and R. Dawkins, *Nature* 371 23 (1994).

14. D. Dennett, *Consciousness Explained* (Allen Lane, London, 1991), p. 202.

15. P. Ormerod, *The Death of Economics* (Faber and Faber, London, 1994), p. 178.

16. B. Grenfell, R. M. May, and H. Tong (eds.), *Phil. Trans. Royal Soc.* A348 (1994).

17. A. N. Refenes, M. Azema-Barac, L. Chen, and S. A. Karoussos, *Neural Computing and Applications* 1 46 (1993); R. G. Hoptroff, *Neural Computing and Applications* 1 59 (1993).

18. There is very little published on this field, for obvious reasons. A good summary can be found in M. Ridley, "The Mathematics of Markets," *The Economist*, October 9, 1993.

19. *The Economist*, October 15, 1994.

20. N. van Kampen, *Stochastic Processes in Physics and Chemistry* (North Holland, Oxford, 1987), p. 320.

21. J. L. Deneubourg, S. Aron, S. Goss, J. M. Pasteels, and G. Duerinck, "Random Behavior, Amplification Processes and Number of Participants: How They Contribute to the Foraging Properties of Ants," in *Evolution, Games and Learning*, D. Farmer A. Lapedes, N. Packard, and B. Wendroff (eds.) (North Holland, Amsterdam, 1986); J. L. Deneubourg, S. Goss, N. Franks, A. Sendova-Franks, C. Detrain, and L. Chretien, in *From Animals to Animats: Proceedings of the First International Conference on Simulation of Adaptive Behaviour* (MIT Press, Cambridge, MA, 1991). See also J. Koza, *Genetic Programming* (MIT Press, Cambridge, MA, 1992), Chapter 12.

22. P. Allen, *Self-Organization and Dissipative Structures. Applications in the Physical and Social Sciences*, W. Schieve and P. Allen (eds.) (University of Texas Press, Austin, 1982), p. 142.

23. Gerard Weisbuch used Swarm, a computer model developed at Santa Fe by Christopher Langton. The computer models the behavior of "agents," whether white blood cells, government firms, or individuals. In this case, the agents are individuals controlled by a neural net. To make it more realistic, a little randomness is also added. The decisions of the agents are made based on their interests and how the information is received. In Weisbuch's simulation, he examined the complicated trade-off between cheap polluting forms of transport and more expensive "green" methods. One of the variables is how long people remember former pollution episodes. Another is price—everybody will use green technology if it is cheap enough. And the model can show when the price differential is large enough to trigger cheating by using cheap polluting cars in a population heavy with greens. "What is important is being in a situation where you get a large gradient of pollution so people can tell bad from good." G. Weisbuch, interview with Roger Highfield, February 1995.

24. K. Mainzer, *Thinking in Complexity* (Springer-Verlag, Berlin, 1994), p. 276.

25. J. Koza, *Genetic Programming* (MIT Press, Cambridge, MA, 1992), Chapter 12.

26. These figures come from the Internet Society in August 1993 (quoted by K. Kelly *Out of Control* [Fourth Estate, London, 1994]).

27. Within hours of Hubble's sending back images of the Levy-Shoemaker comet collision with Jupiter, NASA had posted them on the World Wide Web. Thousands of people immediately logged on to view the photographs, to the extent of almost clogging up the network.

28. M. Butler, introduction to M. Shelley, *Frankenstein or The Modern Prometheus (The 1818 text)* (Oxford University Press, Oxford, 1994), p. xviii. Butler refers

to how Shelley knew of the debate between two surgeons, John Abernethy and William Lawrence, over the vitalist issue. Butler admits that the suggestion that Victor Frankenstein partly caricatures Abernethy remains guesswork. However, she builds a powerful case, pointing out that both Percy and Mary Shelley knew the debaters and that Lawrence was a close friend. For her novel's third edition in 1831, Mary Shelley made Frankenstein a more religious and sympathetic character, while depicting his scientific education as somehow dangerous. This is the version modern readers have generally read, and which has evolved into the Frankenstein myth of the mad scientist and his awful creature.

29. "Modern literature offers us no Unities, so I have turned to science and am trying to complete a four-decker novel whose form is based on the relativity proposition." See A. Friedman and C. Donley, *Einstein as Myth and Muse* (Cambridge University Press, Cambridge, 1990), p. 86.

30. Ibid., p. 107.

31. The picture given by the play of the personalities and the code-breaking effort is fictional, in the opinion of Turing's former colleague Peter Hilton. Nor was Turing an embarrassment because of overt homosexuality. "It was not an issue," Hilton told the American Association for the Advancement of Science, February 1995.

32. C. Jencks, *The Architecture of the Jumping Universe* (Academy Editions, London, 1995).

33. P. Muldoon, *The Annals of Chile* (Faber and Faber, London, 1994), p. 10.

34. R. Littell, *The Visiting Professor* (Faber and Faber, London, 1993), p. 8.

35. T. Stoppard, interview with Roger Highfield, April 1994.

36. Ibid.

37. T. Stoppard, *Arcadia* (Faber and Faber, London, 1993), p. 47.

38. W. Latham, interview with Roger Highfield, September 1994. See also S. Todd, W. Latham, and P. Hughes, *The Journal of Visualisation and Computer Animation* 2 98 (1991); W. Latham and S. Todd, IBM UKSC Report No 248, "Mutator, a Subjective Interface for Evolution of Computer Sculptures," and S. Todd and W. Latham, *Evolutionary Art and Computers* (Academic Press, London, 1992).

39. K. Sims, interview with the authors, August 1994. See also K. Sims, *Computer Graphics* 25 319 (1991), K. Sims, in *Artificial Life IV*, R. Brooks and P. Maes (eds.) (MIT Press, Cambridge, MA, 1994), p. 28 and "Evolving Virtual Creatures," *Siggraph* (1994).

40. S. Thaler, paper presented to the World Congress of Neural Nets 1993 in Portland, Oregon.

41. M. Markus, "Hallucinations: Their Formation on the Cortex Can Be Simulated by a Computer," in *Caos & Meta-Psicologia*, C. Dias and L. Ribeiro (eds.) (Fenda, Lisbon, 1994), p. 65. Zebra stripes have been found in experimental brain maps and can be understood in terms of the dimensionality of the self-organizing feature map generated by the neural network. Stripes arise if the input data vectors to the network have a dimensionality higher than that of the map itself (see T. Kohonen, *Self-Organization and Associative Memory* [Springer, Berlin, 1989], p. 156).

42. M. Bedau, "Philosophical Aspects of Artificial Life," in *Towards a Practice of Autonomous Systems*, F. Varela and P. Bourgine (eds.) (Bradford/MIT Press, Cambridge, MA, 1991), p. 494.

43. L. Adleman, *Science* 266 1021 (1994).

44. R. Feynman, in *Miniaturization*, D. Gilbert (ed.) (Reinhold, New York, 1961), p. 282.

45. One way to speed the process has been found by Thomas Meade of the California Institute of Technology. His laboratory uses single stranded scraps of DNA as the basis of a highly sensitive sensor. These strands will only bind to a matching strand, providing extremely high sensitivity to those complementary strands. These binding events can be detected directly because Meade has found that electrons are able to pass down the double strand 10,000 times more easily than down the single strand. He envisages a microchip containing thousands of these single-strand DNA probes, every one sensing a different genetic code. Only when a complementary strand of DNA docks with the appropriate probe on the chip will a telltale signal occur. T. Meade, interview with Roger Highfield, February 1995.

46. DNA experiments have been proposed to solve the "SAT" problem of computer science. R. Lipton, *Science* 268 542 (1995). Adleman has also given some thought to constructing DNA computers and concluded that there is cause for optimism. L. Adleman, "On Constructing a Molecular Computer," Draft, 1995.

APPENDIX

1. N. Fox, American Association for the Advancement of Science, February 1994. Using a specially designed electroencephalogram cap, the brain wave activity of babies has been mapped by Nathan Fox of the University of Maryland. He studied the differences in the early personality of infants and to see how brain wave activity spreads between left and right sides of the brain. Fox found that infants who show particular right-sided, right-frontal patterns of electrical activity during the first two years of life are more likely to be fearful, inhibited, and, later in life, shy and timid. The frontal lobe of the brain is involved in the regulation of emotions in adults and infants. Fox believes that the centers at work in the right hemisphere of the brain may be linked with fear, withdrawal, fleeing, and freezing behaviors.

2. An interdisciplinary team at the University of Chicago used PET to highlight how alcohol acts on the brain. In a double-blind trial of small amounts of alcohol, they found metabolic differences between those who enjoyed alcohol and those who did not. "Those who preferred alcohol appeared to have more activity in the left side of the brain, and in particular the temporal lobe," said Malcolm Cooper, director of Chicago's PET Center. The scan also revealed a boost of metabolic activity in the speech area, suggesting why drunks talk too much; metabolic depression of regions of the cerebellum, which coordinates movement, the reason that drunks stagger; and metabolic depression of the limbic

system, a region that controls primitive responses such as sexual arousal and violence. This accounts for why drunks are boisterous. M. Cooper, J. Metz, H. de Wit, and J. Mukherjee, *The Journal of Nuclear Medicine* 34 798 (1993), and M. Cooper, interview with Roger Highfield, February 1995.

3. This feature leads to the detector's only responding when two photons simultaneously hit it.
4. S. Zeki, *Sci Amer* 267 43 (1992).
5. S. Zeki, *A Vision of the Brain* (Blackwell Scientific Publications, Oxford, 1993), p. 227.
6. S. Zeki, S. Shipp, and R. Fracowiack, *Nature*, 3 August 1989.
7. S. L. Ogawa, L. M. Lee, A. R. Kay, and D. W. Tank, *Proc Natl Acad Sci USA* 87 9868 (1990).
8. The brain uses anaerobic oxidative metabolism when it is in urgent need of more energy, rather than relying on oxygen from the bloodstream. The reason for this peculiar and counterintuitive behavior is unclear.
9. B. Shaywitz, S. Shaywitz, K. Pugh, R. Constable, P. Skudlarski, R. Fulbright, R. Bronen, J. Fletcher, D. Shankweller, L. Katz, and J. Gore, *Nature* 373 607 (1995).
10. X. Hu, American Association for the Advancement of Science, February 1994.
11. S. Zeki, interview with Roger Highfield, January 1995.
12. R. Salmelin, R. Hari, O. Lounasmaa, and M. Sams, *Nature* 368 463 (1994).

GLOSSARY

Adaptation   Any change in the structure or function of an organism that allows it to survive and reproduce more effectively in its environment.

Afferent   Conducting information inward; said of fibers in the nervous system that bring messages to the brain, and of the transfer of signals toward an individual **neuron**. Opposite of **efferent**.

Algorithm   A procedure or series of steps that can be used to solve a computable problem. In computer science, it describes the logical sequence of operations to be carried out by a **software** program. Not all mathematical problems have computable solutions.

Algorithmic complexity   A measure of the complexity of a problem given by the size of the smallest program that computes it or a complete description of it. Simpler things require smaller programs.

Amino acids   The molecular building blocks of **proteins**.

Animats   Artificial animals comprised of software and hardware.

Arithmetic   A branch of mathematics concerned with the study of numbers and their properties.

Artificial intelligence   A branch of computer science whose goal is the design of machines that have attributes associated with human intelligence, such as learning, reasoning, vision, understanding speech, and, ultimately, consciousness.

Artificial life   A field of study that aims to discover the essential nature and universal features of "life": not only life as we currently know it, but life *as it could be*, whether on earth, within computers, or elsewhere, and in whatever shape or form that it may be found or made within our universe.

Artificial neural network   A kind of computer model, loosely inspired by the **neural network** structure of the brain, and consisting of interconnected processing units that send signals to one another and turn on or off depending on the sum of their incoming signals. Artificial neural networks may be composed of either computer software or hardware or both.

Assembly language   A language that is usually the lowest level available on a computer and, in its simplest form, bears a one-to-one relationship to the **machine code** that issues instructions directly to a microprocessor, but has additional facilities. It is converted into machine code by means of an assembler program.

Association   The ability to relate different pieces of information. An associative memory is one that, given only partially complete input, can associate the corresponding complete set of stored data. **Artificial neural networks** display this property.

Attractor   A way to describe the long-term behavior of a system. Equilibrium and steady states correspond to fixed point attractors, periodic states to limit-cycle attractors and chaotic states to **strange attractors**.

Autocatalysis   Catalysis of a reaction by one of its own products.

Automata theory   The mathematical study of machines and their capabilities for solving problems by means of **algorithms**.

Axon   The long fiber extending from a **neuron** that carries a signal to other neurons.

Belousov-Zhabotinski reaction   A chemical reaction named after two Russian scientists that displays a remarkable wealth of self-organizing features.

Bifurcation   A branch, when there are two distinct choices available to a system. In the case of a chemical-clock reaction, a bifurcation can mark the concentration of a reactant beyond which the periodic color changes of the chemical clock occur.

Binary classifier   Of **genetic algorithms**, when a solution to a classification problem is expressed as a binary string of information. Each bit corresponds to the presence (1) or absence (0) of a particular property in the same way that a chromosome contains genes corresponding to the presence of a protein in the body.

Binary logic   A system in which variables can adopt one of only two different values, usually taken to be zero or unity.

Bit   Contraction of "binary digit." A bit is the smallest unit of information in a binary number system. The value of a bit is usually referred to as a one or zero.

Brusselator   A simplified theoretical model of a chemical reaction showing self-organizing features—like regular color changes in time.

Byte   A group of binary digits eight bits long, which forms the basic unit of computer memory.

Catalyst   A substance able to accelerate a chemical reaction, yet left chemically unchanged by the process.

Capacitor   A device for storing an electrical charge.

Cell   A discrete, membrane-bound portion of living matter; the smallest unit capable of an independent existence.

Cellular automaton   A computer program or piece of hardware consisting of a regular lattice or array of cells. Each cell is assigned a set of instructions by means of an **algorithm** that tells it how to respond to the behavior of adjacent cells as the automaton advances from one discrete time step to the next. Cellular automata are inherently parallel computing devices.

Central processing unit (CPU)   The main component of a computer that performs the actual execution of instructions.

Cerebral cortex   The outer layer (grey matter) of the cerebral hemispheres, evolution's most recent addition to the nervous system.

Chaos   The term used to describe unpredictable and apparently random behavior in **dynamical systems**.

Chromosome   A long strand of **nucleic acid**, usually DNA, containing thousands of genes, in a protective package. There are twenty-three pairs in all human cells, except eggs and sperm.

Cognitive science   The study of intelligence, embracing various academic disciplines: linguistics, experimental psychology, computer science, philosophy, neuroscience.

Compiler   A computer program that translates a set of instructions written in a high-level language such as Fortran into a form (machine code) that can be understood directly by the computer.

Complexity   The study of the behavior of macroscopic collections of simple units (e.g., atoms, molecules, bits, neurons) that are endowed with the potential to evolve in time.

Computability   In mathematics, the property of being calculable on the basis of an **algorithm**, and hence by a computer (or **universal Turing machine**). The natural numbers underpinning mathematics comprise two groups, the **computable** and the noncomputable numbers; there are infinitely many more noncomputable than computable numbers.

Computable numbers   A number that can be computed by an individual **Turing machine**.

Computation   A calculation (usually of numbers) performed by means of an **algorithm**.

Computer   A device that operates on data (input) according to specified instructions (program), usually contained within the computer and which produces results as output. In a digital computer, all tasks are carried out by numbers encoded as binary signals—such as ones and zeros—in contrast to an analog machine which operates using continuously variable signals.

Concentration gradient   Change in the concentration of a substance from one area to another.

Connectionism   Another name for the study of **artificial neural networks**, derived from the myriad connections between processing elements in a network.

Content addressable memory   Capability of recalling an item from a memory using only incomplete aspects of the memorized item, rather than using an explicit address for accessing.

Cost function   In complex optimization problems, it measures how good any particular solution is—the lower (or higher) the value of this function, the better the solution. Also called a "fitness" function.

Determinism   The doctrine that events are completely determined by previous causes rather than being affected by free will or random factors.

Differential equations   Equations involving the instantaneous rate of change of certain quantities with respect to others. Newton's equations of motion, for example, are differential equations linking the force experienced by a body to the instantaneous rate of change of its velocity with time.

Dissipative structure   An organized state of matter arising beyond the first **bifurcation** point when a system is maintained far from **thermodynamic equilibrium**.

DNA (deoxyribonucleic acid)   A very large **nucleic acid** molecule carrying the genetic blueprint for the design and assembly of proteins, the basic building blocks of life.

Dopamine   A neurotransmitter that transmits impulses between **neurons** within the brain.

Dynamical systems   General term for systems whose properties change with time.

Efferent   Of neurons, conducting signals away from a given cell.

Emergent property   A global property of a complex system that consists of many interacting subunits. For example, consciousness is an emergent property of the many neurons in a human brain.

Entropy   A quantity that determines a system's capacity to evolve irreversibly in time. Loosely speaking, we may also think of entropy as measuring the degree of randomness or disorder in a system.

Enzyme   A biological **catalyst**, usually composed of a large protein molecule which accelerates essential chemical reactions in living cells.

Epilepsy   A disorder of brain function characterized by sporadic recurrence of seizure caused by avalance discharges of large numbers of neurons.

Equilibrium   In **thermodynamics**, the final state of time evolution at which all capacity for change is spent. Equilibrium thermodynamics is concerned exclusively with the properties of such static states.

Eugenics (Greek, "well born")   The study of ways in which the physical and mental quality of a people can be controlled and improved by selective breeding.

Evolution   General term for the unfolding of behavior with the passage of time. In biology, the Darwinian theory according to which higher forms of life have arisen out of lower forms with the passage of time.

Evolutionarily stable strategy   An assemblage of behavioral or physical characters (collectively, a **game theoretic** strategy) of a population that is resistant to replacement by any forms bearing new traits, because the new traits will not be capable of successful reproduction.

Excitable medium   A medium—such as the soup of chemicals in the Belousov-Zhabotinski reaction—that changes its state when subjected to a stimulus exceeding a certain threshold level. After excitation, such a medium becomes refractory; that is, it is unresponsive to further stimuli for a period of time.

Expert system   Computer program that uses a direct encoding of human knowledge to help solve complex problems, such as diagnosing an illness or interpreting the law. Also called a knowledge-based system.

Feature detector   A group of neurons that becomes active only if a particular feature is present in the sensory input.

Feedback   A general term for the mechanism whereby the consequences of an ongoing process become factors in modifying or changing that process. The original process is reinforced in positive feedback and suppressed in negative feedback.

Feedforward networks   Networks whose architectures are such that the neurons can be divided into layers, with the neural activities in one layer only being able to influence the activity in later (not earlier) layers. Also called multilayer perceptrons.

Ferromagnetic material   Material that acquires strong magnetism when placed in an external magnetic field. Examples are iron, cobalt, nickel, and their alloys.

Fitness landscape   A landscape representing the fitness measure or **cost function** of a problem, such as that in the traveling salesmen problem, **spin glasses**, or the reproductive capability of a real or virtual organism.

Flop   A floating point operation in a computer—that is, an operation that can be applied to floating point numbers whose decimal points can be in any position. A computer's numerical processing power can be measured in flops—the number of floating point operations it can carry out per second.

Fluid mechanics   The study of the macroscopic behavior of flowing fluids (liquids and gases).

Formalism   A philosophical doctrine espoused by the German mathematician David Hilbert according to which the symbols contained in mathematical statements possess a structure with useful applications. Should be compared with **intuitionism**, **logicism**, and **Platonism**.

Fractal geometry   From Latin *fractus*, "broken." The geometry used to describe an irregular pattern. Fractals display the characteristic of self-similarity, an unending series of motifs within motifs repeated at all length scales.

Frustration   A term used in the physics of **spin glasses** and also applied to interactions in some types of **neural networks**, indicating the tendency for conflicting demands to be placed on spin or neuronal interactions. In the former case, this arises due to the existence of both **ferromagnetic** and antiferromagnetic interactions between spins; in the latter, owing to excitatory and inhibitory inputs experienced by neurons.

Fuzzy logic   In mathematics and computing, a form of knowledge representation suitable for intrinsically imprecise notions such as "hot" and "cold," or "good" and "bad," which may depend on their context.

Gaia hypothesis   The proposal that the earth's living and nonliving components form an inseparable whole that is regulated and kept adapted for life by complex **feedback** processes.

Game theory   A branch of mathematics that deals with strategic problems such as those that arise in business, commerce, evolution, and warfare, by assuming that the contestants involved invariably try to maximize personal gain.

Gene   A unit of heredity composed of the **DNA molecule**, responsible for passing on specific characteristics from parents to offspring.

Genetic algorithm   An adaptive computational method that seeks good solutions to complex (typically **NP** hard **problems** using ideas based on Darwinian evolution).

Genetic code   The sequence of chemical building blocks of DNA (bases) that spells out the instructions for making **amino acids**.

Genotype   The genetic constitution of an individual.

Geological time   The time scale embracing the history of the earth, from its physical origin to the present day.

Geometry   The branch of mathematics concerned with the properties of space.

Giga   Prefix signifying multiplication by 1,000,000,000.

Glucose   A sugar present in blood that is the source of energy for the body.

Gödel's theorem   One of the most important theorems in mathematical logic, which states that it is impossible to reduce mathematics to a finite set of axioms and rules.

GOFAI   Good Old-Fashioned **Artificial Intelligence**, a top-down approach that tries to compartmentalize intelligence into discrete "modules" dealing with specific types of knowledge and information, such as perception, planning, and executing actions. Each of these modules is equipped with explicit models of the external world, and they are supposed to interact by means of logical rules within an "inference engine" to produce intelligent behavior. **Expert systems** are a product of the GOFAI philosophy.

Ground state   The state of lowest energy in a system.

Halting problem   The result, found by Alan Turing, that it is impossible to devise a mechanical procedure, or **algorithm**, that can decide whether an arbitrary computer program will halt after a finite number of steps. The reason is due to the existence of **uncomputable numbers**, and can ultimately be traced back to **Gödel's theorem**.

Hardware   The physical parts of a computer system, usually consisting of mechanical, electronic, and optical components.

Hebbian learning rule   A rule according to which a synaptic connection is strengthened whenever the two neurons involved fire simultaneously. Used in both biological and artificial neural networks.

Hemisphere   One side, left or right, of the **cerebral cortex**.

High-level language   In computing, a programming language designed to suit the needs of the programmer: it is independent of the internal **machine code** of any particular computer.

Homunculus   The "little inner man" conjectured by the ancients to reside in the head, to observe the environment via the senses, and to respond appropriately. The so-called Cartesian theater of the mind would be played out before him.

Hydrodynamics   The science of macroscopic fluid behavior, synonymous with **fluid mechanics**.

Hypercycle   Proposed scenario for the origin of self-replicating molecular systems. These template-instructed replicating cycles involve catalytic feedback loops wherein molecule A begets molecule B which begets molecule C which begets molecule A and so on.

Integrated circuit   The "silicon chip," a small piece of semiconductor material such as silicon that is fabricated so that it contains a complete electronic circuit.

Intractable   A problem so computationally demanding that the time required to solve it is prohibitively long. Such problems, which are not solvable in polynomial time, are said to be in the class **NP**.

Intuitionism   The philosophical doctrine that mathematics cannot deal with the properties of most infinite sets, and that only those statements that can be proved by finite methods can be justifiably asserted.

Irrational number   A number that cannot be expressed as an exact fraction. One example is *pi*. Most irrational numbers are also **uncomputable**.

Irreversibility   The one-way time evolution of a real system, giving rise to an arrow of time.

Iteration   A method of solving a problem by performing the same steps repeatedly until a certain condition is satisfied.

Life   Said of a system capable of undergoing Darwinian style evolution by **natural selection**.

Limit cycle attractor   An **attractor** describing regular (periodic or quasi-periodic) temporal behavior, for instance, in a chemical clock undergoing regular color changes.

Linear equation   A relationship between two variables that, when plotted on Cartesian axes, produces a straight-line graph.

Logic gate   One of the basic logical components used in an electronic device. For example, the output of an AND gate will only be "true" if all its inputs are "true" and the output of an OR gate will be "true" if any of its inputs are "true."

Logicism   The doctrine that all mathematics can be deduced from logic. Frege's attempt to achieve this was brought down by Russell's paradox, leading to the rise of **formalism** and **intuitionism**.

Machine code   The sequence of binary patterns that is executed by the hardware of a computer; the set of instructions that a computer's **central processing unit** can understand and obey directly, without any translation.

Mainframe   A now old-fashioned term for a large computer used for heavy duty processing. Because of the exponential increase in computing power over recent years, most mainframes are less powerful than today's desktop computers and workstations.

Measurement problem   In quantum theory, the problem of accounting for the outcome of experimental measurements on microscopic systems.

Mega   Prefix denoting multiplication by a million.

Memory   In computer science, the space within a computer where information is stored while being actively processed.

Microscopic   A term used to describe tiny dimensions, compared with the everyday or macroscopic dimensions of the world that can be directly perceived with the senses. The distinctions come into their own when contrasting a description of the world couched in terms of individual atoms and molecules with one that relies on global properties of vast numbers of atoms and molecules.

Microtubule   Tiny hollow tube made of the tubulin protein found in almost all cells with a nucleus. Microtubules help to define the shape of a cell by forming a scaffolding that is part of the cytoskeleton.

MIPS   Acronym for million instructions per second. One measure of the speed of a computer processor.

Molecular biology   The study of the molecular basis of life, including the biochemistry of molecules such as **DNA** and **RNA**.

Morphogenesis   The evolution of form in living things.

Mutation   A change in **genes** produced by a chance or deliberate change in the **DNA** that makes up the hereditary material of an organism.

Nanotechnology   The building of devices on a molecular scale.

Natural selection   The process whereby gene frequencies in a population change through certain individuals producing more descendants than others because they are better able to survive and reproduce in their environment. The accumulated effect of natural selection is to produce **adaptations**. Can be used also for certain artificial systems.

Neural network   The brain's actual interconnected mesh of **neurons**.

Neuron   The nerve cell that is the fundamental signaling unit of the nervous system.

Neurotransmitter   A molecule that diffuses across a **synapse** and thus transmits signals between nerve cells.

Nonequilibrium   The state of a macroscopic system that has not attained **thermodynamic equilibrium** and thus still has the capacity to change with time.

Nonlinear   The mathematical property of combining in a more complicated way than simple addition. Nonlinear behavior is typical of the real world and means in a qualitative sense "getting more than you bargained for" unlike linear systems, which produce no surprises. For example, **dissipative** nonlinear systems are capable of exhibiting **self-organization** and **chaos**.

NP-complete   The hardest of all **NP problems**.

NP problem   A problem whose solution requires a time that depends exponentially on its size (something to the power of N, where N represents the size of the problem). These problems are called **intractable** because the time required to solve them rapidly becomes unbounded as the size of the problem grows. Even the raw power of a computer has little effect. These problems, which are not solvable in polynomial time, are said to be in the class NP.

Nucleic acid   Complex organic acid made up of a long chain of units called nucleotides. The two types, known as **DNA** and **RNA**, form the basis of heredity.

Number theory   The abstract study of the structure of number systems and the properties of positive integers (whole numbers).

Optimization   The refinement process used to find the best solution to a problem. To solve an optimization problem computationally, one writes a program in such a way as to maximize or minimize the value of a cost function.

Optoelectronics   Branch of electronics concerned with the development of devices that respond not only to electrons but also photons.

Organic chemistry　Branch of chemistry that deals with carbon-containing compounds.

Parallelism　Leading-edge computer technique that allows more than one computation at the same time by using several different processing units running concurrently. The most extreme form is massive parallelism, in which thousands of individual processors work in unison on different parts of a single giant problem. Concurrent processing (time sharing) is achieved on serial machines by making the single CPU switch between several tasks.

Perceptron　A simple computational model of a biological neuron comprising some input channels, a processing element, and a single output. Each input value is multiplied by a channel weight, summed by the processor, passed through a nonlinear filter and put into the output channel.

Phase change　A change in the physical state of a substance from solid to liquid, liquid to gas, or solid to gas.

Phenotype　The overall attributes of an organism arising due to the interaction of its **genotype** with the environment.

Pixel　Acronym for a picture element, a single dot on a computer screen (the three-dimensional equivalent is called a voxel).

Platonism　A philosophical doctrine according to which mathematical objects exist in advance of and independently of our knowledge of them and of any physical instantiation of them, and that our insight into mathematical truth is achieved by the construction of proofs. Should be compared with **logicism, intuitionism,** and **formalism.**

Polymer　A chemical substance formed by the chain-like linkage of simpler molecules.

Postsynaptic　Pertaining to the receiving side of a **synapse**; by contrast, the presynaptic pertains to the transmitting side of the synapse.

Program　The set of instructions that implements an **algorithm** that can be understood by a computer.

Protein　A class of large molecules found in living organisms, consisting of strings of amino acids folded into complex but well-defined three-dimensional structures.

Punctuated equilibrium　Evolutionary theory developed to explain discontinuities in the fossil record.

Quantum computer　A hitherto theoretical device that could compute various things that are uncomputable using a **universal Turing machine.**

Quantum mechanics　The mechanics that rules the microscopic world, when energy changes occur in abrupt, tiny quantum jumps.

RAM　Acronym for Random Access Memory. A **memory** device used in almost all modern computers.

Random number　One of a series of numbers having no detectable pattern. No **universal Turing machine** (i.e., **computer**) can generate truly random numbers, since these are **uncomputable**; though a universal **quantum computer** should in principle be able to achieve this.

Rational number　Any number that can be expressed as an exact fraction, and is therefore computable.

Reaction   In chemistry, the coming together of atoms or molecules with the result that a chemical change takes place.

Real number   Any of the **rational** or **irrational numbers**.

Reductionism   A doctrine according to which complex phenomena can be explained in terms of something simpler. In particular, atomistic reductionism contends that macroscopic phenomena can be explained in terms of the properties of atoms and molecules.

Reynolds number   A number used in **fluid mechanics** that indicates whether a fluid flow in a particular situation will be smooth or turbulent.

RISC   Acronym for Reduced Instruction Set Computer. A microprocessor that carries out fewer instructions than traditional microprocessors so that it can work more quickly.

RNA (ribonucleic acid)   The genetic material used to translate DNA into **proteins**. In some organisms, it can also be the principal genetic material.

ROM   Acronym for Read Only Memory. Unlike **RAM**, ROM chips can only be read and not written to.

Scalar   A quantity with magnitude but no direction. Examples include mass and temperature.

Scanning tunneling microscope   A microscope that produces a magnified image by moving a tiny tungsten probe across a specimen. By the magnitude of the flow of charge between tip and specimen, the contours of the surface can be determined.

Search space   The variation in a **cost function** is best envisaged as a landscape of potential solutions where the height of each feature is a measure of its cost. This undulating landscape is sometimes called the search space.

Self-organization   The spontaneous emergence of nonequilibrium structural organization on a macroscopic level due to collective interactions between a large number of simple, usually microscopic, objects.

Self-organized criticality   A generic pattern of self-organized nonequilibrium behavior in which there are characteristic long-range temporal and spatial regularities.

Simulated annealing   A method for searching a landscape of possibilities for the best solution to a complex problem, as expressed by the lowest (or highest) point on that landscape. The name comes from the process of annealing, during which a material is first heated and then slowly cooled to enhance ductility and strength. During annealing, the component atoms of the metal are allowed to settle into a lower energy, more stable arrangement than prior to the process.

Software   The programs executed by a computer system, as distinct from the **hardware**.

Soma   The main body of a neuron. Computationally, it acts approximately like a linear threshold filter.

Speech recognition   Any technique by which a computer can understand ordinary speech.

Spin glass   Magnetic material (typically an alloy) whose atomic magnets are involved in a mixture of **ferromagnetic** and antiferromagnetic interactions causing **frustration**, so that not all the constraints necessary to minimize the system's overall energy can be simultaneously satisfied. There are exponentially many stable states, and finding the global ground state is an NP-hard optimization problem. Spin glasses share many properties in common with associative Hopfield networks.

Statistical mechanics   The discipline that attempts to relate the properties of macroscopic systems to their atomic and molecular constituents.

Steady state   A nonequilibrium state that does not change with time.

Strange attractor   An **attractor** that has a fractal (fractional) dimension: describes chaotic dynamics in dissipative dynamical systems.

Supercomputer   The fastest, largest, and most powerful type of computer.

Synapse   The junction between two nerve cells across which a nerve impulse is transmitted.

Teleology   The study (or doctrine) of final causes, particularly in relation to design or purpose in nature.

Tera   A prefix indicating one million million.

Thermodynamics   The science of heat and work.

Transistor   A solid-state electronic component made of a semiconductor with three or more electrodes that can regulate a current passing through it. Used as an amplifier, oscillator, or switch.

Turing machine   In modern terminology, a computer program.

Uncertainty principle   The **quantum mechanical** principle that says it is meaningless to speak of a particle's position, momentum, or other parameters, except as the result of measurements. It gives a theoretical limit to the precision with which a particle's momentum and position can be measured simultaneously: the more accurately one is determined, the more uncertainty there is in another.

Uncomputable number   A number that cannot be computed by any **Turing machine**, including the universal one.

Universal Turing Machine   In modern terminology, a general-purpose programmable computer, which can perform any computable sequence of operations.

Vector   Any quantity with a magnitude and a direction. Examples are velocity and acceleration.

Virtual reality   Advanced form of three-dimensional computer graphical simulation in which a participant is plunged into an artificial multimedia environment.

Virus   A short stretch of computer code that can copy itself into one or more larger "host" computer programs when it is activated. In biology, a virus is a scrap of genetic material, usually wrapped in a protein overcoat, that needs a host cell to reproduce, and which can also evolve.

Wave function   The central quantity in quantum theory that is used to calculate the probability of an event occurring—for instance, an atom emitting a photon—when a measurement is made.

BIBLIOGRAPHY

CHAPTER 1

Appleyard, B., *Understanding the Present* (Pan Books, London, 1992).

Cohen, J., and I. Stewart, *The Collapse of Chaos* (Viking, London, 1994).

Coveney, P., and R. Highfield, *The Arrow of Time* (W. H. Allen, London, 1990; HarperCollins, London, 1991; Ballantine, New York, 1991).

Crick, F., *The Astonishing Hypothesis* (Simon & Schuster, London, 1994).

Darwin, C., *The Origin of Species by Means of Natural Selection* (Penguin, Harmondsworth, 1968).

Emmeche, C., *The Garden in the Machine: The Emerging Science of Artificial Life* (Princeton University Press, Princeton, 1994).

Langton, C., C. Taylor, J. D. Farmer, and S. Rasmussen (eds.), *Artificial Life II* (Addison-Wesley, Redwood City, CA, 1992).

Lewin, R., *Complexity* (Macmillan, New York, 1992).

Lewontin, R. C., *The Doctrine of DNA* (Penguin, London, 1993).

Peitgen, H.-O., and P. Richter, *The Beauty of Fractals* (Springer-Verlag, Berlin, 1986).

Prigogine, I., *From Being to Becoming: Time and Complexity in the Physical Sciences* (Freeman, San Francisco, 1980).

Ridley, M., *The Red Queen* (Penguin, London, 1993).

Vonneumann, N., *John von Neumann, as Seen By His Brother* (© N. Vonneumann, 1396 Lindsay Lane, Meadowbrook, PA 19046-1833, 1992).

Waldrop, M., *Complexity* (Simon & Schuster, New York, 1992).

Weinberg, S., *The First Three Minutes* (André Deutsch, London, 1977).

CHAPTER 2

Barrow, J., *Theories of Everything: The Quest for Ultimate Explanation* (Oxford University Press, 1991).

———, *Pi in the Sky* (Oxford University Press, 1992).

Borowski, E. J., and J. M. Borwein, *Dictionary of Mathematics* (Collins, London, 1989).

Casti, J. L., *Searching for Certainty: What Science Can Know About the Future* (Scribners, London, 1992).

Chaitin, G. J., *Algorithmic Information Theory* (Cambridge University Press, 1987).

———, *Information-theoretic Incompleteness* (World Scientific, Singapore, 1992).

———, *Information, Randomness and Incompleteness* (World Scientific, Singapore, 2nd ed., 1990).

———, in *The New Scientist Guide to Chaos*, N. Hall, ed. (Penguin, London, 1991).

Cornwell, J. (ed.), *Nature's Imagination* (Oxford University Press, Oxford, 1995).

Coveney, P., and R. Highfield, *The Arrow of Time* (W. H. Allen, London, 1990).

Davies, P. C. W., *The Cosmic Blueprint* (Heinemann, 1987).

———, *The Mind of God* (Simon & Schuster, New York, 1992).

Davis, M. (ed.), *The Undecidable* (Raven, New York, 1965).

Garey, M., and D. S. Johnson, *Computers and Intractability* (W. H. Freeman, New York, 1979).

Herken, R., *The Universal Turing Machine* (Oxford University Press, Oxford, 1988).

Hodges, A., *Alan Turing: The Enigma* (Vintage, London, 1983).

Hofstadter, D., *Gödel, Escher Bach: An Eternal Golden Braid* (Harper & Row, New York, 1979).

Hughes, P., and G. Brecht, *Vicious Circles and Infinity: An Anthology of Paradoxes* (Penguin, London, 1978).

Kline, M., *Mathematical Thought from Ancient to Modern Times* (Oxford University Press, 1972).

Lakatos, I., *Proofs and Refutations* (Cambridge University Press, 1976).

Penrose, R., *The Emperor's New Mind* (Oxford University Press, 1989).

———, *Shadows of the Mind* (Oxford University Press, 1994).

Popper, K., *Conjectures and Refutations* (Routledge and Kegan Paul, London, 1963).

———, *The Logic of Scientific Discovery* (Hutchinson, London, 1972).

———, *Realism and the Aim of Science* (Hutchinson, London, 1993).

Reid, C., *Hilbert* (George Allen and Unwin, London, 1970).

Rucker, R., *Infinity and the Mind* (Bantam Books, London, 1982).

———, *Mind Tools* (Houghton-Mifflin, Boston, 1987).

Webb, J., *Mechanism, Mentalism, and Metamathematics* (Reidel, Dordrecht, 1980).

Weinberg, S., *Dreams of a Final Theory* (Viking, London, 1994).

CHAPTER 3

Augarten, S., *Bit by Bit: An Illustrated History of Computers* (Ticknor and Fields, New York, 1984).

Babbage, C., *Passages from The Life of a Philosopher* (Longman, Green, Longman, Roberts, & Green, 1864).

Barrow, J., *Pi in the Sky* (Penguin, London, 1992).

———, *Theories of Everything* (Oxford University Press, 1991).

Boole, G., *The Claims of Science, Especially as Founded in Its Relations to Human Nature* (Taylor and Walton, London, 1851).

———, *An Investigation of the Laws of Thought* (Dover, New York, 1958).

———, *The Mathematical Analysis of Logic, Being an Essay Towards a Calculus of Deductive Reasoning* (Macmillan, Barclay and Macmillan, Cambridge, 1847).

Campbell-Kelly, M. (ed.), *The Works of Charles Babbage* (Pickering & Chatto, London, 1989).

Chandy, K. M., and S. Taylor, *An Introduction to Parallel Processing* (Jones and Bartlett, Boston, 1992).

Coveney, P., and R. Highfield, *The Arrow of Time* (W. H. Allen, London, 1990; Ballantine, New York, 1991).

Donnelly, J., *A Modern Difference Engine: Software Simulators for Charles Babbage's Difference Engine No. 2* (Armstrong Publishing Co., 1992).

Garey, M., and D. S. Johnson, *Computers and Intractability* (W. H. Freeman, New York, 1979).

Goldstine, H., *The Computer from Pascal to von Neumann* (Princeton University Press, Princeton, 1972).

Heims, S., *John Von Neumann and Norbert Wiener: From Mathematics to the Technologies of Life and Death* (MIT Press, Cambridge, MA, 1980).

Hillis, D., *The Connection Machine* (MIT Press, Cambridge, MA, 1989).

Hillis, D., and B. Boghosian, "Parallel Scientific Computation," *Science* 261, 856 (1993).

Hodges, A., *Alan Turing: The Enigma* (Burnett Books, London, 1983).

Hyman, A., *Charles Babbage, Pioneer of the Computer* (Princeton University Press, Princeton, 1982).

Kahn, D., *The Codebreakers: The Story of Secret Writing* (Macmillan, London, 1967).

Kaufmann, W. J., and L. L. Smarr, *Supercomputing and the Transformation of Science* (Scientific American Library, New York, 1993).

Klir, G. J., and T. A. Folger, *Fuzzy Sets, Uncertainty and Information* (Prentice-Hall, Englewood Cliffs, NJ, 1992).

Kosko, B., *Fuzzy Thinking: The New Science of Fuzzy Logic* (HarperCollins, New York, 1994).

Lindgren, M., *Glory and Failure: The Difference Engines of Johann Müller, Charles Babbage and Georg and Edvard Scheutz* (MIT Press, Cambridge, MA, 1990).

MacHale, D., *George Boole: His Life and Work* (Boole Press, Dublin, 1985).

Macrae, N., *John von Neumann* (Pantheon Books, New York, 1992).

Moseley, M., *Irascible Genius: A Life of Charles Babbage, Inventor* (Hutchinson, London, 1964).

Penrose, R., *The Emperor's New Mind* (Oxford University Press, 1989).

Rheingold, H., *Tools for Thought* (Prentice-Hall, Englewood Cliffs, NJ, 1986).

Swade, D., *Charles Babbage and His Calculating Engines* (Science Museum, London, 1991).

Swade, D., and J. Palfreman, *The Dream Machine: Exploring the Computer Age* (BBC Books, 1991).

Thinking Machines Corporation, *Connection Machine CM-200 Series Technical Summary* (Thinking Machines Corporation, Cambridge, MA, 1991).

Trew, A., and G. Wilson (eds.), *Past, Present, Parallel: A Survey of Available Parallel Computing Systems* (Springer-Verlag, Berlin, 1991).

von Neumann, J., *The Computer and the Brain* (Yale University Press, New Haven, 1958).

Zurek, W., *Complexity, Entropy and the Physics of Information* (Santa Fe Institute Series, Vol. 8, Addison-Wesley, Redwood City, CA, 1991).

CHAPTER 4

Baxter, R., *Exactly Solved Models in Statistical Mechanics* (Academic Press, London, 1982).

Coveney, P., and R. Highfield, *The Arrow of Time* (W. H. Allen, London, 1990).

Doolen, G. (ed.), *Lattice Gas Methods for Partial Differential Equations*, Vol. IV in the Santa Fe Institute Studies in the Sciences of Complexity (Addison-Wesley, Redwood City, CA, 1990).

Fischer, K. H., and J. A. Hertz, *Spin Glasses* (Cambridge University Press, 1991).

Hillis, W. D., "The Connection Machine: A Computer Architecture Based on Cellular Automata," *Physica* 10D, 213 (1984).

Hofstadter, D., *Gödel, Escher Bach: An Eternal Golden Braid* (Harper & Row, New York, 1979).

Mezard, M., G. Parisi, and M. Virasoro, *Spin Glass Theory and Beyond* (World Scientific, Singapore, 1987).

Minsky, M., *Computation: Finite and Infinite Machines* (Prentice-Hall, Englewood Cliffs, NJ, 1967).

Mydosh, J. *Spin Glass Physics: An Experimental Approach* (Taylor and Francis, London, 1993).

Sherrington, D., "Magnets, Microchips and Memories: From Spin Glasses to the Brain," *Speculations in Science and Technology* 14, 316 (1991).

———, "Complexity Due to Disorder and Frustration," in *1989 Lectures on Complex Systems*, E. Jen (ed.) (Addison-Wesley, Reading, MA, 1990).

Toffoli, T., and N. Margolus, *Cellular Automata Machines* (MIT Press, Cambridge, MA, 1987).

von Neumann, J., *The Computer and the Brain* (Yale University Press, New Haven, 1958).

Wolfram, S., "Cellular Automata as Models of Complexity," *Nature* 311, 419 (1984).

Wolfram, S. (ed.), *Theory and Applications of Cellular Automata* (World Scientific, Singapore, 1986).

Wuensche, A., and M. J. Lesser, *The Global Dynamics of Cellular Automata: An Atlas of Basin of Attraction Fields of One-Dimensional Cellular Automata* (Addison-Wesley, Reading, MA, 1992).

CHAPTER 5

Aleksander, I., and H. Morton, *An Introduction to Neural Computing* (Chapman and Hall, London, 1990).

Amari, S., and A. Arbib, *Competition and Cooperation in Neural Nets*, Lecture Notes in Biomathematics, Vol. 45 (Springer-Verlag, Berlin, 1982).

Amit, D., *Modelling Brain Function* (Cambridge University Press, 1989).

Beale, R., and T. Jackson, *Neural Computing* (Adam Hilger, Bristol, 1990).

Broadbent, D. (ed.), *The Simulation of Human Intelligence* (Blackwell, Oxford, 1993).

Davis, L., *Genetic Algorithms and Simulated Annealing* (Pitman, London, 1987).

————, *Handbook of Genetic Algorithms* (van Nostrand-Rheinhold, New York, 1991).

Dreyfus, H. L., *What Computers Still Can't Do: The Limits of Artificial Intelligence* (MIT Press, Cambridge, MA, 1992).

Freeman, J. A., and D. M. Skapura, *Neural Networks: Algorithms, Applications and Programming Techniques* (Addison-Wesley, Reading, MA, 1991).

Goldberg, D., *Genetic Algorithms in Search, Optimization and Machine Learning* (Addison-Wesley, Reading, MA, 1989).

Grossberg, S., *Neural Networks and Natural Intelligence* (MIT Bradford Press, Cambridge, MA, 1988).

Hebb, D. O., *The Organization of Behavior* (Wiley, New York, 1949).

Hertz, J. A., A. Krogh, and R. G. Palmer, *Introduction to the Theory of Neural Computation* (Addison-Wesley, Reading, MA, 1991).

Hofstadter, D., *Gödel, Escher Bach: An Eternal Golden Braid* (Harper & Row, New York, 1979).

Hofstadter, D., and D. Dennett (eds.), *The Mind's I* (Basic Books, New York, 1981).

Holland, J., *Adaptation in Natural and Artificial Systems* (University of Michigan Press, Ann Arbor, 1975).

Holland, J., K. J. Holyoak, R. E. Nisbett, and P. R. Thagard, *Induction: Processes of Inference, Learning and Discovery* (MIT Press, Cambridge, MA, 1986).

Kohonen, T., *Self-Organisation and Associative Memory* (Springer-Verlag, Berlin, 1989).

Koza, J. R., *Genetic Programming: On the Programming of Computers by Means of Natural Selection* (MIT Press, Cambridge, MA, 1992).

Koza, J. R., and J. P. Rice, *Genetic Programming: The Movie* (MIT Press, Cambridge, MA, 1992).

Minsky, M., *Computation: Finite and Infinite Machines* (Prentice-Hall, Englewood Cliffs, NJ, 1967).

————, *The Society of Mind* (Simon & Schuster, New York, 1986).

Minsky, M., and S. Papert, *Perceptrons* (MIT Press, Cambridge, MA, 1969).

Penrose, R., *The Emperor's New Mind* (Oxford University Press, 1989).

Rosenblatt, F., *Principles of Neurodynamics* (Spartan Books, Washington, D.C., 1962).

Rumelhart, D. E., and J. L. McClelland (eds.), *Parallel Distributed Processing: Explorations in the Microstructure of Cognition*, Vol. I and II (MIT Press, Cambridge, MA, 1986).

von Neumann, J., *The Computer and the Brain* (Yale University Press, New Haven, 1958).

Whitley, L. D., and J. D. Schaffer, *ICOGANN-92 International Workshop on Combina-*

tions of Genetic Algorithms and Neural Networks (IEEE Computer Society Press, Los Alamitos, CA, 1992).

CHAPTER 6

Babloyantz, A., *Molecules, Dynamics and Life* (Wiley-Interscience, New York, 1986).

Brush, S., *The Kind of Motion We Call Heat* (North-Holland, Amsterdam, 1976).

Casti, J. L., *Searching for Certainty* (Scribners, London, 1992).

Coveney, P., "The Second Law of Thermodynamics: Entropy, Irreversibility and Dynamics," *Nature* 333, 409 (1988).

———, "L'irréversibilité du Temps," *La Recherche* 20, 190 (1989).

Coveney, P., and R. Highfield, *The Arrow of Time* (W. H. Allen, London, 1990).

Doolen, G. (ed.), *Lattice Gas Methods for Partial Differential Equations* (Addison-Wesley, Redwood City, CA, 1990).

Gaynord, R. J., and P. R. Wellin, *Computer Simulations with* Mathematica: *Explorations in Complex Physical and Biological Systems* (Springer-Verlag, Berlin, 1995).

Glansdorff, P., and I. Prigogine, *Thermodynamic Theory of Structure, Stability and Fluctuations* (Wiley, New York, 1971).

Gleick, J., *Chaos* (Sphere, London, 1988).

Gray, P., and S. Scott, *Chemical Oscillations and Instabilities* (Oxford University Press, 1990).

de Groot, S., and P. Mazur, *Non-Equilibrium Thermodynamics* (Dover, New York, 1984).

Hall, N. (ed.), *The New Scientist Guide to Chaos* (Penguin Books, Harmondsworth, 1991).

Lorenz, E., *The Essence of Chaos* (UCL Press, London, 1993).

Mandelbrot, B., *The Fractal Geometry of Nature* (W. H. Freeman, New York, 1977).

Nicolis, G., and I. Prigogine, *Exploring Complexity* (W. H. Freeman, New York, 1989).

———, *Self-Organisation in Non-Equilibrium Systems* (Wiley-Interscience, New York, 1977).

Prigogine, I., *From Being to Becoming* (W. H. Freeman, New York, 1980).

Ruelle, D., *Chaotic Evolution and Strange Attractors* (Cambridge University Press, 1989).

———, *Chance and Chaos* (Princeton University Press, Princeton, 1991).

Scott, S., *Chemical Chaos* (Oxford University Press, 1992).

Stewart, I., and M. Golubitsky, *Fearful Symmetry* (Penguin, London, 1992).

Wolfram, S. (ed.), *Theory and Applications of Cellular Automata* (World Scientific, Singapore, 1986).

Zurek, W. (ed.), *Complexity, Entropy and the Physics of Information* (Addison-Wesley, Redwood City, CA, 1988).

CHAPTER 7

Alberts, B., D. Bray, J. Lewis, M. Raff, K. Roberts, and J. Watson, *The Molecular Biology of the Cell*, 2nd ed. (Garland, New York, 1989).

Axelrod, R., *The Evolution of Co-operation* (Penguin Books, London, 1990).

Babloyantz, A., *Molecules, Dynamics and Life* (Wiley-Interscience, New York, 1986).

Ball, P., *Designing the Molecular World* (Princeton University Press, Princeton, 1994).

Bendall, D. (ed.), *Evolution from Molecules to Men* (Cambridge University Press, 1982).

Coveney, P. V., and R. Highfield, *The Arrow of Time* (W. H. Allen, London, 1990).

Cramer, F., *Chaos und Ordnung. Die komplexe Struktur des Lebendingen* (Deutsche-Verlags-Anstalt, Stuttgart, 1988).

Crick, F., *Life Itself* (Macdonald, London, 1981).

Cronin, H., *The Ant and the Peacock* (Cambridge University Press, 1991).

Dawkins, R., *River Out of Eden* (Weidenfeld & Nicolson, London, 1995).

———, *The Selfish Gene*, 2nd ed. (Oxford University Press, 1989).

Deamer, D., and G. Fleischaker, *Origins of Life: The Central Concepts* (Jones and Bartlett, Boston, 1994).

Eigen, M., *Steps Towards Life* (Oxford University Press, 1992).

Eigen, M., and P. Schuster, *The Hypercycle. A Principle of Natural Self-Organization* (Springer-Verlag, New York, 1979).

Fleischaker, G., S. Colonna, and P. L. Luisi (eds.), *Self-Production of Supramolecular Structures*, Proceedings of a NATO Advanced Research Workshop held in Maratea, Italy, September 1993 (Kluwer, Dordrecht, 1994).

Goodwin, B., *How the Leopard Changed Its Spots* (Weidenfeld & Nicholson, London, 1994).

Gould, S. J., *Ever Since Darwin* (Penguin, Harmondsworth, 1980).

———, *The Panda's Thumb* (Penguin, Harmondsworth, 1980).

———, *Wonderful Life* (Hutchinson Radius, London, 1980).

Grene, M. (ed.), *Dimensions of Darwinism* (Cambridge University Press, 1983).

Kauffman, S. A., *The Origins of Order: Self-Organization and Selection in Evolution* (Oxford University Press, 1993).

Langton, C. (ed.), *Artificial Life: Proceedings of an Interdisciplinary Workshop on the Synthesis and Simulation of Living Systems* (Addison-Wesley, Redwood City, CA, 1989).

Langton, C., C. Taylor, J. D. Farmer, and S. Rasmussen (eds.), *Artificial Life II: Proceedings of the Workshop on Artificial Life held February, 1990, in Santa Fe, New Mexico* (Addison-Wesley, Redwood City, CA, 1992).

Lewontin, R. C., *The Doctrine of DNA* (Penguin, London, 1993).

Lovelock, J. E., *Gaia* (Oxford University Press, 1979).

Maynard Smith, J., *Did Darwin Get It Right?* (Penguin Books, London, 1993).

Orgel, L., *The Origins of Life* (Wiley, New York, 1973).

Peacocke, A., *An Introduction to the Physical Chemistry of Biological Organisation* (Oxford University Press, 1983).

Ridley, M., *The Red Queen: Sex and the Evolution of Human Nature* (Viking Penguin, London, 1993).

Schrödinger, E., *What Is Life?* (Cambridge University Press, 1944).

Sigmund, K., *Games of Life* (Oxford University Press, 1993).

Varela, F., *Principles of Biological Autonomy* (North-Holland, Amsterdam, 1979).

Varela, F., and W. Stein (eds.), *Thinking About Biology* (Addison-Wesley, Redwood City, CA, 1992).

CHAPTER 8

Bendall, D. (ed.), *Evolution from Molecules to Men* (Cambridge University Press, 1982).

Brooks, R., and P. Maes (eds.), *Artificial Life IV* (MIT Press, Cambridge, MA, 1994).

Dawkins, R., *The Selfish Gene*, 2nd ed. (Oxford University Press, 1990).

Eigen, M., *Steps Towards Life* (Oxford University Press, 1992).

Emmeche, C., *The Garden in the Machine: The Emerging Science of Artificial Life* (Princeton University Press, Princeton, 1994).

Goodwin, B., *How the Leopard Changed Its Spots* (Weidenfeld & Nicholson, London, 1994).

Holland, J. H., *Adaptation in Natural and Artificial Systems: An Introductory Analysis with Applications to Biology, Control, and Artificial Intelligence*, Revised ed. (MIT Press, Cambridge, MA, 1992).

——, "Genetic Algorithms," *Sci Amer* 267, 44–50 (1992).

Koza, J., *Genetic Programming: On the Programming of Computers by Means of Natural Selection* (MIT Press, Cambridge, MA, 1992).

Langton, C. (ed.), *Artificial Life: Proceedings of an Interdisciplinary Workshop on the Synthesis and Simulation of Living Systems* (Addison-Wesley, Redwood City, CA, 1989).

Langton, C., C. Taylor, J. D. Farmer, and S. Rasmussen (eds.), *Artificial Life II: Proceedings of the Workshop on Artificial Life held February 1990, in Santa Fe, New Mexico* (Addison-Wesley, Redwood City, CA, 1992).

Langton, C. (ed.), *Artificial Life III* (Addison-Wesley, Redwood City, CA, 1989).

Levy, S., *Artificial Life* (Pantheon, New York, 1992).

Michalski, R. S., J. G. Carbonell, and T. M. Mitchell, *Machine Learning: An Artificial Intelligence Approach*, Vol. I and II (Morgan Kaufman, Los Altos, 1986).

Pagels, H. R., *The Dreams of Reason: The Computer and the Rise of the Sciences of Complexity* (Bantam, New York, 1988).

Prusinkiewicz, P., and A. Lindenmayer, *The Algorithmic Beauty of Plants* (Springer-Verlag, Berlin, 1990).

Ridley, M., *The Red Queen* (Viking, London, 1993).

Sober, E., *The Nature of Selection: Evolutionary Theory in Philosophical Focus* (University of Chicago Press, Chicago, 1984).

Thompson, D., *On Growth and Form* (Cambridge University Press, 1942).

CHAPTER 9

Blakemore, C., *The Mind Machine* (BBC Books, London, 1991, and Penguin Books, 1994).

Blakemore, C., and S. Greenfield (eds.), *Mindwaves* (Blackwell, Oxford, 1987).

Boden, M., *The Creative Mind* (Sphere Books, London, 1992).

——, *Artificial Intelligence and Natural Man* (MIT Press, London, 1987).

——, *The Philosophy of Artificial Intelligence* (Oxford University Press, 1990).

Boden, M., A. Bundy, R. M. Needham (eds.), "From Artificial Intelligence to the Mind," *Philosophical Transactions of the Royal Society* 349 (15 October 1994).

Changeux, J.-P., *Neuronal Man* (Pantheon, New York, 1985).

Chomsky, N., *Barriers* (MIT Press, Cambridge, MA, 1986).

————, *Language and Mind* (Harcourt Brace and World, 1968).

————, *Reflections on Language* (Pantheon, New York, 1975).

Churchland, P., *Neurophilosophy: Toward a Unified Science of the Mind/Brain* (MIT Press, London 1986).

Churchland, P., and T. Sejnowski, *The Computational Brain* (MIT Press, Cambridge, MA, 1992).

Cotterill, R., *No Ghost in the Machine* (Heinemann, London, 1989).

Crick, F., *The Astonishing Hypothesis* (Simon & Schuster, London, 1994).

Damasio, H., and A. Damasio, *Lesion Analysis in Neuropsychology* (Oxford University Press, 1989).

Dennett, D., *Consciousness Explained* (Penguin, London, 1991).

Denton, D., *The Pinnacle of Life* (Cassels, Allen & Unwin, London, and Harper-Collins, San Francisco, 1993).

Dreyfus, H. L., *What Computers Still Can't Do: A Critique of Artificial Reason* (MIT Press, London, 1993).

Edelman, G., *Neural Darwinism: The Theory of Neuronal Group Selection* (Oxford University Press, 1989).

————, *Bright Air, Brilliant Fire* (The Penguin Press, London, 1992).

Goldberg, D., *Genetic Algorithms in Search, Optimization and Machine Learning* (Addison-Wesley, Reading, MA, 1989).

Grossberg, S., *Neural Networks and Natural Intelligence* (MIT Bradford Press, Cambridge, MA, 1988).

Hanson, S. J., and C. R. Olson (eds.), *Connectionist Modeling and Brain Function: The Developing Interface* (MIT Press, Cambridge, MA, 1990).

Hebb, D. O., *Organization of Behavior* (Wiley, New York, 1949).

Heidegger, M., *Basic Writings* (Harper & Row, New York, 1977).

Hofstadter, D., *Gödel, Escher Bach: An Eternal Golden Braid* (Harper & Row, New York, 1979).

Hofstadter, D., and D. Dennett (eds.), *The Mind's I* (Basic Books, New York, 1981).

Hubel, D. H., *Eye, Brain and Vision* (Scientific American Library, W. H. Freeman, 1988).

Kohonen, T., *Self-Organization and Associative Memory*, 3rd ed. (Springer-Verlag, Berlin, 1989).

Lewontin, R. C., *The Doctrine of DNA: Biology as Ideology* (Penguin, London, 1993).

Marcel, A. J., and E. Bisiach (eds.), *Consciousness in Contemporary Science* (Oxford University Press, 1992).

Marr, D., *Vision* (Freeman, New York, 1982).

Minsky, M., *The Society of Mind* (Simon & Schuster, New York, 1986).

Moravec, H., *Mind Children: The Future of Robot and Human Intelligence* (Harvard University Press, Cambridge, MA, 1988).

Morris, R. G. M. (ed.), *Parallel Distributed Processing: Implications for Psychology and Neuroscience* (Oxford University Press, 1989).

Penrose, R., *The Emperor's New Mind* (Oxford University Press, 1989).

————, *Shadows of the Mind* (Oxford University Press, 1994).

Pinker, S., *The Language Instinct* (Morrow, New York, 1994).

Posner, M. I., and M. E. Raichle, *Images of Mind* (Scientific American Library, W. H. Freeman, 1994).

Rose, S., *The Making of Memory* (Bantam Press, London, 1992).
Rumelhart, D. E., and J. L. McClelland (eds.), *Parallel Distributed Processing*, Vol. 1 and 2 (MIT Press, London, 1986).
Skinner, B. F., *About Behaviorism* (Penguin, London, 1993).
Traub, R. D., and R. Miles, *Neuronal Networks of the Hippocampus* (Cambridge University Press, 1991).
Wills, C., *The Runaway Brain* (HarperCollins, London, 1994).
Wolpert, L., *The Triumph of the Embryo* (Oxford University Press, Oxford, 1991).
Zeki, S., *A Vision of the Brain* (Blackwell Scientific Publications, Oxford, 1993).

CHAPTER 10

Anderson, P. W., K. Arrow, and D. Pines, *The Economy as an Evolving Complex System* (Santa Fe Institute Studies in the Sciences of Complexity, Vol. 5, Addison-Wesley, Redwood City, CA, 1989).
Arthur, B., *Sci Amer*, 92–99 (February 1990).
Axelrod, R., *The Evolution of Cooperation* (Penguin, London, 1990).
Coveney, P., and R. Highfield, *The Arrow of Time* (HarperCollins, London, 1991).
Crichton, M., *Jurassic Park* (Century, London, 1992).
Dawkins, R., *The Selfish Gene* (Oxford University Press, Oxford, 1989).
Dennett, D. C., *Brainstorms* (Bradford Books, Montgomery, VT, 1978).
Dennett, D., *Darwin's Dangerous Idea: Evolution and the Meaning of Life* (Simon & Schuster, New York, 1995).
Gell-Mann, M., *The Quark and the Jaguar* (Little, Brown, London and Boston, 1994).
Hofstadter, D., *Gödel, Escher Bach: An Eternal Golden Braid* (Penguin Books, London, 1979).
Hofstadter, D., and D. Dennett, *The Mind's I: Fantasies and Reflections on Self and Soul* (Penguin Books, London, 1982).
Jencks, C., *The Architecture of the Jumping Universe* (Academy Editions, London, 1995).
Kelly, K., *Out of Control* (Fourth Estate, London, 1994).
Kelves, D. J., and L. Hood (eds.), *The Code of Codes: Scientific and Social Issues in the Human Genome Project* (Harvard University Press, Cambridge, MA, 1992).
Lewontin, R. C., *The Doctrine of DNA* (Penguin, London, 1993).
Mainzer, M., *Thinking in Complexity* (Springer-Verlag, Berlin, 1994).
Ormerod, P., *The Death of Economics* (Faber and Faber, London, 1994).
Prigogine, I., and M. Sanglier (eds.), *The Laws of Nature and Human Conduct* (Brussels Task Force of Research Information and Study on Science, Brussels, 1985).
Todd, S., and W. Latham, *Evolutionary Art and Computers* (Academic Press, London, 1992).

APPENDIX

Posner, M., and M. Raichle, *Images of Mind* (Scientific American Library, New York, 1994).
Zeki, S. A., *A vision of the Brain* (Blackwell Scientific Publications, Oxford, 1993).

INDEX

NOTE: Italicized page numbers refer to photos and drawings.